Phylogenetics of Bees

Editors

Rustem Abuzarovich Ilyasov
Doctor in Division of Life Sciences
Major of Biological Sciences, and Convergence
Research Center for Insect Vectors
Incheon National University
Songdo-dong, Incheon
South Korea
and
Ufa Scientific Center
Institute of Biochemistry and Genetics
Russian Academy of Sciences
Ufa
Russia

Hyung Wook Kwon
Professor in Division of Life Sciences
Major of Biological Sciences, and Convergence
Research Center for Insect Vectors
Incheon National University
Songdo-dong, Incheon
South Korea

CRC Press
Taylor & Francis Group
Boca Raton   London   New York

CRC Press is an imprint of the
Taylor & Francis Group, an **informa** business

A SCIENCE PUBLISHERS BOOK

Cover credit: Cover illustration reproduced by kind courtesy of the first editor, Rustem Abuzarovich Ilyasov.

CRC Press
Taylor & Francis Group
6000 Broken Sound Parkway NW, Suite 300
Boca Raton, FL 33487-2742

First issued in paperback 2021

© 2020 by Taylor & Francis Group, LLC
CRC Press is an imprint of Taylor & Francis Group, an Informa business

No claim to original U.S. Government works

Version Date: 20190924

ISBN-13: 978-1-03-208181-6 (pbk)
ISBN-13: 978-1-138-50423-3 (hbk)

Library of Congress Cataloging-in-Publication Data

Names: Ilyasov, Rustem Abuzarovich, 1980- editor. | Kwon, Hyung Wook, 1968-
  editor.
Title: Phylogenetics of bees / editors: Rustem Abuzarovich Ilyasov, Hyung
  Wook Kwon.
Description: Boca Raton, FL : CRC Press, [2020] | Includes bibliographical
  references and index.
Identifiers: LCCN 2019037641 | ISBN 9781138504233 (hardcover)
Subjects: LCSH: Bees--Phylogeny. | Bees--Variation.
Classification: LCC QL563 .P495 2020 | DDC 595.79/9--dc23
LC record available at https://lccn.loc.gov/2019037641

Visit the Taylor & Francis Web site at
http://www.taylorandfrancis.com

and the CRC Press Web site at
http://www.crcpress.com

Preface

The first idea of writing this book came about after reading the books of Ruttner "Biogeography and Taxonomy of Honeybees" (1988), Michener "The Bees of the World" (2000), Kipyatkov "Life Cycles in Social Insects Behaviour, Ecology And Evolution" (2006), and Hepburn "Honeybees of Asia" (2011). Meanwhile, I met with Professor Hyung Wook Kwon who invited me to a postdoctoral position at Incheon National University, South Korea. In Korea, we are working very hard and met many times with local beekeepers and scientists at conferences of The Korean Beekeeping Society. Compared with western honeybees, eastern honeybees have received very thin coverage on honey bee taxonomy, phylogenetics, physiology, biochemistry, genetics, and pathology. Especially, the eastern honey, bee *Apis cerana,* has no clearly defined subspecies like the western honey, bee *Apis mellifera.* In order to address this disparity, we decided to write this book, which combines the efforts of western and eastern teams of scientists. The book contains works not only about phylogenetics but also about current aspects of adaptivity and expansion to the New World. We fervently hope that this collation will provide a stimuli to broaden the base of the development of taxonomy and phylogenetics of the Asian and European honeybees. We also hope that it will stimulate further research, both empirical and theoretical, in this important and promising field. We wish to thank all contributing authors as well as the referees for their collaboration and help. Also, I appreciate my scientific advisors Professor Alexey Nikolenko of the Institute of Biochemistry and Genetics of Ufa Federal Research Center of Russian Academy of Sciences and Professor Hyung Wook Kwon of the Incheon National University for their help in editing and for the constructive discussions on basic ideas in this book.

Rustem Abuzarovich Ilyasov
Incheon National University
Songdo-dong, Incheon, Yeonsu-gu
September 24, 2019

Acknowledgments

We are grateful to the CRC Press publisher, Russian Government Assignments (RGA) (the Grant No. No. AAAA-A16-116020350026-0), Russian Found of Basic Research (RFBR) (the Grant No. 17-44-020648 Povolzhye), the National Research Foundation of Korea (NRF), the Korea government (MSIP) (the Grant No. 2016R1A2B3011742), the Cooperative Research Program for Agriculture Science and Technology Development (RDA) (the Grant No. PJ012526), and the Postdoctoral Fellowships in the Incheon National University (2017–2019) for kindly supporting both editors– Sc.D. R.A. Ilyasov and Prof. H.W. Kwon.

Contents

Introduction

Hymenoptera is a large order of insects, comprising the sawflies, wasps, bees, and ants. Over 150,000 living and over 2,000 extinct species of Hymenoptera have been described (Mayhew 2007, Janke et al. 2013, Aguiar et al. 2013).

Honey bees are flying eusocial insects closely related to wasps and ants, and are known for their role in pollination and, in the case of the best-known bee species, the western honey bee, for producing honey and beeswax. Bees are a monophyletic lineage within the superfamily Apoidea and are presently considered a clade, called Anthophila. There are over 16,000 known species of bees in seven recognized biological families (Michener 2000, Danforth et al. 2006). They are found on every continent except Antarctica, in every habitat on the planet that contains insect-pollinated flowering plants.

Some species, including honey bees, bumble bees, and stingless bees live socially in colonies. Bees are adapted for nectar and pollen nutrition, the former primarily as an energy source, and the latter for protein and other nutrients. Most pollen is used as food for larvae. Bee pollination is important both ecologically and commercially. The decline in wild bees has increased the value of pollination by commercially managed hives of honey bees.

The size of the bees ranges from tiny stingless bee species whose worker bees are less than two millimeters long to *Megachile pluto*, the largest species of the leafcutter bee, whose females can attain a length of forty millimeters. The most common bees in the Northern Hemisphere are the Halictidae or sweat bees, but they are small and often mistaken for wasps or hoverfly species. Vertebrate predators of bees include birds, such as bee-eaters; insect predators include beewolves and dragonflies.

Human beekeeping or apiculture has been practiced for a thousand years, since at least the times of Ancient Egypt and Ancient Greece. Apart from honey and pollination, honey bees produce beeswax, royal jelly, and propolis. Bees have appeared in mythology and folklore, through all phases of art and literature, from ancient times to the present day, though primarily focused in the Northern Hemisphere, where beekeeping is far more common.

Some big questions in the study of honey bees are their evolution and phylogenetics. The resolution hinges on understanding the phylogenetic relationships of the four bee subgroups: orchid bees, bumble bees, stingless bees, and honey bees. These problems are studied by three methods, based on morphological data, nuclear DNA (nDNA) characters, and mitochondrial DNA (mtDNA) features. The fact that no consensus is reached reflects the complexity of the challenges, and shows the need for additional research.

Honey bees compose a single genus *Apis* in the family Apidae. All *Apis* honey bee species are similar in morphology, social biology, nest architecture, foraging behavior, and the use by foragers of a complex "dance" to signal direction and distance to food sources. *Apis mellifera* is one of the best-studied insects in the world, though many basic questions about the biology of this species remain unanswered. However, the similarities among honey bee species do not agree with the diversity of behavior and ecology found among the Asian species. Even the number of *Apis* species is not known with any certainty.

Honey bees of *Apis* species are grouped into three lineages: the cavity-nesting bees, *A. mellifera*, *Apis cerana*, and *Apis koschevnikovi*; the dwarf bees, *Apis florea* and *Apis andreniformis*; and the giant bees of the *Apis dorsata* group. Each of these lineages occurs over a wide range of habitats and climates. Each of the three lineages includes numerous morphologically and ecologically differentiated populations, ranging from allopatric subspecies to reproductively isolated sympatric species. Many populations, such as those on islands in the Japanese, Indonesian, Malaysian, and Philippine archipelagos, are geographically isolated.

The nomenclature and systematics of *Apis* are complex. Maa (1953) recognized 24 honey bee species and divided them into three genera. Subsequent researchers tended to ignore Maa's classification and recognized four species of honey bees, one corresponding to each of Maa's genera or subgenera: *A. florea*, the dwarf bee; *A. dorsata*, the giant bee; *A. cerana*, the eastern cavity-nesting bee; *A. mellifera*, the western cavity-nesting bee.

Sakai et al. (1986), based on quantitative cladistic analysis, evaluate 23 species and subspecies (or races) from the entire natural range of the genus. Their data matrix includes morphological, behavioral, and biochemical characters and they present both cladistic and phenetic analyses. However, their cladistic analysis is difficult to evaluate, because they did not indicate how the characters were polarized or how coding decisions were made for continuously varying characters.

A review on the Asian bees has led to the "rediscovery" of two of Maa's species, *A. andreniformis* (Wu and Kuang 1987, Otis 1991), and *A. koschevnikovi* (*A. vechti*) (Koeniger et al. 1988, Tingek et al. 1988). Another new species *Apis nigrocincta* (Hadisoesilo and Otis 1996) and *Apis nuluensis* (Tingek et al. 1996) have recently been discovered from Sulawesi and Borneo, respectively, and their ranges have not yet been adequately determined.

Among the four species of the genus *Apis* found in Asia, *A. cerana* F., *A. dorsata* F., and *A. florea* F. are sympatric through much of their ranges. *A. mellifera*, the European bee, is exotic to Asia and its range is allopatric. Ruttner (1987) has recently reviewed the geographic distribution of *A. cerana*, which is found in a very wide area comprising mostly southern and eastern Asia, and its range extends from Afghanistan in the west to the Philippines in the east, and in the north from Ussuri to Java and Timor in the south. Thus, *A. cerana* is found not only in tropical and subtropical regions of Asia, but also in cooler climates, such as Siberia, north China, and the higher altitudes of the central Asian mountains (Koeniger et al. 1976b). Michener (1974) reported the wide distribution of *A. cerana* in Southeast Asia, extending from Sri Lanka and India to Japan and Southeast to the Maluku Islands. Peng et al. (1989)

and Otis (1991) discussed the distribution of *A. cerana* in China and considered the bees in China to be a variety of *A. indica* and different from the Japanese variety.

The honey bees *A. mellifera* L. were distributed across all Europe, Africa, and *A. cerana* F. across all Asia, which were divided into many subspecies and currently have been expanded into all regions and continents, excluding only Antarctica (Ruttner 1988). Now, 31 subspecies of *A. mellifera* are recognized by international scientists (Ruttner 1988, Engel 1999, Sheppard and Meixner 2003, Chen, C. et al. 2016). They have colonized the extensive area, characterized by a variety of climate and vegetation, over thousands of years, diversifying through natural selection into well-characterized populations that have been identified as subspecies, distinguishable firstly on a morphological and ethological basis, and more recently through molecular biology studies (De la Rúa et al. 2005, Meixner et al. 2013). After the introduction of multivariate statistical methods, which were substantially developed by Daly and Balling (1978) and Ruttner et al. (1978), they were applied to the genus *Apis* by Ruttner (1988). The current classification of honey bees stems from multivariate methods of analysis originally advanced by DuPraw (1965a) and substantially developed by Ruttner (1988) and Daly (1992). Ruttner (1987) distinguished four different races of *A. cerana*; namely, *Apis cerana cerana*, *Apis cerana himalaya*, *Apis cerana indica*, and *Apis cerana japonica*, based on statistical methods, including principal component analysis, discriminant analysis, and cluster analysis. In the recent past, attempts have been made to identify different races of *A. cerana* and *A. mellifera* by multivariate analysis using computer-assisted standard statistical methods (Verma et al. 1994, Hepburn et al. 2001, Kandemir et al. 2005, Radloff et al. 2005a, Takahashi et al. 2007, Tan et al. 2008, Özkan Koca et al. 2009, Abrol 2013).

Based on an extensive sample collection and multivariate analyses, Ruttner et al. (1978) hypothesized that north-eastern Africa and the Near East might be the center of origin of *A. mellifera*. They proposed that the species invaded Africa and Europe in three distinct branches, a South and Central African branch A, a North African and West European branch M, and a North Mediterranean branch C. This classification was further refined by the addition of a fourth evolutionary branch, called O, which included the Near and Middle Eastern subspecies (Ruttner 1988). Mitochondrial DNA variations have been used to confirm the presence of three lineages A, M, C in Africa, western Europe, and southeastern Europe (Smith and Brown 1988, Cornuet and Garnery 1991), and the existence of a fourth mitochondrial lineage O in the Middle East has also been confirmed (Arias and Sheppard 1996, Palmer et al. 2000).

The same general structure of the honey bee subspecies also emerged from microsatellite surveys (Estoup et al. 1995). The main discrepancy provided by such molecular markers was a clear genetic disruption between the branches M and A in the Liberian Peninsula (Smith 1991a, Franck et al. 1998). Variation in morphological and mitochondrial characters may provide information about different time scales. Mitochondrial non-coding sequences evolve rapidly, but they are not directly affected by natural selection. In contrast, morphological characters may respond to selective pressures imposed by environmental conditions (Hepburn and Radloff 1997). Due to maternal inheritance without recombination, the mtDNA is passed intact from generation to generation along maternal lineages. Thus, the mitochondrial data provides information on biogeographic patterns resulting from mutation, migration, and genetic drift, while morphology responds to current environmental conditions.

MtDNA haplotypes have proven to be a useful tool for unraveling the population structure of honey bees (Hall and Smith 1991, Smith 1991b). Even though *A. cerana* has been reported to have plenty of geographic variations among its populations (Maa 1953, Peng et al. 1989, Ruttner 1988), studies of the mtDNA of *A. cerana* and other Asian honey bees are very few in comparison with *A. mellifera* (Cornuet and Garnery 1991, Sihanuntavong et al. 1999, Ferreira et al. 2009). Studies on the diversity of morphological (Ruttner 1988, Radloff et al. 2005a) and mitochondrial haplotypes (Hepburn et al. 2001, Smith 2002) across the range of *A. cerana* are suggestive of its high biological diversity. Recent whole-genome studies of *A. mellifera* (Weinstock et al. 2006, Wallberg et al. 2014) and *A. cerana* (Park et al. 2015, Diao et al. 2018, Yokoi et al. 2018) shed light on their genetic variability, evolution, and adaptation. The assessment of genetic variability among natural populations of honey bees is the first step towards the genetic improvement of the species. Biological and ecological differences between different geographic races and ecotypes of honey bees provide an excellent opportunity for their genetic improvement by selection and breeding.

In this book "Phylogenetics of bees", authors from different countries of the world have reviewed and summarized published research on the origin and evolution of colony in Apidae (Chapter 1), on phylogenies of Asian honey bees (Chapter 2), on the origin of the European bees and their intraspecific biodiversity (Chapter 3), on the classic taxonomy of Asian and European bees (Chapter 4), the genetic diversity of honey bee *A. mellifera* in Siberia (Chapter 5), on the current drivers of taxonomic biodiversity loss in Asian and European bees (Chapter 6), on the loss of taxonomic biodiversity of honey bees *A. mellifera* and *A. cerana* in Russia (Chapter 7), on the breeding of better and healthy honey bees as the only way to save the native biodiversity (Chapter 8), on the threats and importance of honey bee populations that spread beyond their native range, the case study of Latin America (Chapter 9), and on the history of honey bees in North America (Chapter 10).

CHAPTER 1

The Origin and Evolution of the Colony in Apidae

Eskov, E.K.

Introduction

In the group of aculeate Hymenoptera (*Aculeata*), the bees represent one of the largest superfamilies of the Hymenoptera formed in the process of the coupled evolution with flowering plants. Since the flourishing of angiosperms occurred in the Late Cretaceous period, the appearance of bees probably dates to this epoch (Michener 1944, 1965, 1975, Brothers 1975, Rasnitsin 1980a,b, Dietz 1986).

The number of bee species progressively increased at the end of the Oligocene—the beginning of the Miocene (Michener 1975, 1979, Rasnitsin 1980a,b). Currently, the superfamily includes 11 families, uniting 520 genera. The greatest taxonomic diversity is in the Neotropical, Neo-, and Palearctic realms, where 315, 260, and 243 genera of bees, respectively, are known. Their lowest representation is in the Australian region, where only 18 genera are found (Radchenko and Pesenko 1994).

Bees have developed care for their offspring. The expression of this instinct depends on the way of life. By this feature, bees are divided into solitary living, social, and parasitic. Most species of the modern superfamily of bees are represented by solitary living insects. Various forms of social organization were found in some representatives of the families Apidae, Bombidae, Halictidae, Anthophoridae, and Euglossiedae.

In the order of the Hymenoptera, social insects are found not only among bees. Fully social species are represented by the superfamily of Formicoidea (ants). Among the other 35 orders of insects, only the representatives of Isoptera (termites) lead a social way of life. Unlike the Hymenoptera, this order is completely represented by socially living species.

Russian State Agrarian Correspondence University, 19 Komsomolskaya street, Apt. 16, Balashiha, Moscow region, 143903, Russia; ekeskov@yandex.ru

The presence among bees of the variety of transitional forms from typical solitary living to highly organized socials allows us to trace the general patterns of the origin and sociality development. The main directions that it could have developed can be traced by comparing the advantages in the struggle for the existence and social relations of modern species of bees that differ in their way of life.

It is accepted to distinguish several levels of the social organization. Eosociality is considered to be the rudimentary (initial) form of social organization. This is expressed by the fact that the female offspring of the female-foundress can provide her with some assistance in the life support in the temporary colony. When it breaks up, the generation of female flies away and establishes its own nests. Temporary colonies also form subsocial species. A similar organization is characteristic of eosocial species. Highly organized eusocial species are those that constantly live in colonies or at least two generations of the female-foundress live with her in one nest. Adult individuals participate in the construction and protection of the nest, the delivery of food, and the feeding of the female-foundress offspring. Along with female-foundress, her daughters can also participate in the reproduction of the offspring.

Nest constructions and trophic provision of offspring

Solitary bees. The nest construction of solitary bees depends on the place of settlement. Most often, they use different cavities in plants, rocks, or the ground. Bees subject these cavities, used for sheltering offspring, to some processing (deepening, expansion, cleaning). Many species that settle in the ground dig out passages and cavities by themselves, some build nests in open spaces (stones, tree trunks, soil surfaces, etc.). The construction of nests is carried out by females (female-foundresses) that are ready to perform reproductive functions. The choice of the place of settlement, materials for the construction of shelters, and ways to process them belong to species-specific features.

Among a wide range of nest structures, the most primitive ones are those where food stocks and developing individuals are in one common cavity. That is typical, for example, for the bee *Metallinella atrocaerulea* (Fig. 1.1 A). It settles in ready shallow cavities in wood that has one inlet. The female fills the nest cavity with food (pollen mixed with nectar); ovipositing 4–12 eggs in it, and then seals the entrance with a plug of crushed leaves. The emergence of larvae and their development does not occur simultaneously. The first to complete the larval stages and to pupae are individuals that emerged from the eggs ovipositioned at the base of the nest at beginning of the settling, the latter—at the upper part from the side facing the plug (Radchenko 1978). There is a competition between the larvae due to the localization of all of them in the same trophic substrate, where the intensity of consumption depends on their age. This prevents the appearance of consolidation among developing individuals.

However, in the superfamily of bees, the use of a common nesting chamber by developing individuals has a limited occurrence. Along with *M. atrocaerulea*, this form of offspring reproduction was found in the family of Anthophoridae in representatives of the genus *Allodapini* (Radchenko and Pesenko 1994). The insignificant representation of species with this form of nesting in the rapidly

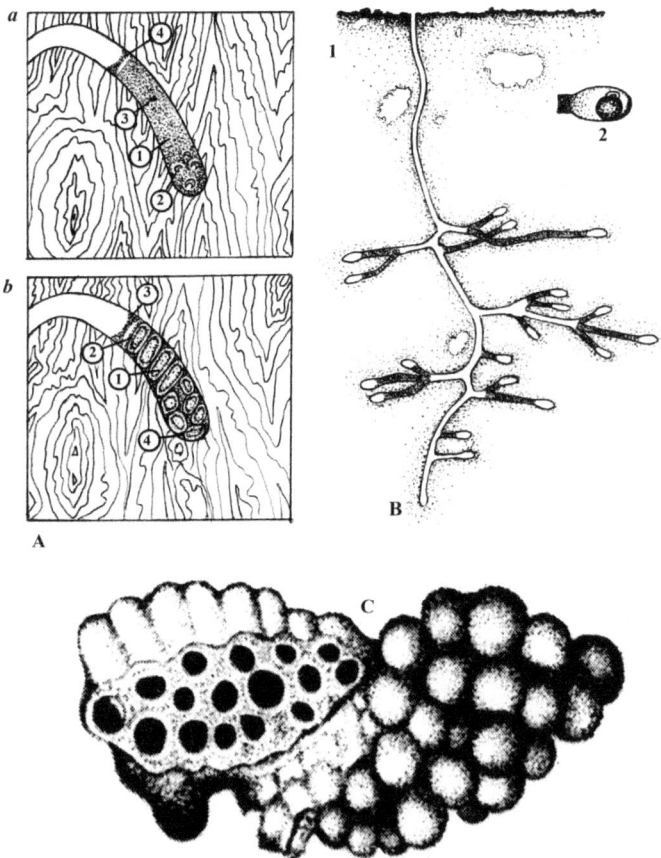

Figure 1.1: Nest structures of solitary living bees: A–a nest of ***Metallinella atrocaerulea*** with developing bees: a–bees at the egg stage and larvae (1: pollen, 2: larva, 3: egg, 4: plug); b–stage of pupae (1: cocoon, 2: excrement, 3: plug, 4: formed cocoon) (Radchenko 1978); B–***Nomioides minutissimus*** (1) and one of the sealed cells with food and egg (2) (Radchenko 1979); C–ground honeycomb of the bee ***Halictus quadricinctus*** (Blagoveshchenskaya 1983).

progressing superfamily of bees is evidently associated with the incipience of rather perfect instincts of caring for offspring based on the use of individual isolation of developing offspring.

In bees, natural selection obviously favored the acquisition of instincts responsible for nesting behavior, providing individual isolation to the developing offspring. This was fully satisfied with the construction of cells where only one individual could develop. It was also specially supplied with food once or periodically along with development. In such a situation, the cell turned into an elementary unit of nesting construction (Fig. 1.1 B, C).

The choice of the settlement location and the materials used for the construction of the cells differ in diverse species of bees widely. Many types of Megachilidaes construct cylindrical or cells close to its form from pieces of leaves or petals of flowers. Differences in the species determine the size of the body of insects, which affects the size of the cells.

The discrepancy between the large diameter of the nest cavity and the biological needs of the individuals developing in them is corrected by an increase in the thickness of their walls. However, the reduction in the diameter of the cells constructed in narrow nest cavities is reflected in the size decrease of the developing bees.

Therefore, improving the nesting behavior of Megachilidaes was accompanied by the acquisition of the ability to control the size of the cavity choice for settlement. Its successful choice is achieved by reducing energy costs for the construction of cells by reducing the thickness of their walls under optimal conditions for the development. Thus, with the possibility of choice, leafcutter bees (*Megachile rotundata*) prefer a diameter of tunnels limited to 6–6.5 mm for settling (Pesenko 1982).

The number of cells constructed in the nest cavity depends on its depth and is limited by the reproductive potential of the females-foundresses. The reproduction of the leafcutter bee is limited to about 40 offspring. Within a day the female can oviposition no more than two eggs. This limits the breeding rate of offspring (Stephen et al. 1969).

The leafcutter bee fills each cell with the pollen mass, lays an egg on it, and then closes (seals) it with the cover. Especially for it, a bee spends up to 6 pieces of leaves of the same plants from which the cell is built. Construction of the nest, that can contain 1–20 cells, is completed with the construction of the plug. The bee spends 8–50 pieces of leaves on it. A small cavity usually remains unoccupied between the plug and the closest cell cover (Stephen and Torchio 1961, Gerber and Klostermeyer 1972). The plug performs a protective role, reducing the negative impact of external environmental factors, and prevents penetration into the nest of other insects (Stephen et al. 1969).

The ratio between females and males that develop in the same nest depends on the size of the cells, their localization in the nest cavity, and the food supply. In addition, the females-foundress oviposition the fertilized eggs from which the females develop into the cells appeared at the beginning of the nest construction, and the unfertilized ones—closer to the exit. Developing females are supplied with food more abundantly than males who leave the nest before females (Stephen et al. 1969, Pesenko 1982). For females, it takes a longer time than for males to develop from egg to adult stage.

The use of plant leaves by Megachilidaes and other species of solitary bees for nesting is conditioned by their easy accessibility, but is limited with low strength and high moisture permeability. This was due to the acquisition by the bees of the tools that contributed to increasing the protection of developing offspring from the adverse effects of the physical and biotic factors of the environment. On this path, instincts and morph physiological adaptations have been developed that ensure the expansion of the diversity of places of settlement. Thus, bees of the genus *Heriades* (Osmiini) and *Megachile* cover the inner surfaces of the cells with resinous substances. This results in an increase in the waterproofing of the cells and protection from the penetration of parasites in it (Tsuneki 1970, Maciel 1976).

In the natural environment, there are objects used by some bees that without additional processing meet the needs of developing individuals. For example, *Osmia rufonirta* and *O. bicolor* (Megachilidae) settle in empty shells of snails in order to use each of them for one developing individual. To do this, the bee firstly cleans the shell, and then brings the food into it, and after ovipositioning, the bee seals the egg with the paste made from chewed leaves. Further care for the offspring is expressed in the sheltering of shells from competitors and enemies, as well as from overheating by the sun's rays. *O. rufonirta* usually rolls the shell from an open surface into a thicket of grassy vegetation, and *O. bicolor* digs it into the soil (Grozdanić 1971).

The strengthening of competition for suitable ground nesting places could affect the acquisition of instincts and morph physiological adaptations for settlement in the ground. First, bees could be attracted by cavities into the dead parts of the root system that emerged to the surface, as well as tunnels and depressions left by soil invertebrates in the ground. So, *Megachile albocincta* has adapted to use earthworms' tunnels on the surface for nesting. As a material for cells building, the bee uses pieces of leaves. The construction of the first cell begins at the depth of the tunnel, and the last one—at its exit. This behavior is similar to the leafcutter bee that settles in terrestrial cavities (Fabre 1963).

The increase in the adaptability of bees to settle in the ground developed in the direction of growing independence from the availability of suitable cavities and environmental factors that threaten the developing offspring. In this way, bees have acquired devices for building in the ground passages and cavities. The form of settlement when the female-foundress digs out a small hole in the ground where it is possible to build at least one cell belongs to the simplest one. Such nesting behavior is characteristic of one species of Colletidaes—*Colletes daviesanus*, settling on sandy, forest, or limestone unshaded hillsides of the southern exposures. In the excavated holes, the bee constructs up to four cells singly or in small rows (Scheloske 1974). The representative of another Megachilidae family—*Megachile japonica* constructs up to eight nests containing only 1–2 cells (Maeta 1979). Each of them is dug out with a cavity about two cm in depth.

The complication of nest behavior of bees settling in the ground was expressed in the acquisition of instincts for the construction of tunnels, labyrinths, and brood chambers. Relatively simple nests of this type are built by the bee *Cantridini aethyctera* (Anthophoridae). It settles in dry forest soil, digs in an almost vertical tunnel with a depth of 8–14 cm, to which 3–6 cells are adjoined (Vilson and Frankie 1977). Side branches from the entrance tunnel to horizontally located brood cells are constructed by bees *Punurginus atriceps* and *P. occidentalis* (Andrenidae) (Rust 1976) and *Nomioides minutissimus* (Halictidae) (Radchenko 1979). In *Atigochlorella edentata* (Halictidae), a bunch-like cluster of cells strengthened on ground supports have a remote resemblance to honeycombs (Eickwort and Eickwort 1973). Closer to them is the ground nest of the *Halictus* bee (*Halictus guadricinctus*). This bee has the nest of a convex or spiral curved structure, measuring approximately 6×12 cm that is formed by several tens of closely spaced cells (Blagoveshchenskaya 1983).

A threat to the developing offspring of bees is represented by a hydro factor in many environmental situations. The development of means of protection against it involves the acquisition of instincts of waterproofing the contents of cells. Bees use plant extracts or their own secrets as the moisture-proof substances. Using the pitch of woody plants, the bee *Melmoma tourea* (Anthophoridae) veneer the walls of brood cells. The surface of the cells is covered with an oral secret in *Nomia melanderi* (Halictidae) settling in alkaline soil (Batra 1970). The secret of the Dufurov gland, which is a complex mixture of liquid triglycerides, is used to cover the cells of the *Anthophora* bee *Antophora antope* (Batra 1980) and *A. abrupta* (Norden et al. 1980).

High reliability of cell content isolation is provided by wax. It is produced in bees by the glands of the body's integument. The use of wax or wax-like substances was found in solitary living Andrenes *Andrena ovatula* (Wafa et al. 1972), Anthophoras *Anthophora urbana* (Mayer and Johansen 1976), proteolines *Ptilothrix bombiformis* (Rust 1980), etc. These substances usually cover only cell walls and the plug is usually constructed from the same substrate the nest is built with. For example, *Andrena urbana* seals the cell with a soil plug (Mayer and Johansen 1976).

Nesting on open surfaces was caused by the absence of suitable cavities, morphological devices for processing natural substrates, and/or their physical properties (high strength, humidity, etc.). Widespread use of such building materials as small stones fastened with dust and moistened by oral secretions was seen in bees *Chalicodoma pyrenaica* (Fabre 1963), *Hoplitis anthocopoides* (Eickwort and Eickwort 1973), *Osmia anthocopoides* (Radchenko and Pesenko 1994), etc. The female firstly supplies each of the alternately constructed cells with food, then ovipositions an egg in, and seals it. The material used to seal the cell does not contain small stones, as they could prevent the release of individuals reaching the imago stage. Usually, the construction of such nests contains a little more than ten cells and it is completed by the construction of an additional shelter above them.

Eo- and subsocial species. Primitive social insects include bumble bees (*Bombus*), represented in the insects' fauna of about 250 species. In bumble bees living in the Palearctic, the overwintered female-foundress during the spring-summer season creates a colony that breaks down in late summer or early autumn. Burrows of rodents, tree hollows, crevices of rocks, etc., are used as places of settlement of bumble bees. The bumble bee nest is covered with various heat-insulating materials that were found nearby, such as dry moss, straw, leaves, etc. A wax-like shell can be used for waterproofing the nest (Eskov 1992, Eskov and Dolgov 1982). When settling on trees species living in the tropics, they use alive and/or dead leaves of plants to shelter the nests (Janzen 1971).

The seasonal cycle of the development of the bumble bee colony begins with the search for a suitable nesting place for the fertilized female where she constructs a cell for feed and a brood chamber for offspring. Wax is used as a building material for bumble bees that is usually mixed with pollen. This building material is used many times in its own nests and when settling in abandoned nests of other colonies (Alford 1971). It is also possible to use other wax-like substances as a building material, for example, plasticine, found in the nest cavity. If there are small containers resembling feed cells, bumble bees can finish building them and store their food reserves there (Dolgov 1982).

The construction of the brood cell by the female-foundress is preceded by the detachment of the wax base. The female forms a pollen ball moistened with nectar and rings it with a small wax roller on it (Dolgov 1982). The females of *B. pratorum*, *B. humilis*, *B. pascuorum*, and others make holes in the pollen ball where one egg is ovipositioned (Alford 1975). The female *B. hortorum* ovipositions eggs at different levels of the pollen ball (Alford 1971). Females of *B. balteatus* and

B. polaris oviposition eggs in pollen balls, fill them with a pollen mixture, and then cover with a waxy membrane (Alford 1971, 1975, Richards 1973). The construction of separate brood cells where the female ovipositions one single egg is characteristic for *B. rufocinctus* (Hoobs 1965).

Different types of bumble bees oviposition 8–16 eggs in the first brood cell. Their number varies also among females of the same species. A different time is required for the female-foundress to complete the first ovipositioning cycle. Under favorable weather conditions, *B. agrorum* spends about three days to oviposition eggs (Free and Butler 1959, Alford 1971, 1975, Dolgov 1982).

Emerged larvae systematically receive food (nectar and pollen). It is brought into the cell through the pollen pocket or temporary holes made in the cell (Alford 1971, Sakagami and Zucchi 1965, Sakagami 1976). As the larva grows, the female repeatedly rebuilds the brood cell, completing the waxy membrane. Subsequent brood cells are usually placed randomly on a common base, constructed at the initial stages of nest construction. With the emergence of working individuals, the colony protection from the adverse effects of biotic and abiotic environmental factors and the reproduction of offspring increases.

Eusocial species. In highly organized social types of bees, nest structures differ mainly in the use of building materials, the orientation of honeycombs, and the location of fodder stocks. According to the features of these species, the greatest differences are found between stingless bees and representatives of the genus *Apis*.

Stingless bees. A large group of stingless bees (Meliponinae) permanently lives in colonies where structural organization and size can significantly vary. For the place of settlement, stingless bees usually use various natural shelters, such as trees or ground cavities, and also inhabited or abandoned termites' mound-building (Wille and Michener 1973). Some species settle in trees (Darchen 1969).

Regardless of the place of settlement, bees isolate the nest from the outside environment with a multi-layered shell using wax, pitch, clay, wood, and other materials (Koeniger 1976c). The entrance to the nest is in the form of a slit or tunnel that in Meliponen is made of clay, and in *Trigona*—of resin (Brian 1986). To protect against penetration of robbers and parasites into the nest, the bee guards the entrance during the day, and at night closes it with a cork constructed from the material used for the building of the tunnel entrance. Some species protect the entrance from the ants with sticky substances (Ihering 1903, Rooley and Michener 1969). In the ground in the lower part of the nest, a drainage pipe can be constructed to remove water (Brian 1986, Michener 1974). Some wax balls may be in the nest as a reserved building material (Brian 1952).

The cells for brood and food, as a rule, differ in shape and location in the nest. The largest cells are constructed to store the reserves of honey and pollen. They are placed on the periphery of the nest. In *Melipona beehen,* its diameter reaches 2–3 cm (Darchen and Delage-Darchen 1975). The main part of the nest cavity is occupied by relatively small cells for brood. They are strengthened by supports. The nest material is used for these constructions. In some species, the brood cells are arranged in groups, forming bunches. For most species, it is characteristic to have the horizontal distribution of the brood cells forming one-sided honeycombs (Fig. 1.2). The construction of vertical two-sided honeycombs was found in *Dactylurina staudingeri*. Working individuals and drones develop in cells of honeycombs. Cells for the reproduction of large females (queens) are located at the bottom of the honeycomb or in the inner surface of the shell of the nest (Michener 1974).

Working individuals are engaged in the construction of the nest cavity and cells. They also fill the cells with food (a mixture of pollen and nectar), adding to it the secretion of hypo- or propharyngeal glands. The queen first consumes part of this food, and then ovipositions an egg. After that, bees seal the cell, not worrying about the trophic supply of the developing individual (Darchen and Delage-Darchen 1975, Sommeijer et al. 1982).

Bees of the genus *Apis*. The genus *Apis* includes three subgenera; each of them unites two species that have similarities in lifestyle, the specificity of the settlement places, and the design of nesting structures. The most evolutionarily advanced subgenus combines *A. mellifera* and *A. cerana*. The intermediate position is taken by the species *A. dorsata* and *A. laboriosa* (subgenus Megapis). The most primitive species are *A. florea* and *A. andreniformis* that form a subgenus (*Micrapis*). The most widespread is *A. mellifera*. Other species live mainly in the tropical forests of Southeast Asia. Only *A. cerana* inhabit the Palearctic (in North-east China, Japan, and Southern Primorye), and *A. laboriosa* lives in the Himalayas, settling at an altitude of 3–4 km above sea level.

Representatives of *Megapis* and *Micrapis* construct only one honeycomb, attached to a tree branch, the bottom surface of a rock, or other aboveground objects. In bees *Micrapis,* the view of at least a small part of the sky from a honeycomb is a limiting factor when choosing a place of settlement. By its polarization, bees

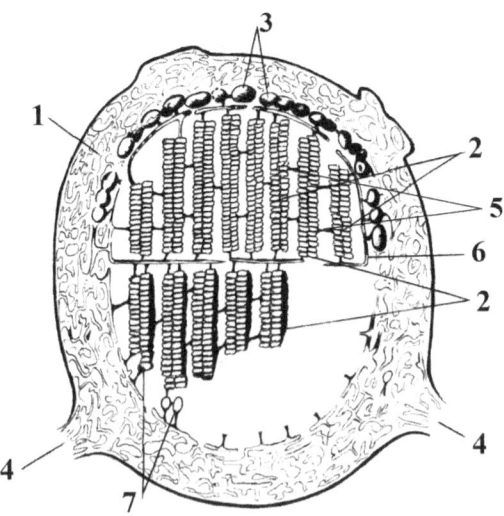

Figure 1.2: The nest of ***Dactylurina staudinderi***: 1–nest shelter, 2–brood cells, 3–food reserves, 4–entrance to the nest, 5–supporting partitions, 6–drainage tubes (Michener 1974).

determine the position of the sun; this is used in the system of spatial orientation. This is not required for the other four species of bees of the genus *Apis* (Lindauer 1971, Eskov 1979, 1992).

Bees start the construction of honeycombs with the place of its attachment and complete it later in the process of mastering the new place of settlement. *A. florea* usually settles on the rocks sided to the south in the summer, and to the east in winter (Dutton and Simpson 1977). Its honeycombs usually have a shape close to a semicircle or an ellipse with one side area of about 500 cm². The dimensions of the cells differ depending on their purpose. Relatively small cells with a diameter of about 2.5 mm, occupying approximately 80% of the honeycomb surface, are used for the development of working individuals. Larger cells are used for the developing drones. In the upper part of the honeycomb, there are large cells used for storing honey and flower pollen. On the lower part of the honeycomb, bees build several cells of an acorn-form shape where queens develop. When settling on tree branches, nest protection from ants is achieved by rings constructing made of sticky resinous plant extracts (Butler 1969, Lindauer and Kerr 1960).

Bees of the subgenus Megapis when settling on trees build honeycombs at a height of up to 20 m, and less often up to 40–80 m above the surface of the soil. The honeycomb can hang under the weight of honey close to the ground, but usually does not come into contact with it. The clustering of nests in one place is characteristic of these bees. More than 150 colonies can settle on one large tree (Butani 1950), although there are also single nests. The minimum diameter of the branches used to build the honeycombs is 12 cm. However, most often bees prefer branches with a diameter of about 30 cm (Morse and Laigo 1969). After the construction is completed, the honeycomb becomes a truncated ellipse, where one side area can reach 0.5–1.5 cm² (Kaiser 1976).

The largest cells (their depth can exceed 8 cm) localized at the top of the honeycomb are used as food stocks (Morse and Laigo 1969, Singh 1962). Cells used for the development of working individuals and drones are under the food tier. In this part, the thickness of the honeycombs decreases to 35–40 mm. Ellipsoidal queen cells are located in the lowest part of the honeycombs (Morse and Laigo 1969).

Protection of honeycombs from the direct influence of biotic and abiotic factors is provided by bees forming a multilayer covering with their bodies. This is formed mainly by young bees, hanging upside down, clinging to each other with their legs (Roepke 1930). The size of such a nest shelter, which usually hangs over the lower border of the honeycomb, depends on the number of bees in the colony that reaches the maximum before the sociotomy (Morse and Laigo 1969).

The bees *A. mellifera* and *A. cerana* differ from the bees Micro- and Megapis in the adaptation to the settlement in shelters (hollows of trees, clefts of rocks, and other cavities), protecting from the influence of unfavorable physical and biotic factors. When it is possible to choose the place of settlement, the honey bee prefers cavities in wood with a volume of 70 ± 103 cm³, located at an altitude of 8–19 m (Petrov 1983). This indicates that they have instincts of evaluating the suitability of a potential dwelling with the biological needs of the colony.

The nests of Indian and honey bees form vertical two-sided honeycombs, whose number and shape depend on the configuration of the nest cavity. One large

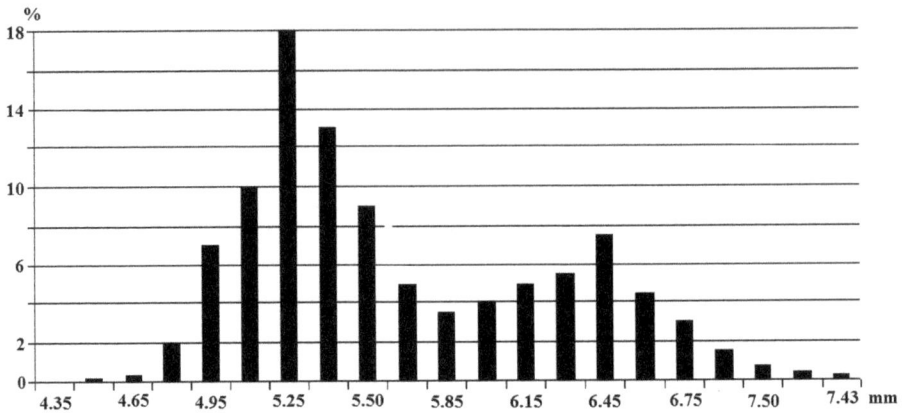

Figure 1.3: Percentage distribution of 43,279 cells (y-axis), differing on the distance between opposite angles (x-axis); honeycombs are built up by bees in free nesting spaces.

honeycomb can be enough for a nest of a honey bee that has settled in a narrow cavity. In representatives of this species, as well as other bees of the genus *Apis*, the upper part of the honeycomb is usually used as food stock. The brood cells are placed below, and the acorn-form queen cells on the periphery. At different times, all honeycomb cells can be used to store food, except queen cells.

The diameter of the cells (the distance between opposite angles) in a honey bee varies from 4.1 to 7.7 mm. The largest frequency of occurrence (about 60%) has cells with a diameter of 5.1–5.4 and 6.5–7.0 mm (Fig. 1.3). The first of them is used for reproduction of working individuals, the others for drones. The tendency of cells enlargement can be traced at the initial stages of honeycomb construction during the period of a new dwelling mastering (Eskov and Eskova 2001, 2012).

In bees of the genus *Apis*, the closeness of cells and the formation of honeycombs made by them were favored by saving the building material and maximizing the use of the nest structure. Probably, the bees built bunch-shaped clusters of cells in the initial phases of the development of this building instinct that are typical for some *Halictus* and relatively low-ordered stingless bees. One-sided honeycombs are distinguished by higher ordering. They are widespread in stingless bees and found in some social types of *Halictus*. At the same time, the needs of relatively small *Halictus* colonies for the reproduction of offspring are satisfied by the presence of only one honeycomb. However, this is not enough for large colonies of stingless bees. Therefore, they had a need for constructing a multi-tiered nesting structure containing several honeycombs connected by a plurality of supports and bridges. They provide an increase in the strength of the brood nest zone, but it is associated with additional costs of the building material and the restriction of the useful volume of the nest cavity. Cells of one-sided honeycombs of stingless bees are unsuitable as food stocks. For these purposes, large cells are used that are tuned outside the brood zone.

The transition from horizontal to vertical placement of cells is associated with a modification of building instincts, which could be motivated by adaptation to life in shelters. With the development of the instinct of two-sided honeycombs cells construction, the selection favored the convergence of the separated common

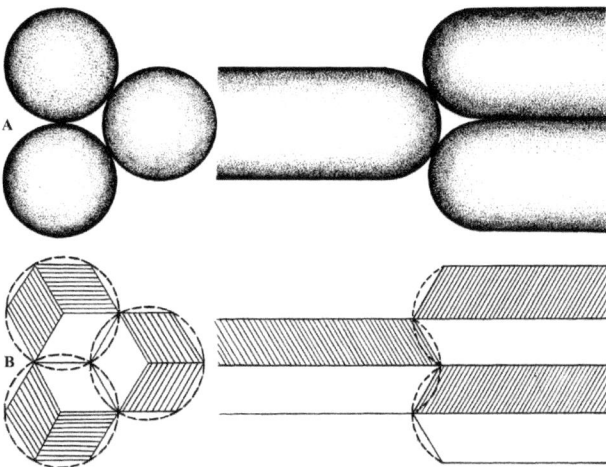

Figure 1.4: Schematic representation of the transformation of rounded cells into hexahedral cells, due to their compaction (A), and the resulting transformation of oval bottoms into trapezoidal (B).

adjacent walls. Partial overlapping of their bases on opposite sides of cells is associated with the convergence of cells. Each of the three converging cells on one side of the honeycomb formed a concave base for the cell on its opposite side. Each of the three converging cells on one side of the honeycomb formed a concave base for the cell on its opposite side. The maximum convergence of cells led to the transformation of their originally rounded shape into hexagonal, and the combined parts of the bases into rhombs. Each of the three rhombs forming the base of the cell began to be included into three adjacent cells on the opposite side, thus maximizing the convergence of the cells of bilateral honeycombs, saving the building material (wax), and providing a relatively high strength (Fig. 1.4).

Morphofunctional differentiation of the bee colony

The size and the members of the colony. Colonies of eo- and subsocial species include a small number of adults and developing individuals. In the nests of *Lasioglossum* sp. (Halictidae) during the intensive development of the colony, there may be 20–25 adult females, approximately the same number of developing individuals at the larval and pupal stage, and about 10 males (Houston 1970). *Exoneurella eremophilia* and *Exoneurella setosa* (Anthophoridae) reproduce males constantly during the period of the colony's existence, but they do not stay for long in the nest. The maximum number of adult females in the nests of these species can reach 20 individuals, and a little more than 40 cells in the brood (Houston 1977).

The number of bumble bee colonies has a pronounced tendency to increase with the migration of various species from zones of cold and temperate climate to the subtropics and tropics. Short summers in the upper latitudes limit or completely prevent the raising of workers. Therefore, with the move to the cold climate zones, bumble bees come closer to the typical solitary living insects in their way of life (Malyshev 1928, Richards 1953). In contrast, in the tropics, the raising of working individuals, mature females, and males is possible throughout the year (Michener and Amir 1977).

In seasonal colonies of species inhabiting the temperate climate, by the time of maximum development, the number of adults is usually limited to a few tens, and to hundreds in the tropics. Among adults in tropical species, the ratio between mature females and working individuals is maintained at about 1:10 (Ihering 1903). In seasonal colonies, females are generally reproduced more than males. For example, during years with warm weather, by the end of summer in *B. agrorum* colonies, the ratio between the reproduction of females and males corresponds to about 0.6:1, and at the early onset of cold weather, changes to 1:1.6 (Brian 1951). The number of males also increases in the case of the female-foundress loss. In *B. terricola* and *B. melanopygus*, the ratio between females and males can be as high as 1:6 in size, and 1:2.9 in biomass (Owen et al. 1980). The quantity of males in the colony increases corresponding to the increase of the period between the onset of ovipositioning by the female-foundress and the working individuals of haploid eggs (Duchateau et al. 2004).

High variability is characterized by interspecific differences in stingless bees. The smallest number of them is several hundred individuals. It is characteristic for the colonies of *Melipona quadrifasciata*, *M. scutellaris*, and *Trigona silvestris*. One to several thousands is characteristic for colonies of *T. capitata*, *T. mombuca*, *T. testaceicornis*, *T. droryana*, and *T. postica*. Fifty to 180 thousand adults can contain colonies of *T. rufrus* (Lindauer and Kerr 1960). The ratio between the number of queens and working individuals in them is usually in the range of 1:3–1:4, and can exceed these limits depending on the availability of food for colonies (Darchen and Delage-Darchen 1975). In *M. beichei* colonies, when lacking food, this ratio increases up to 1:9, and with an abundant food supply, it can reach 1.3:1 (Darchen 1973).

In the colonies of the honey bee, the workers usually have the largest representation. Their quantity is subject to pronounced seasonal variability depending on the external temperature and productivity of the forage place. In zones with a temperate and cold climate, the maximum numbers of worker bees reach 30–70 thousand individuals in the summertime, and the minimum is at the end of wintering and the resumption of flying activity. Drones are the temporary members of the bee colony. They emerge in April–June and are expelled from colonies in late summer or early autumn. Drones can remain in the nests for the winter only in queenless colonies.

The honey bee is characterized by a monogynous organization of the colony. The presence of two or more adult queens in a colony is possible only during the period of sociotomy, when the separation of the new colony resulting with swarming is delayed due to unfavorable weather. The participation of two queens in the carrying of reproductive function is possible in colonies that replace the queens. However, their life together is not usually long.

Life expectancy and reproductive potential of females. In social insects, females (females-foundresses) performing reproductive functions usually differ from their female offspring, at least from those that appear in the initial phases of the development of the new colony. These females in social insects most often perform auxiliary functions and do not directly contribute to the reproduction of their species. This is most often associated with their sexual underdevelopment.

Females-foundresses. The lifespan of females-foundresses and their female offspring varies, and depends on the level of the social organization of the bees. Females-foundresses of eo- and subsocial bees have a relatively low life expectancy. In species that have the lowest level of social organization, the female-foundress may not survive to see her offspring achieve the adult stage. Unequal trophic support at the stage of the larvae is the cause of differences in the development of reproductive organs. As a result, sexually mature females are engaged in the reproduction of offspring, and underdeveloped females perform subsidiary functions. This form of communality is known in the bees of *Exoneurella eremophilia* and *E. setosa* (Anthophoridae), settling in the core of the stems of herbaceous plants. Females-foundresses in these bees do not always survive until the appearance of the first underdeveloped females. When they reach the imago they remain in the nest and after the emergence of mature females, they participate in the protection of the nest and the trophic support of the generation of their sisters (Houston 1977).

In higher-organized social species of *Ceratina*, for example, in *Ceratina japonica*, *C. flavipes*, and *C. iwatai*, the females-foundress' care of developing offspring helps to reduce its chance of death approximately from 50 to 3–19% (developing individuals can exist without females-foundresses in the case of their deaths). Strengthening the relationship between females-foundresses and offspring of *Ceratina* is sometimes expressed (in about 10% of cases) in joint overwintering, and afterwards the females-foundresses start to perform reproductive functions (Sakagami and Maeta 1977). The increase in the life expectancy of females-foundresses, related to the increased interrelation with their offspring, is even more pronounced in some species of *Halictus* bee. At *Lasioglossum* (*Dialictus*) *umbripenne* the female-foundress lives about a year with the first generation of her offspring, performing the function of working individuals. During this time, she is not replaced by her daughters (Wille and Orozco 1970).

In one-year-old colonies of bumble bees, working individuals always develop in the first brood cell constructed by the female-foundress, their number in *B. agrorum* and *B. humilis* reaches 8 (Alford 1970), and 16 in *B. hortorum* (Alford 1971). In the next generations, mature females develop. Reproduction of males takes place at the end of the development season of bumble bee colonies. The lifespan of females-foundresses is limited to one year, and workers and males die when the cold comes. Females-foundresses live much longer in perennial colonies inhabiting the tropics. However, in these colonies, several females of different ages may simultaneously oviposition fertilized eggs (Sakagami and Zuchi 1965, Michener and Amir 1977).

The activity of reproductive function in bumble bees' females-foundresses increases in the process of development of their colonies. In the first brood cell, the female *B. agrorum* usually ovipositions 8 eggs within three days. This is followed by a reproductive pause that ends after the emergence of first working individuals, and after that new brood cells are built in the nest (Alford 1970). With the increase in the number of adults, the reproductive activity of the female can reach 12 eggs per day

(Brian 1951, 1952). The reproductive potential of perennial tropical colonies is much higher than that of annuals. Within a day, the female *B. transversalis* can oviposition up to 300 eggs (Michener 1979).

The females (queens) of bees of the genus *Apis* are distinguished by high fertility and lifespan. In different races of the honey bee, the queen can oviposition from a few hundred to three thousand eggs during the day, which, along with genetic differences, depends on the ecological situation and the number of working individuals in the colonies (Eskov 1995, Taber 1980). The reproductive activity of the queens decreases as they age. The life expectancy of queens is limited to 4–6 years, and the maximum, according to the observations of Betts (Eskov 2013), can reach 8 years. The death of the queens most often happens during the wintering period. The probability of death increases as they age.

Worker individuals. Despite the level of social organization in bees, the underdeveloped females, as a rule, act as workers. They appear in seasonal colonies at the initial stages of their development. In social *Halictus*, workers live about 30 days (Wille and Orozco 1960). In some types of *Halictus,* small underdeveloped females appear along with large, mature females in the first reproduction. They replace the female-foundress in the case of her death. In the colonies of *Exoneurella eremophilia* and *E. setosa*, the representation of mature females can reach 75 percent. They perform a reproductive function with the female-foundress in the colony (Houston 1977).

Males usually develop from eggs ovipositioned by unfertilized worker bees. In some species, the contribution of working individuals to the reproduction of males is of dominant importance. Thus, in the colonies of bumble bees, the proportion of males developing from eggs ovipositioned by working individuals can reach 90%, and 95% in the *Trigona* (Lin and Michener 1972). The behavior of working individuals' ovipositioning eggs differs from the behavior of females-foundresses. Workers are characterized by randomly ovipositioning several eggs in one brood cell. In this process, workers often occupy other cells, eat eggs, and replace them with their own ones (Katayama 1973). Females are obligatorily similarly related to brood cell constructed and occupied with eggs ovipositioned by working individuals (Garofalo 1978).

The life expectancy of worker bumble bees depends on the forms of participation in the life of the colony. Bumble bees engaged in the delivery of feed live, as a rule, fewer than workers who perform intra-nest work. In colonies of species inhabiting temperate climates, the average life expectancy of workers is limited to about 25 days, and the maximum lifespan reaches 69 days (Brian 1952). This is similar to tropical species of *B. morio*, where foragers live an average of 36 days, and individuals who specialize in performing intra-nest work live twice longer, about 73 days (Garofalo 1978).

Among bees of the genus *Apis*, working individuals begin to perform reproductive functions in the absence of the queen. In a queenless colony of a honey bee, up to 25% of ovulating bees can be found (Perepelova 1928), while representatives of Megapis have only one (Velthuis et al. 1971). The number of eggs ovipositioning by the ovulating working individual in *A. mellifera* varies from 19 to 32 (Perepelova 1928).

The working bee spends 17–251 seconds (Haydak 1969) for egg ovipositioning, and the queen about 10 seconds. Convergent similarities of working bees and bumble bees engaged in ovipositioning eggs are expressed in disorder (chaotic) behavior.

The lifespan of workers of honey bees is subject to seasonal variability. Bees actively participating in the replenishment of food reserves of spring-summer generations live 30–40 days, and wintering ones up to 6–7 months (Eskov 1995). The maximum life expectancy of bees of spring-summer generations can reach 89 days, and 304 days for wintering. Some bees can survive up to 307–396 days in queenless colonies (Maurizio 1958). The concentration of carbohydrate food consumed by worker bees has a significant effect on their life expectancy. The increase in water content in it leads to a reduction in life expectancy (Eskov 1992, 1995).

Determination of gender. In bees, gender differentiation is mainly associated with a cyclic replacement of haplo- and diploidy. In most known species, females of bees develop from fertilized eggs and males from unfertilized eggs. If you do not take into account the mutation process, then the kinship coefficient between female-foundress and her daughters in a case of single pairing approaches 1:4, and in the case of pairing with two males (polyandry) increases up to 1:2, with a triple up to 5:12, and tenfold up to 3:10.

The probability of mating females with one or some more males is related to species-specific features, but may depend on the environmental situation. Polyandry is not widely spread among bumble bees. In most of their species, females mate with males at the end of summer and beginning of autumn. However, females in *B. hypnorum* may have a 2-3-fold pairing (Roseler 1973). The honey bee is characterized by polyandry. Queens of this species can mate during the mating season with 17 drones under favorable weather conditions, otherwise, they remain unfertilized.

In the case of arrhenotoky, common in groups of honey bees living on the Eurasian continent, colonies with unfertilized queens are eliminated. The greatest probability of survival without fertilized queens is in the colonies of the species characterized by thelytoky. It was first discovered in the bees of *A. mellifera capensis* inhabiting the Cape Province of Africa (Onions 1912). The colonies of these bees usually contain about 20% of individuals with enlarged ovarioles. Among these initially working bees, there are individuals that oviposition unfertilized eggs for 28–42 days, from which females hatch and develop. In cape bees, the thelytoky is stimulated by the loss of the ovulating female. If there is a thelytoky in the colony, the drones will develop from eggs ovipositioned by working individuals. These bees freely interbreed with European bee races. When crossing *A. mellifera capensis* with *A. m. carnica*, the thelytoky dominates in metis, and in the case of crossing with *A. m. ligustica*, this feature turns into a recessive state (Ruttner 1977).

The development of drones from fertilized eggs is discovered in the honey and Indian bees. Bees usually destroy such drones in the early stages of ontogeny. Nevertheless, sometimes they develop up to the imago stages and can occupy up to 30% of place in the cells using the working individuals' reproduction (Woyke 1980a). Adult diploid drones differ in appearance from haploid ones, having at the same time similarities with females and males (mosaic gynandromorphism). The increase in the size of certain parts of the body can be due to polyploidy (Woyke 1980b). Gynandromorphism in diploid *Melipona* drones is expressed in the presence of similarities with females in the structure of the exoskeleton, wings, and eyes (Kerr 1974).

The seasonal sequence of reproduction of sexually mature females and males. In one-year-old bee colonies, the sequence of reproduction of mature females and males, their time of flight from the nest, and mating is determined by the hereditary program acquired in the process of adaptation to typical living conditions. For example, in *Osmia lignaria*, females complete development and leave the nest before males, whose size decreases from the first to the last generations. This is caused by deterioration in trophic support (Torchio and Tepedino 1980). In bees of *Lasioglossum (Dialictus) umbripenne*, reproductive periods of sexually mature offspring are associated with the cycles of drought, but the seasonal cycle of development of colonies is completed by the reproduction of drones (Wille and Orozco 1960). For the second half of the summer, the reproduction of sexually mature offspring takes place in colonies of bumble bees living in temperate climate zones. In colonies of tropical species, uninterrupted reproduction of males and females happens throughout the year. There the female can live for more than one year. Nevertheless, new colonies are always based on the same female (Michener and Amir 1977).

In the colonies of the subsocial species, where the females are completely incapable of independent existence, the cyclicity of the rearing of sexually mature offspring is usually associated with certain phases of colony development. For example, in a young colony of *Trigona nebulata komiensis*, which contains about 200 individuals at the beginning of the development of a new nest, males are usually absent. Males appear in colonies in some periods of times when the number of individuals increases. In colonies, including 1–2 thousand adults, the representation of males may reach 7 percent (Roger 1969). In the colonies of the honey bee at the beginning of the active season, drones appear firstly among the sexually mature individuals, and only then queens. In addition, the reproduction of drones is characteristic of all or most of the overwintered bee colonies, and the queens are reproduced in small numbers only by those colonies where bees are preparing for sociotomy. In the bee colony that has found itself without a queen, the reproduction of queens is possible at any time of the annual cycle if there are developing larvae of working bees in the nest.

Polymorphism and polyethism. The development of functional differentiation (polyethism) in bee colonies is associated with an intensification of morph physiological differentiations (polymorphism) and the acquisition of functional specialization. However, functionally significant changes in species with a high level of social development were occurring only among females. They diverged on working individuals and ovulating females (queens).

Females. In the eo- and subsocial species, females-foundresses and their assistants almost or completely do not differ from each other in appearance. There are no differences in seasonal generations of females of Australian *Halictus* of the genus *Lasiglossum*, although they have developed collective nest protection and mutual construction activities (Knerer and Schwarz 1976). The ovulating females of the genus *Evylaeus* practically do not differ in size from the individuals performing the functions of workers (Knerer 1980).

The pronounced differentiation of females according to morphometric features, body weight, and nest behavior are traced in *Eulaema nigrita*. In these bees, small females remain in the nest and perform the function of working individuals, while relatively large ones fly away (Zucchi et al. 1969). There are similar situations in *Lasioglossum* (*Dialictus*) *umbripenne* (Wille and Orozco 1960), *Exoneurella eremophilia*, and *E. setosa* (Houston 1977), where the first generation of relatively small females participates in foraging, building, and nest protection, although many of them (about 50%) have developed ovaries, can mate and oviposition eggs in the presence of the female-foundress in the nest. Small females rarely participate in ovipositioning.

Reproduction of physiologically underdeveloped workers at the initial phases of the development of colonies is mainly due to a lack of trophic support in the larval stage. In most biological situations, underdeveloped females cannot mate, but this does not deprive them of the opportunity to assist the female-foundress in the reproduction of sexually mature offspring. The ratio of adult workers and developing individuals influence reproduction in the colony of sexually mature and underdeveloped females. Their representation decreases with the increase in the number of adults participating in the trophic supply of larvae. That can be clearly seen, for example, in annual bumble bee colonies (Povrean 1971).

The sequence of transitional forms reproduced in one-year colonies is a result of the differentiated supply of food to the larvae from the beginning to the completion of their development. However, their functional differentiation may not have a strict connection with the size of individuals. Regardless of the size in *B.* (*Fevidobombus*) *morio*, the specialization is expressed by the fact that one part of the individuals is engaged in foraging, the other performs intra-nest work. Depending on the biological situation along with them, some females can participate in the delivery of food, heating the brood, guarding the nest (Garofalo 1978). In *B. agrorum*, large bumble bees usually bring pollen and nectar into the nest, and small ones only bring nectar (Brian 1952).

Among eusocial species, the presence of transitional forms between the largest queens (Darchen and Delage-Darchen 1975) and working individuals is widespread in stingless bees. They have the sizes of working individuals, increasing, respectively, with the growth of their quantities in colonies, which is determined by the improvement of trophic brood breeding (Imperatriz-Fonseca 1976). Along with the size of the body, the queen differs from the working individuals in a complex of morph physiological features. The worker individuals are stronger in the eyes and mandibles. The specific features of worker bees include the presence of a specialized device for collecting and transporting pollen on their third pair of legs. The queen differs from the working individuals in the rather developed ovaries, as well as the structure of the abdominal nervous system. In the queen, the fourth to the seventh segments of the abdominal ganglia are displaced to the anterior part of the body, and in the workers to the rear (Darchen and Delage-Darchen 1975).

For a large set of characters, the queen and the worker of the *Apis* bees are different. The phenotype of females at the stage of the larva is modified by a diet. The larvae of workers in *A. mellifera* are supplied with royal jelly (the secret of the hypopharyngeal and mandibular glands of adult bees) only during the first two days, and the queens till the end of the larval stage. As a result, the queen at the beginning of the imaginal stage is approximately twice as big by mass than the working individuals. The queen differs from working bees by the absence of wax glands, morphological structures for collecting pollen, the eyes and the oral apparatus are less developed, but the reproductive organs are hypertrophied. European groups of honey bees have about 200 ovarioles in the ovary of the queens, workers of *A. mellifera ligustica* have 12 on average (Chaud-Netto and Bueno 1979), and *A. m. capensis* have 19.6 (Velthuis et al. 1971).

The presence of an underdeveloped reproductive system in worker bees allows them, under certain conditions, to participate in the reproduction of males (*A. m. mellifera*) or females (*A. m. capensis*). Ovulating working individuals in breaks between the ovipositioning can engage in the construction of honeycombs, feeding larvae, foraging, and other intra- and out-of-nest work.

Polyethism in colonies of eusocial species is not associated with strict morph physiological differentiation. Usually, the functions change with age. Among the bees of spring-summer generations, young workers aged between 3 and 12 days are usually involved in the brood rearing. At this time, they have actively functioning hypopharyngeal glands. In 11–13 days, the development of wax glands is intensified in bees, and this allows them to actively participate in the construction of honeycombs. In 17–21 days, bees start getting engaged in food delivery. Young foragers can participate in the protection of the home, localizing at the entrance in the flying hole.

The marked age differentiation of the performed function can be modified under the influence of various extreme factors. For example, in a colony consisting only of young bees, their participation in foraging can begin at 5–8-day-old. In contrast, the bees of the autumn generation, having lived for several months, participate in the brood rearing. So, the age dependence of polyethism is observed only in the presence of bees of different ages in the colony.

Males. Unlike females, males do not participate in the life support of their colonies. In seasonal colonies, they often leave their nests after reaching imago and puberty. The assumption of the presence in the colonies of some Australian *Halictus Lasioglossum* (*Chilalictus*) sp. males involved in the protection of the nest (Houston 1970) was not confirmed (Knerer and Schwarz 1978). These males, distinguished by the presence of an enlarged head and mandibles, resemble representatives of the termites. The emergence of such males in *Halictus* is caused by a lack of trophic support in the larval stage. These males do not leave their nests because their wings and vision are underdeveloped (Knerer and Schwarz 1976). Perhaps drones in the nests of *Melipona Schwarziana quadripunctata* can take part in the processing of nectar, which occurs during its consumption (Brian 1986).

The morph physiological and morphometric variability of drones is mainly associated with their trophic supply. The size of the cells is important here. For example, in *Halictus*, the enlargement of drones occurs in the case of development in cells for the reproduction of females, which are abundantly supplied with food (Knerer and Schwarz 1978). In *Melipona* the size of drones developing in queens cells increases by approximately one-third (Inperatriz-Fonseca 1976). In the honey bee, drones usually develop in enlarged cells. However, in some biological situations, drones sometimes develop in relatively small cells of a working type that leads to their underdevelopment—a 1.8-fold decrease in body weight (Eskov 1995).

In the nests of the honey bee, young drones come into the tactile contacts with the working individuals during the first 4–5 days of the imaginal stage, and receive food regurgitated from the crop from them. With age, the attractiveness of drones decreases for worker bees (Jaycox 1961). Being needed in their care, young drones are usually located in the center of the nest and old ones are on its periphery. The first flights of drones from the hive start at the age of 4–8 to 14 days, and mass flights during mating season from 16-day age (Park 1923, Butler 1969, Muszynska 1979). In bees inhabiting cold and temperate climates zones, the expulsion of drones from the nest occurs before the beginning of wintering.

Mechanisms of social consolidation

The dominance of ovulating females. Signs of dominance or pronounced mutual aggression are usually not observed in the colonial populations of bees, although each ovulating female cares mainly about her offspring. Conflicts are possible only when capturing other's cells, and this is often accompanied by the removal of other females' offspring from them. Competition for nesting sites becomes more acute with an increase in the population density of the colony. In the small communal nests of the *Halictus* of the *Conanthalictus dicksoni* and *C. conanthi*, one nest may be occupied simultaneously by 2–3 females similar in appearance (Rosen and Meginley 1976). Similarly, polymorphism and polyethism in the ovulating females of the *Lasioglossum* sp., co-participants in settling their settlements (Knerer and Schwarz 1976) were not detected.

In colonies of social types, females usually differentiate into dominant and subordinate. Dominant females are mainly engaged in the reproduction of males and other females, and subordinates are engaged in trophic support of the colony, construction, and protection of the nest. The means used by females to achieve a dominant position can be conditionally divided into ethological and physiological ones.

Ethological domination. The hierarchy of females in bee colonies evidently developed on the inequality of their trophic support in the larval stage. In eo- and subsocial one-year colonies of bees, the first generations of females are not adequately provided with food, which is reflected in their underdevelopment. Compared to the female founder, they are usually less in mass and poorer in the development of the reproductive organs. This is due to the fulfillment of the auxiliary work in the nest, although working individuals are capable of ovulation and often participate in the reproduction of males.

Possessing an advantage in strength, female founders and/or sexually mature females counteract the reproduction of offspring by underdeveloped females and working individuals. Large mature females, competing in the struggle for reproduction with underdeveloped females, destroy the eggs oviposited by them and use the freed cells to reproduce their own offspring. This limits the contribution to the population of underdeveloped females.

Among bees with different levels of social organization, it is a widespread phenomenon when dominant females eat eggs oviposited by working individuals. This behavior is known in the colonies of primitive *Halictus Lasioglossum zephyrum* (Michener and Brothers 1974), in more socially organized advanced bumble bees (Free 1970, Garofalo 1978), and in eusocial species of bees. In *Melipona*, the process of eating eggs oviposited by the working individuals by sexually mature ovulating females (queens) has become one of the acts of the realization of the reproductive instinct. Eggs of workers began to be used by the queens as a trophic substrate, which compensates for the costs of ovulation.

Another form of cannibalism developed in colonies of bees of the genus *Apis*. The number and age of larvae eaten by worker bees mostly depend on the productivity of the feed area used by bee colonies and their physiological status. Indian bees feed up to about 95% of the brood before sealing during periods of abundant planting of nectar and the production of pollen. At moderate productivity of the food area, the number of larvae that survive to the pupa stage decreases to 50%, and while fasting the reproduction of the working individuals is suspended, although the queen ovipositions eggs (Woyke 1976).

The activity of cannibalism in the honey bee is subject to seasonal variability. Seventy five to eighty percent of bees develop from the egg stage to the imago in spring, up to 80–90% in summer, and 50–75% in autumn. Usually, most workers destroy developing drones. The exception is colonies that have lost their queens. In such colonies, even in autumn, the consumption of developing drones is about 25% (Woyke 1977).

The eating of brood by Indian and honey bees is not related to the dominance of the queens. However, this is associated with maintaining the homeostasis of the colony and regulating its development in accordance with the environmental situation. The regulative role of cannibalism is expressed in the fact that worker bees, while eating part of the brood, provide an improved trophic supply to the remaining larvae and queen. In contrast, in the one-year-old dominance colonies, the female-foundress can delay the reproduction of males for a time and increase the reproduction of females.

Tactile contacts and trophallaxis. Singly living species of bees tend to have an indifferent attitude towards their offspring. The offspring of the female-foundress also often do not have contact with each other and with her. The origin of sociality is associated with the acquisition of the instinct of caring for developing offspring by the female-foundress, which can be expressed in its protection. Strengthening the connection of the female-foundress with her offspring is expressed in the periodic supply of larvae with food. The primitive methods of feeding larvae include regurgitation of the contents of crops on them, which occurs in certain species of *Ceratinas* that use common brood cells for offspring placing (Sakagami 1960). The *Ceratinas C. japonica* and *C. flavipes*, nestling in the stems of plants, are forced to destroy the barrier between the sealed cells to feed the larvae. Probably, the beginning of defecation of larvae is the meaningful signal for females. The female after unsealing the cell removes the excrement, and then seals it again (Sakagami and Maeta 1977).

Tactile interaction of female bumble bees with developing offspring often precedes the delivery of food to the brood cells. This is expressed by the fact that the adult individual firstly senses the larvae with antennae and only then supplies it with food (Sakagami and Zucchi 1965, Sakagami 1976). Nevertheless, this limits tactile interaction in the bumble bee colony. Specific forms of interaction of adult individuals are known in *B. agrorum*, where foragers that return to the nest can attack the passive individuals in it. Sometimes bumble bees in the nest attack foragers returning to the nest and lick the nectar regurgitated by them (Brian 1952). Usually, bumble bees consume pollen from the feed cells and very seldom take it directly from the feet (from pollen baskets) of foragers (Sakagami and Zucchi 1965).

By means of trophallaxis, females-foundresses of some species of carpenter bees living in South Africa interact with their offspring (Watmough 1974, Bonelli 1977). The female-foundress of *Xylocopa combusta* often enters into trophic contact with young adult individuals (Watmough 1974). In the *Lasioglossum zephyrum*, foragers returning to the nest firstly contact the female-foundress, and then release from their fodder burden. Their tactile interaction is usually maintained in the process of progressing to the feeding cell (Breed and Gamboa 1977).

The trophallaxis and tactile interaction in the colonies of bees of the genus *Apis* reach high order and perfection. They even have small groups of workers, isolated from the colony, that exchange food. The frequency of trophic contacts increases with the number of bees in the group. By radioactive labeling of feed obtained by a solitary-living bee, it is established that it is distributed in a group of 20 bees in 6 hours (Skirkevicius 1986). Even faster, but at different rates, food and other substances are distributed among different polyethical groups in the nest of the bee colony. The greatest activity of food exchange is in bee-foragers, and the smallest in those that perform intra-nesting work (Nixon and Ribbands 1952).

The amount of feed distributed during trophic contacts depends on the individual characteristics of the donor bees. In total, about 7% of bees transmit up to 90% of the contents of crops in the process of trophallaxis, about 50% from 50 to 60%, and about 40% within 5 percent. (Duo et al. 1975). The largest number of contents of crops workers (an average of 60%) transfer to queens (Pen et al. 1975).

Worker bees do not always supply drones with food. In the active period of the colony's life, approximately 10% of drones consume food from honey cells. In the process of trophallaxis with worker bees, only 50% of drones satisfy their nutritional needs. Moreover, in the process of trophic contact, the bee can give the drones up to 95% of the contents of the crop (Duo et al. 1975).

In the process of trophic and tactile contacts between members of the bee colony, pheromones that fulfill the regulatory function are distributed. At present, about 30 varieties of different pheromones are found in the honey bee, differing in the mechanisms and results of physiological effects on the organism of adults and developing bees. Royal jelly that is distributed by the bees in contact with the queen, has especially high activity and a wide spectrum of action. After contacting it, the bees become activated and within minutes enter into trophic or tactile contact with other bees (Seeley 1979, Ferguson and Free 1980). The initiators of contacts are usually bees who left the surroundings of the queen (Butler 1980, Skirkevicius 1986).

Acoustic and electrical communication signals. Honey bees communicate through a multichannel system implemented in complex instincts, where realization is stimulated by bees. The mechanisms of generation and perception of communication are multifunctional. In species phylogeny, acoustical and other forms of communication promoted consolidation and transformation of the bee colony into an evolutionary biological unit. The broad ecological valency of the honey bee is to a great extent associated with the development of an efficient, reliable communication system. The generation of specific sound signals is an additional function of the flight apparatus. This mechanism, based on the vibration of thoracic segments by indirect flight muscles, allows the bee to broadly vary the spectral structure of the signal as it is delivered (Eskov 1969). The body covers can accumulate an electrostatic charge, thereby generating an electric field.

Low-frequency acoustic and electric fields are perceived by fast-adapting *trichodes* sensilla located between the facetted eyes and the occipital suture (Fig. 1.5). The frequency, intensity, and duration of a signal causing hair vibrations are coded by the bipolar neuron of the receptor into the quantity and repetition rate of less frequent action potentials (Fig. 1.6). They are generated by the neuron only while the hair moves (deflects). The hair fixed in a deflected position does not give rise to action potentials and provides no response to electric or acoustic fields (Eskov 1975).

Intra-nest signaling includes acoustic communication via air and substrate. Signals that are mainly air-borne are received by *trichodes* sensilla; substrate vibrations are perceived by subgenual organs. The match of signal transmission and perception mechanism obviates transition of energy losses between different media, thus facilitating the communication.

The air channel is used in the communication of dancers with the bees that are mobilized. The relatively small intensity of dancer signals is apparently necessary and sufficient for reliable communication. This is further improved by adjusting the frequency-amplitude spectra of acoustic signals to avoid interference (Eskov 1972). Probably, the impact of air-borne vibrations on the phonoreceptor hairs is enhanced by the electric field created by the charged body of the dancer (Eskov 1974, 2018).

Signaling via substrate is typical only of queens competing in the period of sociotomy. Queens as other bees generate sounds by the flight apparatus, but differ in that during signaling, they press their vibrating thorax to the substrate (comb). These vibrations differentially affect the queens and the worker bees. Queens get excited and go up the intensity gradient to meet their rival, while workers freeze, which favors the unhindered approach of the queens. Furthermore, queens generate signals with major components in noise-free ranges (Eskov 2018).

Specific suppression of the motility of worker bees is caused by queen signals or their simulation, and by sinusoidal vibrations of the substrate. However, bees are

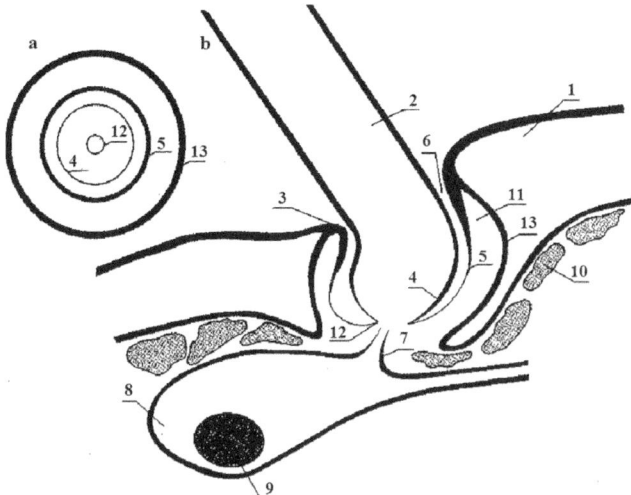

Figure 1.5: Microstructure of the *trichodes* fast-adapting sensilla of the honey bee: a–cross-section through the central part of the cuticular capsule; b–longitudinal cut (1: cuticle, 2: proximal part of hair, 3: neck, 4: base, 5: articular membrane, 6: entrance to cuticular capsule, 7: dendrite, 8: body of neuron, 9: nucleus, 10: accompanying cells, 11: capsule cavity, 12: intra-hair canal, 13: capsule wall).

markedly activated by briefly pulsed vibration, which are associated with invasion of burglars into the nest. Activation is attended with the intensification of sounds generated by adult bees; the duration of the acoustic reaction of the colony to vibro-stimulation depends on its physiological condition (Eskov 1992). This is intended to scare off animals that try to enter the nest.

The ability of bees to estimate the distance, memorize, and reproduce during the dance the coordinates of the target (food, water, new settlement) is unrelated to transfer of experience by training/learning. The entire system of communication in the honey bee colony is implemented in complex instincts, the realization of which is stimulated by the bees' needs that vary depending on physiological state and ecological situation. In species phylogeny, acoustical and other forms of communication promoted consolidation and transformation of the bee colony into an evolving biological unit.

The broad ecological valency of the honey bee, acquired in conjunction with the perfection of thermal adaptation, is in great measure associated with the development of an efficient, reliable, and economic communication system. Involvement of only separate bees in the search for and notification about food sources saves the colony energy that is spent on food provision. Natural selection also favored the acquisition of the instinct for finding resettlement sites and notifying the colony of their coordinates during sociotomy (Eskov and Toboev 2011). The energy expenditures of a group of nest scouts are negligible relative to losses that might be suffered if the completely moving colony had to look for a new home. The maximum spectral energy of vibration of honeycombs in the spring-summer period is in the range of 80–165 Hz, or 10–18 dB per octave.

The most intense low-frequency vibrations of honeycomb cells are observed under the influence of acoustic streams generated by fanning bees. The frequency

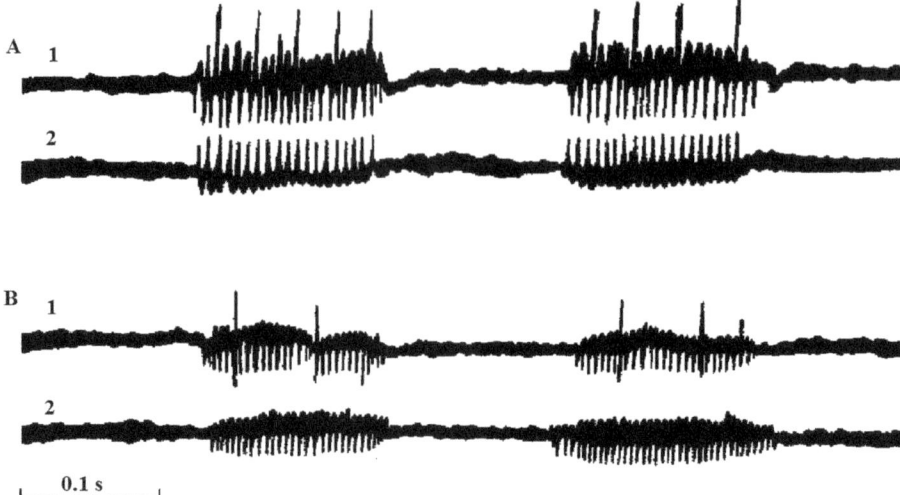

Figure 1.6: Electro-responses (A1, B1) of a rapidly adapting *trichodes* sensilla to sound pulses (A2, B2), differing in frequency and intensity.

and intensity of these vibrations increase as the temperature rises above the value that is optimal for the bee colony. During sociotomy, vibrations generated by conflicting queens are spread over honeycomb cells. High-frequency vibrations of relatively low intensity are generated by the mechanical impacts of adult individuals on the honeycomb cells.

Developing individuals make a small contribution to the vibration of honeycomb cells. Honeycomb vibrations with a disordered temporal structure occur when young bees gnaw through the honeycomb cell walls. Slight vibrations are generated by individuals at the larval stage of development. They periodically turn over in the honeycomb cells, which is accompanied by the generation of vibrations by the friction of the larval body against cell walls. Larval locomotion is stimulated by starvation. The intensity of the sounds generated by the older larvae near the surface reaches 12–17 dB, which is insufficient for their perception through the air. However, the honeycomb vibrations generated by the larvae rubbing against the cell walls can be perceived by the subgenual organs of adult bees, which may attract the nurse bees to the hungry larvae. A similar type of signal communication between larvae and adults is used in communications of paper wasps (Eskov 1979). They generate vibrations by rubbing their mandibles over the cell walls. There is a parallelism in the use of acoustic communication between larvae and adults in remote hymenopteran species-bees and wasps.

Conclusions

In the development of the strategy of nesting behavior, an important role belongs to the acquisition of the instinct responsible for offspring reproduction in individual cells. The possibility of differentiated food supply and regulation of the diet at the larval stage is associated with developing individuals. This excludes intra-species competition between developing individuals. Their morph physiological differentiation is mainly due to the inequality of trophic support, which can vary in weight and chemical composition.

In the process of development of building instincts, selection favored the preservation and improvement of those that ensured the increase of independence from unfavorable factors of the external environment. On this basis, eo- and sub-social species settling in the soil developed devices for constructing the waterproofing of brood cells, thus protecting the feed from waterlogging or drying solitary living, and social species of bees have adapted to use resinous plant extracts as a waterproofing material. Increasing the independence from the habitat was achieved by the development of morph physiological adaptations, which allow one to use their own secretions for waterproofing and/or building cells, among which waxes and wax-like substances have become widespread. In species that do not have specialized wax excretory organs, the use of epicuticle waxes has been obtained. The improvement of wax using mechanisms is associated with the development of waxing glands.

The process of the nesting structure of order was motivated by the need to enlarge the size of the nest that is associated with the increase in the number of members of the colony. In primitive social species with a relatively small number of bees, the brood cells do not have a strict ordering. A special disorder of the nesting design is typical for different types of bumble bees. They, like stingless bees, re-use the building material, which contributes to reducing energy costs for nest structures. However, this is not developed in the honey bee, who reached the highest level of social organization.

In a honey bee, honeycomb cells do not have strict differentiation into food stores and brood cells. This provides the mobility of using limited nesting space for the reproduction of working individuals and drones in the spring and summer periods, and for storing food in the freed brood cells. After the completion of the brood reproduction, food placed in the cells that occupy the most part of honeycombs is necessary for wintering bees. They do not feel the need to migrate in search of food when they aggregate under the influence of cooling in a limited part of the nest (in the food cells). Therefore, with the acquisition of the instinct of settlement in shelters, the construction of vertical two-sided honeycombs and the use of cells for the reproduction of brood and/or the storage of reserves at different times of the annual life cycle is related to the development of zones with a temperate and cold climate.

Social relations in bees are based on the development of the instinct of caring for the offspring of the female-foundress and her first generations that are physiologically underdeveloped females. The underdevelopment of females, causing their partial or complete elimination from reproduction, in eo- and subsocial species, occurs mainly in connection with the restriction of trophic supply of larvae, which can have a cyclic character under the influence of seasonal variability in the productivity of the food place. Underfeeding is their limited direct contribution to the population, to which the first generation of females is subjected. Nevertheless, they contribute to the female-foundress in the reproduction of her offspring, who, due to the abundant trophic support, reaches a normal sexual development.

The acquisition and improvement of the means of dominating of the female-foundress, who contributes to the consolidation of colony members, is associated with a decrease in the contribution to the population of its potential competitors, referred to as assistants. A widespread method of domination, based on eating the female-foundress of eggs ovipositioned by underdeveloped females, eliminates them from reproduction. Using this, the female-foundress or other mature female replenishes the reproduction costs. Physiological dominance, which reaches the highest perfection in eusocial species, is associated with the development of trophallaxis and tactile contacts. They ensure the distribution of pheromones, thereby increasing the interdependence among members of the bee colony.

Colony development in bees is associated with increased interconnectivity and interdependence between adults and developing individuals. Their progressive consolidation, which reaches the highest perfection in eusocial species, has facilitated the transformation of the colony into a biological unit that is individually subject to the action of natural selection and other factors of evolution. The genetic relationship of the offspring of ovulating females and its development under similar conditions, excluding intra-breeding competition, caused the emergence of colony selection. It involves the acquisition of a specific form of reproduction by means of sociotomy, which provides an increase in the number and expansion of the range occupied by eusocial species. With high life expectancy, which is characteristic for queens (Eskov 1975, Darchen and Delage-Darchen 1975), the replacement, and renewal of genotypes occurs in the process of sociotomy of bee colonies.

In the communication system of dancing bees and the bees mobilized by them, the situation required to stimulate foraging is insufficient. The motivation for potential foragers alone is also insufficient. That is why they react indifferently to the reproduction of sounds by model dancing bees. Acoustic coupling between a signaling bee and the bees mobilized by it is ensured when the bees attracted by the dancer accompany is at a distance necessary to ensure an acoustic contact. To obtain more information about the quality and smell of food, the bees that are mobilized by the dancer stop it and then come into trophic contact with it. This is achieved by the generation of a specific acoustic signal.

The ability of bees to assess the covered distance and memorize the coordinates of a motivated search target (a source of food and water or a place for settlement) and reproduce them in dance is not related to experience in the form of learning. The entire system of communications in a honey bee colony is represented by sophisticated instincts, the realization of which is stimulated by the needs of bees, which vary depending on their physiological state and environmental situation.

The acoustic interaction between the dancing bees and the bees mobilized by them developed during the transformation of a bee colony in an evolving biological unit. In this sense, a bee colony is sort of a complex multicellular organism, in which the mobile information of receptor organs is encoded in action potentials. In a bee colony, a similar function is played by the acoustic communication signals, which are used as an unconditioned reflex level. For this reason, bees are indifferent to such signals under conditions that are inadequate for their realization in a bee colony. The coding of acoustic information by a dancing bee into a sequence of acoustic pulses and their effect on the mobilized bees shows a convergent similarity with the coding of sensory stimuli perceived by receptors of organisms of different complexity into a sequence of action potentials.

A bee colony, which unites many biologically independent individuals, has a single regulatory center. The colony is consolidated by the hormones secreted by the queen due to the interaction of worker bees. They play the dominant role in an adequate response to changes in environmental conditions. The discrete structure of the frequency-amplitude spectrum of the acoustic noise produced by a bee colony is apparently generated by the processes of mutual adjustment of the sounds generated by different groups of bees in the similar state. This is a manifestation of the general consistent pattern of auto-synchronization of the processes occurring in the biological systems (Wiener 1963) that do not have a single regulatory center. However, the synchronization of acoustic processes in a bee colony helps to unite its members into a single biological system representing an evolving biological unit.

CHAPTER 2

Phylogenies of Asian Honey Bees

Raffiudin, R.* and *Shullia, N.I.*

Introduction

Numerous biological studies have been devoted to the diverse social Asian honey bees. Some species have vast distributions (i.e., *Apis cerana, Apis dorsata, Apis florea*, and *Apis andreniformis*), while others are endemic or native to small regions of Asia (i.e., *Apis nigrocincta, Apis dorsata binghami, Apis breviligula*, and *Apis nuluensis*) (Maa 1953, Roubik et al. 1985, Ruttner 1988, Hadisoesilo et al. 1995, Hadisoesilo and Otis 1996, Otis 1996, Tingek et al. 1996, Trung et al. 1996, Smith and Hagen 1996, Damus and Otis 1997, Engel 1999, Hadisoesilo et al. 1999, Smith et al. 2000, Sittipraneed et al. 2001, Tanaka et al. 2003, Suka and Tanaka 2005, Arias and Sheppard 2005, Raffiudin and Crozier 2007, Hadisoesilo et al. 2008, Lo et al. 2010, Hepburn and Radloff 2011a,b). This chapter focuses on phylogenetics of the Asian honey bee based on molecular studies, and adds to the recent literature (Cao et al. 2012, Fitriya et al. 2012, Zhao et al. 2014, Nagir et al. 2016, Takahashi et al. 2016, Eimanifar et al. 2017, Okuyama et al. 2017, Takahashi et al. 2017, Wakamiya et al. 2017, Takahashi et al. 2018, Santoso et al. 2018, Shinmura et al. 2018, Shullia et al. 2019, Yang et al. 2019). Inferences regarding the phylogenetics of the Asian honey bees have been discussed. The common mitochondrial genes used as phylogenetic markers discussed in this chapter are cytochrome oxidase subunits 1 and 2 (*COX1/COI* and *COX2/COII*, respectively), *NADH* dehydrogenase 2 (*ND2/nad2*), cytochrome b (*CYTB/CYTB*), and *16S* ribosomal RNA (*16S rRNA/ lsrRNA/rrnL*).

Department of Biology, Faculty of Mathematics and Natural Sciences, Institute Pertanian Bogor (IPB) University, Dramaga Campus, Bogor, 16680, Indonesia; nurulinsani64@gmail.com

* Corresponding author: rika.raffiudin@apps.ipb.ac.id

Phylogenetics based on molecular characteristics of inter- and intraspecies of Asian honey bees were extensively investigated in Smith (1991b) and Smith and Hagen (1996), including that of *A. cerana* in large areas of Asia. They range from mainland Asia (India, Thailand, and Malaysia) to the archipelagoes of Indonesia and the Philippines. A biogeographical analysis of honey bees using restriction fragment length polymorphisms of mitochondrial DNA was described by Smith (1991b). The phenograms revealed that the *A. cerana* intra-population in Borneo, Malaysia, Thailand, Japan, and India had a 0–1.86% range of sequence divergence, while *A. cerana* from the Luzon and Andaman islands had greater divergence rates from the mainland, ranging from 2.89 to 5.62 percent. The same pattern is found in the biogeography of *A. dorsata* on the mainland, except that Asian honey bees from Andaman Island cluster in the same clade as those from the mainland. Smith (1991b) mentioned that the shared haplotypes among *A. cerana* from Borneo, Malaysia, Thailand, Japan, and India were established during the late-middle Pleistocene Era (160,000 years ago). In this era, the sea level was 160–180 m lower than at present, thus the islands of the Sunda Shelf, including Borneo, Java, and Sumatra were joined to the mainland (Heaney 1986). However, the Andaman, Sulawesi, and Luzon islands were isolated by a deep-water channel; therefore, the isolations of populations in those regions diverged from related mainland populations. This condition also affected the distinction of *A. d. binghami* population in Sulawesi from mainland populations (Smith 1991b). *Apis andreniformis* does not occur on Sulawesi nor some other Philippine islands. As mentioned above, during the middle Pleistocene, the sea level was 160–180 m lower than at present; however, the channel between Borneo and Palawan today is 145 m; therefore, the Greater Palawan is connected to Borneo. Sulawesi is separated from Borneo by a deep-water channel in the Makassar Strait, and the Philippines from Palawan by the Sulu Sea. Thus, *A. andreniformis* only occurs in Palawan and not in other islands of the Philippines (Smith et al. 2000, Hepburn and Radloff 2011b). Furthermore, the speciation between *A. andreniformis* and *A. florea* supposedly began during the early Pleistocene, owing to the barrier of the arc in the Arakan that extends southward from near Manipur to southern Myanmar (Hepburn and Radloff 2011b).

The present distribution of the honey bee may have resulted from historical geological motions. The first fossil of a recognizable honey bee was linked to the Oligocene Epoch (Engel 1998). The distribution and speciation of the modern honey bee should be considered to have occurred in the Pleistocene (Smith 1991b, Smith et al. 2000, Hepburn and Radloff 2011b).

Phylogenetics of Asian honey bee interspecies

Nuclear and Mitochondrial Genes. For several decades, the genus *Apis* consisted of four species: *A. mellifera*, *A. cerana*, *A. dorsata*, and *A. florea* (Koeniger 1976b). Explorations in the 1980s added five more *Apis* species, namely, *A. andreniformis* (Wu and Kuang 1987), *A. koschevnikovi* (Tingek et al. 1988), *A. laboriosa* (Sakagami and Matsumura 1980, Underwood 1990), *A. nigrocincta* (Hadisoesilo et al. 1995), and *A. nuluensis* (Tingek et al. 1996).

Arias and Sheppard (2005) analyzed the complete Asian honey bee using *ND2* mitochondrial DNA and the intronic region of Elongation Factor 1-alpha (EF1-α) genes. Their study revealed that the honey bee clusters were in agreement with those of a morphological cladistics analysis by Alexander (1991). The clusters formed were as follows: the giant honey bees *A. dorsata, A. binghami,* and *A. laboriosa);* the dwarf honey bees *A. andreniformis* and *A. florea;* and the cavity-nesting bees *A. mellifera, A. cerana, A. koschevnikovi, A. nuluensis,* and *A. nigrocincta.* However, the cluster of Asian cavity-nesters was found to be paraphyletic. Raffiudin and Crozier (2007) constructed a honey bee molecular phylogeny of all presently known species using Bayesian consensus data from two mitochondrial genes (*COX2* and *16S rRNA*) and a nuclear gene, inositol-1.4.5 triphosphate receptor (*itpr*). The most probable honey bee evolutionary topology was revealed, as shown in Fig. 2.1. This topology was confirmed by the cladistics tree obtained by Alexander (1991) based on common morphological traits among species. The Asian open-nesting honey bees *A. dorsata* and *A. florea* formed a single clade separate from that of the *A. mellifera-*

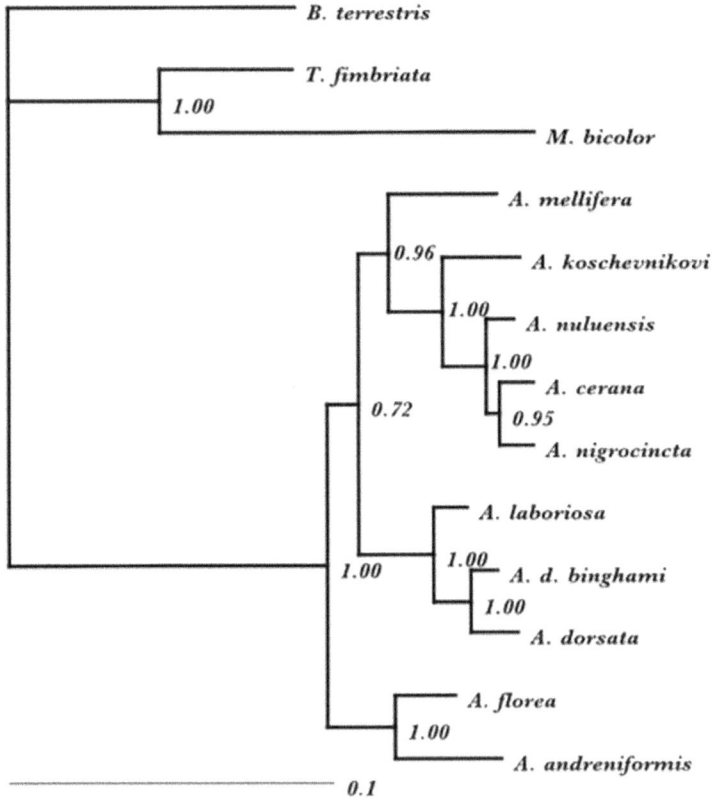

Figure 2.1: Honey bee phylogenetic tree inferred from Bayesian consensus data of two mitochondrial genes (*COX2* and *16S rRNA*) and a nuclear gene, inositol-1.4.5 trisphosphate receptor (*itpr*) (Raffiudin and Crozier 2007).

containing group, and dwarf honey bees are the basal group for the giant and cavity-nesting honey bees. In addition, previously unresolved correlations among the morphological data of the *A. dorsata* groups (Alexander 1991) were resolved.

Lo et al. (2010), using the same nuclear and mitochondrial genes as Raffiudin and Crozier (2007), further added the Philippines giant honey bee *A. breviligula* and the yellow Indian *Apis indica* to the re-constructed phylogenetic tree. Smith and Hagen (1996) and Smith et al. (2000) commenced the recognition of the yellow Indian *A. cerana* as different from Plains Indian *A. cerana* based on the non-coding regions between *COX1* and *COX2*. Lo et al. (2010) suggested that *A. indica*, the Plains Honey bee of south India, was a separate species from *A. cerana* (see further discussion in the section on *A. cerana* below). *Apis nuluensis* and *A. nigrocincta* are two described species of honey bees that are connected with *A. cerana* by a very short branch (Raffiudin and Crozier 2007). This finding is very interesting because both species have restricted distributions and are the results of recent speciation from *A. cerana*. Details of *A. nuluensis* and *A. nigrocincta* are discussed below in the sections on Bornean and Sulawesi honey bees, respectively.

Apis cerana: the greatest distribution and diversity level based on morphometric and molecular analyses

The eastern honey bee *A. cerana* occupies a vast area, ranging from western Afghanistan to the Philippines, and thus has the greatest number of subspecies among the Asian honey bees. *Apis cerana* is mostly distributed in western and north-eastern Asian countries (Afghanistan, Pakistan, Kashmir, China, and Korea) (Ruttner 1988). Moreover, *A. c. cerana* is also distributed in the southern-most areas of Russia (Ussuri krai) and south to northern Vietnam (Engel 1999). The subspecies *A. c. heimifeng* Engel can be found in central China at relatively high elevations and can be differentiated from *A. c. cerana* by the dark brown to black scutellum and metasomal tergum 3 and 4 (Engel 1999). Ruttner (1988) mentioned that *A. cerana* exists in the Himalayan region as the *A. cerana himalaya* race, which had been previously named by Maa (1944) as *A. c. skorikovi*. The name was replaced again by Engel (1999), becoming *A. c. skorikovi* Engel. *Apis cerana japonica* is distributed widely in Japan, except on Hokkaido Island (Hepburn et al. 2001). In south and south east Asia, *A. c. indica* (Maa 1953) is the main subspecies, inhabiting mainland India, Sri Lanka, Myanmar, Thailand, Malaysia, Sumatra, Java, Borneo, Lombok, Bali, Flores, and most of Sulawesi, Timor, and Sabah (Damus and Otis 1997). Engel (1999) described the *A. cerana* from Java as the Far East as Timor as the *A. c. javana* Enderlein subspecies and *A. cerana* from Sumatra Island was identified as *A. c. johni* Skorikov. The Philippine archipelago exhibits another subspecies *A. c. philippina* (Maa 1953). Another mountain honey bee is *A. nuluensis*, found at elevations of 1524–3400 m in two mountainous regions of Sabah, Mount Kinabalu, and the Crocker Range (Tingek et al. 1996). However, based on morphological characteristics, Engel (1999, 2012) recognized *A. nuluensis* as a subspecies of *A. cerana*.

The morphometric analysis of *A. cerana* interpopulations from southeast Asia and Sri Lanka was intensely studied by Damus and Otis (1997) and Radloff et al. (2010). Based on morphometric characteristics, the well-known subspecies *A. c. cerana* from Hong Kong and *A. c. japonica* from Japan were confirmed as different subspecies from others in southeastern Asia and Sri Lanka (Damus and Otis 1997). This finding also clustered at least three morphologically distinct *A. cerana* groups. The first group had a range including Sri Lanka, the Malay Peninsula, Sabah/Palawan, Java, Bali, Lombok, Flores, and central Sulawesi, while the second group ranged in South Sulawesi. Evidence that the South Sulawesi group was distinct from other Malaysian and Indonesian *A. cerana* indicated that it might have been isolated for a long period, accelerating the phenetic divergence (Damus and Otis 1997). The last group of southeast Asian *A. cerana* (Damus and Otis 1997) included samples from Timor. However, Engel (1999) identified the *A. cerana* in Timor as the same as those in Java, belonging to the group of *A. cerana javana*.

Radloff et al. (2010) performed an extensive study of Asian *A. cerana* population structure using morphological characteristics. They found homogeneous morphoclusters, and adding the local sample locations to the clusters revealed six morphoclusters, Northern, Himalayan, Indian plains, Indo-Chinese, Philippine, and Indo-Malayan *cerana*. The Northern *cerana* morphocluster extends from northern Afghanistan, Pakistan, and northwestern India across southern Tibet, northern Myanmar, China, and Korea to far eastern Russia, and Japan (Radloff et al. 2010). The Himalayan *cerana* occurs in northern India and includes northwest, north-east, and some of southern Tibet and Nepal. The Indian plains *cerana* is established in central and southern India and Sri Lanka and is known as *A. c. indica*. The Indo-Chinese *cerana* is located in Myanmar, northern Thailand, Laos, Cambodia, and southern Vietnam. The Philippine *cerana* morphocluster is restricted to the Philippines, whereas the neighboring morphocluster, Indo-Malayan *cerana*, occurs from southern Thailand, through Malaysia up to Indonesia (Radloff et al. 2010). The differentiation of the Philippine *cerana* from the Indo-Malayan *cerana* is supported by the fact that honey bees from Luzon, previously identified as *A. c. philippina*, are very distinct from *A. cerana* from the Indo-Malayan population (Damus and Otis 1997).

Highly diverse molecular analyses

Smith and Hagen (1996) pioneered Asian *A. cerana* molecular-based investigative approaches using the non-coding regions of mitochondrial *COX1* and *COX2* genes to reveal the population structure of this species. They analyzed *A. cerana* samples covering South Asia (Nepal, India, Sri Lanka, and the Andaman Islands), Southeast Asia (Thailand, Malaysia, Indonesia, and the Philippines), and East Asia (Hong Kong, Korea, and Japan). The evolving non-coding regions of *A. cerana* were divided into two major populations, the western group (Indian "plain bees", Sri Lanka, and the Andaman Islands) and the eastern group (South and Southeast Asia, plus India "hill bees"). Furthermore, within the eastern group, two haplotypes were created, the Sundaland group (Peninsular Malaysia, Borneo, Java, Bali, Lombok, Timor, and Flores) and the Philippine group (Luzon, Mindanao, and Sangihe). *Apis cerana* samples from Korea and the Philippines were added to the analysis (Smith et al. 2000), and revealed 41 haplotypes of non-coding regions from 153 colonies. All the haplotypes can be clustered into four major groups, Asian mainland, Sundaland, Palawan, and Luzon-Mindanao (Smith et al. 2000). The Luzon-Mindanao group is in agreement with the Damus and Otis morphometric analysis (1997), in which *A. cerana* from Luzon recognized as *A. c. philippina* was different from both *A. cerana* and *A. nigrocincta*.

In the works of Smith and Hagen (1996) and Smith et al. (2000), the Indian *A. cerana* had a high diversity of haplotypes that were distributed in both western (plain bees) and eastern (hill bees) haplotype groups based on non-coding regions of *COX1/COX2*. A further molecular analysis of *A. cerana* from the plains of India using datasets of two mitochondrial genes (*16S rRNA* and *COX2*) and one nuclear gene (*itpr*) employing Bayesian and maximum parsimony trees (Lo et al. 2010) revealed that the Plains Honey bee collected from Karnataka groups genetically diverged by 2.1% from mainland *A. cerana*. This genetic distance value is similar to the 2.2% between the two established species of *A. cerana* from Sabah Malaysia and *A. nigrocincta*. Therefore, Lo et al. (2010) suggested that *A. indica*, the Plains Honey bee of southern India, is a separate species from *A. cerana*.

To obtain the genetic differences among the "eastern group" defined by Smith and Hagen (1996) or the Northern *cerana* morphocluster derived from Radloff et al. (2010), during the last decade, complete mitogenomes of *A. cerana* from several locations were published, such as *A. cerana* from China (GenBank Acc. No. GQ162109, Tan et al. 2011), the Japanese honey bee *A. c. japonica* (GenBank Acc. No. AP017314, Takahashi et al. 2016), *A. cerana* from Sabah, Borneo (GenBank Acc. No. AP018149, Okuyama et al. 2017), and *A. cerana* from Taiwan (GenBank Acc. No. AP017983 and AP017984, Shinmura et al. 2018). The new findings based on the mitogenome of *A. cerana* from Korea (GenBank Acc. No. AP018431, Ilyasov et al. 2018) found that *A. cerana* from Korea has genetically diverged from the subspecies *A. c. japonica* and the Chinese *A. c. cerana*. Ilyasov et al. (2018) assessed *A. cerana* in Korea using a comparative analysis of these mitogenomes combined with two nuclear genes, *Vitellogenin* precursor (*VG*) and EF1-α, and six morphological characteristics, forewing length and width, cubital index, hind leg length, metatarsal index, and the length of metasomal tergum 3 plus 4. The comprehensive data analysis suggested that *A. cerana* from Korea is a different subspecies compared with *A. c. cerana* from China and *A. c. japonica* from Japan, because of the genetic divergences of 2.57% and 2.58%, respectively (Ilyasov et al. 2018). Those genetic divergences matched those of mtDNA between animals at the subspecies level (0.8%–8.0%), thus *A. cerana* from Korea could be named *Apis cerana koreana*.

In *A. cerana* from China, Tan et al. (2007) revealed nine haplotypes of the CO1/*COX2* non-coding region, which increased the known haplotypes on the Asian mainland, and they were clustered in the Asian mainland group of Smith et al. (2000). In a separate study, Zhao et al. (2014) investigated the phylogeography of the *A. cerana* population in China using the *COX1* gene, and they found 57 new haplotypes from the Chinese mainland and Hainan Island, excluding the 21 previous haplotypes of the GenBank database's sequence. A phylogenetic tree using 78 haplotypes divides the *A. cerana* population into two main group lineages: A and B (Zhao et al. 2014). The A lineage consists of the Indo-China + Hainan + Japan-Korea-Russia sub-lineage + Taiwan sub-lineages, and the B lineage B consists of Indonesia + Indo-Malayan sub-lineages.

The diversity of *A. cerana* mtDNA genes in the Indo-China region has been well explored. Using the same non-coding region as Smith and Hagen (1996) and Smith et al. (2000), six haplotypes were observed in Burma (Smith et al. 2004). The haplotypes of *A. cerana* from the northern and southwestern regions of Burma also clustered in the group from the Asian mainland, while, interestingly, the haplotype form from southeastern Burma clustered in the Sundaland group (Smith et al. 2004).

Using the *16S* RNA gene, Sittipraneed et al. (2001) found at least four haplotypes of *A. cerana* in Thailand. Those haplotypes were grouped based on the sample locations, the north-to-central region (haplotype A), the peninsular Thailand and Phuket region (haplotype B), and Samui Island (haplotype C). Further, Songram et al. (2006), using another mtDNA gene (*ATPase6–8*), consistently found three region-based haplotypes of *A. cerana* populations from northern Thailand, peninsular Thailand, and Samui Island.

Smith (2011) mentioned that more *A. cerana* samples from Sundaland are needed to determine whether there are variations among these honey bees on the three main islands: Sumatra, Java, and Kalimantan. In an attempt to further explore the intraspecies variations of *A. cerana* in Sundaland, the phylogenetic tree, inferred from *COX1* using current and previous *A. cerana COX1* sequences from Indonesia and Malaysia, was constructed (Tanaka et al. 2001a,b, Tanaka et al. 2003, Okuyama et al. 2017) (Fig. 2.2). The phylogenetic tree inferred from a single *COX1* gene clearly separated into two main clades, the Indo-Malayan lineage containing Sabah, Sarawak, and South-East-West Kalimantan, and the Indonesian lineage, containing East-West Java, Bali, Lombok, South Kalimantan, and Central Sulawesi (Fig. 2.2). Most of the variations in *A. cerana* among the two clades and within each clade are derived from the third codon position of the *COX1* gene (Fig. 2.3). Thus, the *COX1* gene result confirmed the *A. cerana* morphometric clusters of Damus and Otis (1997) and Radloff et al. (2010).

The *COX1* gene has been used as a DNA barcode to determine interspecies in animals (Hebert et al. 2003). Using the current *COX1* for the intraspecies phylogenetic analysis of *A. cerana* revealed a high mutation rate that allowed the differentiation of the *A. cerana* from Kalimantan (except South Kalimantan) from those in Java. The genetic distances among *A. cerana* in Sabah, Sarawak, Kinabalu, and East and West Kalimantan are in the 0.000–0.009 range (Table 2.1). Moreover, they are genetically different populations than those of *A. cerana* from Bali, and West and East Java, having genetic differences in the 0.035–0.042 range (Table 2.1). Although in the Pleistocene Era, Sumatra, Kalimantan, and Java were united owing to the lower sea level, compared with the situation at present (Heaney 1986), and high levels of variation occurred among *A. cerana* in these regions. Thus, these genetic variations may have occurred during the separation of the islands. With this

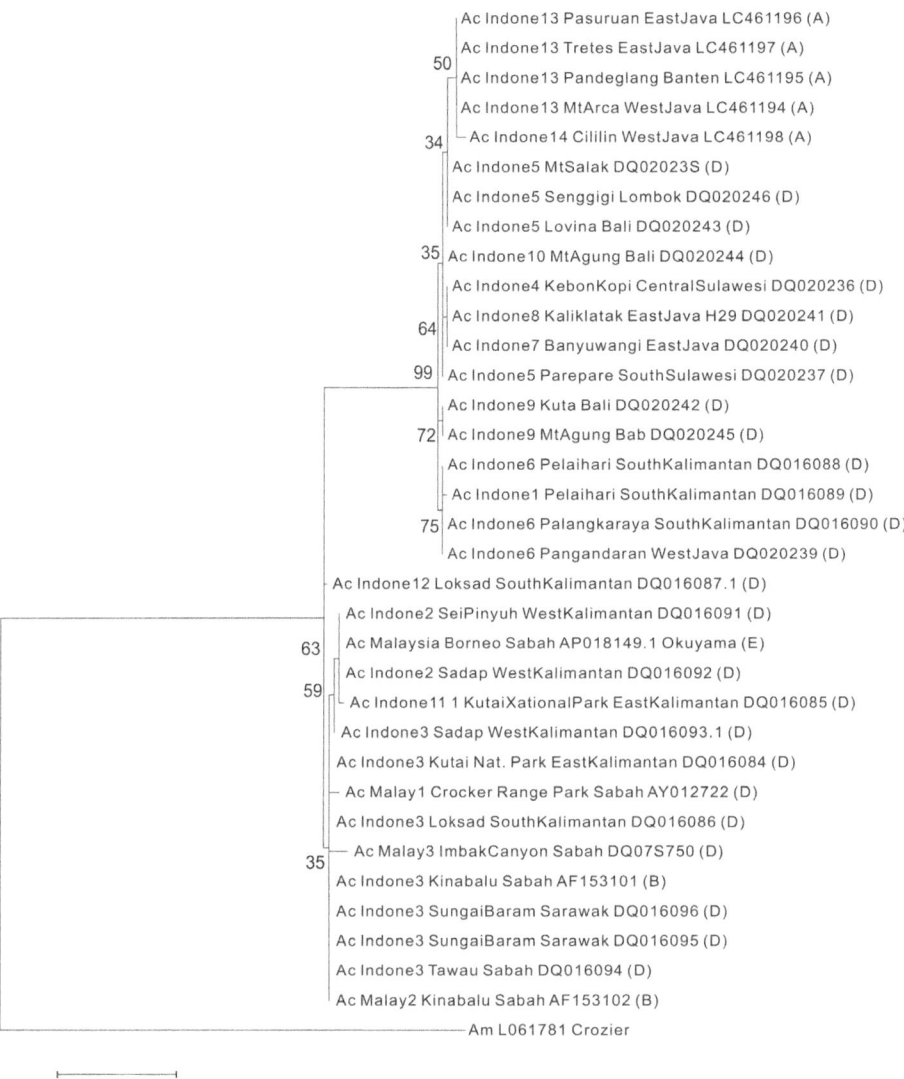

Ac Indone13 Pasuruan EastJava LC461196 (A)
Ac Indone13 Tretes EastJava LC461197 (A)
50
Ac Indone13 Pandeglang Banten LC461195 (A)
Ac Indone13 MtArca WestJava LC461194 (A)
34
Ac Indone14 Cililin WestJava LC461198 (A)
Ac Indone5 MtSalak DQ02023S (D)
Ac Indone5 Senggigi Lombok DQ020246 (D)
Ac Indone5 Lovina Bali DQ020243 (D)
35 Ac Indone10 MtAgung Bali DQ020244 (D)
Ac Indone4 KebonKopi CentralSulawesi DQ020236 (D)
64
Ac Indone8 Kaliklatak EastJava H29 DQ020241 (D)
Ac Indone7 Banyuwangi EastJava DQ020240 (D)
99 Ac Indone5 Parepare SouthSulawesi DQ020237 (D)
Ac Indone9 Kuta Bali DQ020242 (D)
72 Ac Indone9 MtAgung Bab DQ020245 (D)
Ac Indone6 Pelaihari SouthKalimantan DQ016088 (D)
Ac Indone1 Pelaihari SouthKalimantan DQ016089 (D)
75 Ac Indone6 Palangkaraya SouthKalimantan DQ016090 (D)
Ac Indone6 Pangandaran WestJava DQ020239 (D)
Ac Indone12 Loksad SouthKalimantan DQ016087.1 (D)
Ac Indone2 SeiPinyuh WestKalimantan DQ016091 (D)
63 Ac Malaysia Borneo Sabah AP018149.1 Okuyama (E)
Ac Indone2 Sadap WestKalimantan DQ016092 (D)
59
Ac Indone11 1 KutaiXationalPark EastKalimantan DQ016085 (D)
Ac Indone3 Sadap WestKalimantan DQ016093.1 (D)
Ac Indone3 Kutai Nat. Park EastKalimantan DQ016084 (D)
Ac Malay1 Crocker Range Park Sabah AY012722 (D)
Ac Indone3 Loksad SouthKalimantan DQ016086 (D)
35 Ac Malay3 ImbakCanyon Sabah DQ07S750 (D)
Ac Indone3 Kinabalu Sabah AF153101 (B)
Ac Indone3 SungaiBaram Sarawak DQ016096 (D)
Ac Indone3 SungaiBaram Sarawak DQ016095 (D)
Ac Indone3 Tawau Sabah DQ016094 (D)
Ac Malay2 Kinabalu Sabah AF153102 (B)
Am L061781 Crozier

0.05

Figure 2.2: Phylogenetic tree of *Apis cerana* generated based on the *COX1* gene, showing the relationships of samples from Java with samples from other locations in Indo-Malaya using the Maximum-likelihood method with 1,000 bootstraps. Information for OTUs are as follows: GenBank accession numbers, letter in bracket indicates sample from: A = new sequences from current study, B = Tanaka et al. (2001a), C = Tanaka et al. (2001b), D = Tanaka et al. (2003), and E = Okuyama et al. (2017).

data, one can presumably find more variations among *A. cerana* from Lesser Sunda (Lombok, Sumbawa, Sumba, and Flores). Owing to the deep strait separating Bali and Lombok, an isolation barrier in this Wallacea region was created.

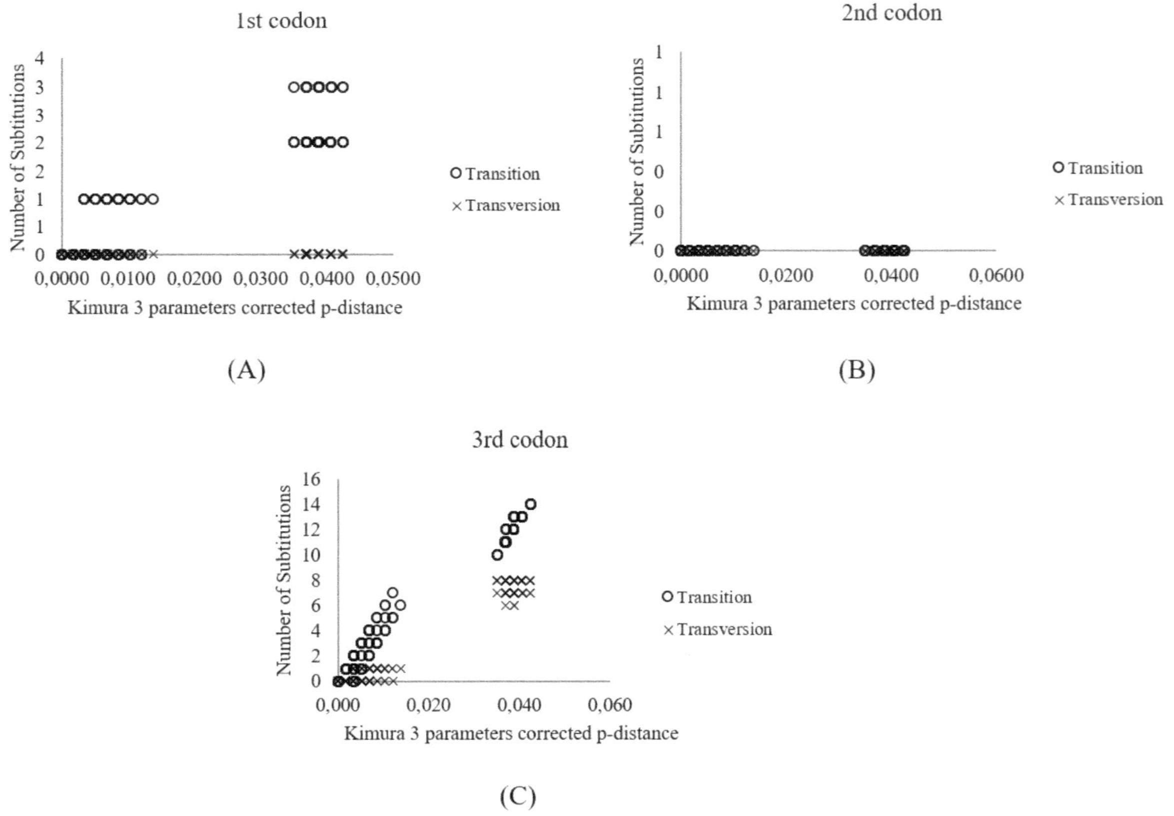

Figure 2.3: Mutational rate of the Indo-Malayan *Apis cerana COX1* gene. (A) 1st codon, (B) 2nd codon, and (C) 3rd codon position. Sample descriptions refer to Figure 2.2.

Table 2.1: Genetic distances based on the *COXI* gene of *Apis cerana* (Ac) from Java and Indo-Malayan. Letter in bracket indicates sample from: A = current study, B = Tanaka et al. (2001a), C = Tanaka et al. (2001b), D = Tanaka et al. (2003), and E = Okuyama et al. (2017). Sample descriptions refer to Figure 2.2.

No.	1*	2	3	4	5	6	7	8	9	10	11	12	13	14	15	16	17	18	19	20	21
1																					
2	0.000																				
3	0.000	0.000																			
4	0.003	0.003	0.003																		
5	0.003	0.003	0.003	0.007																	
6	0.003	0.003	0.003	0.007	0.007																
7	0.003	0.003	0.003	0.009	0.009	0.002															
8	0.005	0.002	0.002	0.005	0.005	0.002	0.003														
9	0.002	0.002	0.000	0.003	0.003	0.003	0.005	0.002													
10	0.000	0.000	0.007	0.010	0.010	0.010	0.012	0.009	0.007												
11	0.007	0.007	0.039	0.041	0.039	0.039	0.041	0.037	0.039	0.042											
12	0.039	0.039	0.039	0.041	0.039	0.039	0.041	0.037	0.039	0.042	0.000										
13	0.039	0.039	0.039	0.041	0.039	0.039	0.041	0.037	0.039	0.042	0.003	0.003									
14	0.039	0.039	0.039	0.041	0.039	0.039	0.041	0.037	0.039	0.042	0.003	0.003	0.000								
15	0.037	0.037	0.037	0.039	0.037	0.037	0.039	0.035	0.037	0.041	0.005	0.005	0.005	0.005							
16	0.039	0.039	0.039	0.041	0.039	0.039	0.041	0.037	0.039	0.042	0.007	0.007	0.007	0.007	0.005						
17	0.037	0.037	0.037	0.039	0.037	0.037	0.039	0.035	0.037	0.041	0.005	0.005	0.005	0.005	0.003	0.002					
18	0.037	0.037	0.037	0.039	0.037	0.037	0.039	0.035	0.037	0.041	0.002	0.002	0.002	0.002	0.003	0.005	0.003				

Table 2.1 contd...

No.	1*	2	3	4	5	6	7	8	9	10	11	12	13	14	15	16	17	18	19	20	21
19	0.037	0.037	0.037	0.039	0.037	0.037	0.039	0.035	0.037	0.041	0.002	0.002	0.002	0.002	0.003	0.005	0.003	0.000			
20	0.039	0.039	0.039	0.041	0.039	0.039	0.041	0.037	0.039	0.042	0.003	0.003	0.007	0.007	0.009	0.010	0.009	0.005	0.005		
21	0.039	0.039	0.039	0.041	0.039	0.039	0.041	0.037	0.039	0.042	0.003	0.003	0.007	0.007	0.009	0.010	0.009	0.005	0.005	0.000	
22	0.037	0.037	0.037	0.039	0.037	0.037	0.039	0.035	0.037	0.041	0.007	0.007	0.010	0.010	0.012	0.014	0.012	0.009	0.009	0.003	0.003

* 1. Ac Indone3 Kutai East Kalimantan DQ016084 (D), 2. Ac Indone3 Kutai East Kalimantan DQ016085 (D), 3. Ac Indone3 Sungai Baram Sarawak DQ016095 (D), 4. Ac Malay1 Crocker Range Park Sabah AY012722 (C), 5. Ac Indone12 Loksad South Kalimantan DQ016087.1 (D), 6. Ac Indone2 Sadap West Kalimantan DQ016092 (D), 7. Ac Indone11 Kutai East Kalimantan DQ016085 (D), 8. Ac Indone3 Sadap West Kalimantan DQ016093.1 (D), 9. Ac Malay2 Kinabalu Sabah AF153102 (B), 10. Ac Malay3 Imbak Canyon Sabah DQ078750 (D), 11. Ac Indone5 Lovina Bali DQ020243 (D), 12. Ac Indone5 Mt Salak DQ020238 (D), 13. Ac Indone7 Banyuwangi East Java DQ020240 (D), 14. Ac Indone4 Kebon Kopi Central Sulawesi DQ020236 (D), 15. Ac Indone9 Mt Agung Bali DQ020245 (D), 16. Ac Indone1 Pelaihari South Kalimantan DQ016089 (D), 17. Ac Indone6 Pangandaran West Java DQ020239 (D), 18. Ac Indone5 Parepare South Sulawesi DQ020237 (D), 19. Ac Indone10 Mt Agung Bali DQ020244 (D), 20. Ac Indone13 Mt Arca West Java LC461194 (A), 21. Ac Indone13 Pasuruan East Java LC461196 (A), 22. Ac Indone14 Cililin West Java LC461198 (A).

The Bornean honey bees

Apis nuluensis. Apis cerana and *A. nuluensis* are sympatric (Tingek et al. 1996), with the distribution of *A. nuluensis* being in the higher elevations (up to 2,000 m) of the Kinabalu and Crocker mountainous regions of Sabah, Borneo, while *A. cerana* is not commonly found at heights greater than 1,500 m above sea level. However, they overlap at 1,500–1,700 m (Koeniger et al. 1996). Arias et al. (1996) analyzed the molecular relationship of the Bornean honey bees—*A. cerana, A. nuluensis,* and *A. koschevnikovi.* A phylogenetic tree constructed using EF1-α and *ND2* genes showed that *A. nuluensis* clustered with *A. cerana.* This indicated that *A. nuluensis* appeared more recently and is a sister species of *A. cerana.* Tanaka et al. (2001a) also determined that *A. nuluensis* is closely related to *A. cerana* based on three mitochondrial genes, *16S* RNA, *COX1*, and *COX2*, and the result corroborated those from Eimanifar et al. (2017). However, when using the whole mitogenome to infer Asian honey bee evolution, Takahashi et al. (2018) found that *A. nuluensis* is not a sister species of *A. cerana.* They showed that *A. nigrocincta* is a sister species of *A. cerana,* which corroborated the results of Raffiudin and Crozier (2007) and Lo et al. (2010).

Furthermore, Takahashi et al. (2002) used the same region as Smith and Hagen (1996), the non-coding *tRNA* leu-*COX2* intergenic region, to analyze the nucleotide variations of the cavity-nesting honey bees, *A. cerana, A. koschevnikovi,* and *A. nuluensis,* in Sabah, Borneo. They reported the first short sequence from Borneo as a new haplotype of *A. nuluensis,* "Sabah short", which consists of 11 nucleotides in the middle of two 11-nucleotide-long stems.

Apis koschevnikovi. Apis cerana and *A. koschevnikovi* also live sympatrically over much of Sundaland, including some regions of Sumatra, the Malay Peninsula, Borneo, and Java (Rinderer et al. 1989, Otis 1996). Recent observations revealed that *A. koschevnikovi* is now rarely seen, except in Borneo, owing to the destruction of its forest habitat (Otis 1991, Hadisoesilo et al. 2008). This Bornean Sabah "Red honey bee" (Ruttner et al. 1989) can be found at elevations from sea level to 1,000 m (Otis 1996).

Otis (1996), based on collections in various museums, found that *A. koschevnikovi* is widely distributed in the Sundaland region of Southeast Asia. The population of this species has likely declined considerably owing to deforestation and conversion of forests to tea, oil palm, rubber, and coconut plantations. Based on our observations, *A. koschevnikovi* lives primarily in forests with densely covered canopies (R. Raffiudin, personal observation). In contrast, *A. cerana* that lives sympatrically with *A. koschevnikovi* is able to live in open, urban, and disturbed areas. Hadisoesilo et al. (2008) agreed that *A. koschevnikovi*'s habitat is the evergreen rain forests of Sundaland.

In 2001, an extensive phylogenetic study of *A. koschevnikovi* was undertaken by Tanaka et al. (2001a), and continued until 2005 (Tanaka et al. 2001b, 2003, Suka and Tanaka 2005). They analyzed the haplotype of *COX1* in *A. koschevnikovi* and compared it to other Bornean honey bees from Indonesia, Malaysia, and Brunei. In total, 15 haplotypes of the *COX1* gene were found in the Bornean *A. koschevnikovi* population, and it was concluded that there were three *A. koschevnikovi* mitochondrial lineages (Suka and Tanaka 2005).

According to our study in South Kalimantan, *A. koschevnikovi* can be found in several districts (Fig. 2.4). In total, 29 colonies of *A. koschevnikovi* were found in five districts, Hulu Sungai Selatan, Hulu Sungai Timur, Balangan, Tanah Bumbu, and Kota Baru, of South Kalimantan, Indonesia (Fig. 2.4). The first three locations are located in the primary forest of Meratus Mountain, which provides a natural evergreen habitat for this species. A high-altitude population *A. koschevnikovi* was found on Laut Island and Sungai Bali Island in Kota Baru, which has primary forests. The 685-bp *COX1* genes of *A. koschevnikovi* (Suka and Tanaka 2005) from the North Kalimantan regions appeared in cluster 1 of the South Kalimantan *A. koschevnikovi* and five new haplotypes (HT16-20) were discovered (Fig. 2.5). This finding added to the 15 haplotypes of Tanaka et al. (2001a, 2001b, 2003) and Suka and Tanaka (2005); therefore, HT1-20 represent the haplotypes. There are two other clusters, 2 and 3 (Fig. 2.5). Thus, the existence of three mitochondrial lineages, as described by Tanaka et al. (2003) and later by Suka and Tanaka (2005), was confirmed.

The genetic distance between the lineage of cluster 1, which contains the current new *A. koschevnikovi* haplotypes from South Kalimantan (HT16-20), and other haplotypes from Borneo ranged from 0.001 to 0.009 (Table 2.2). This lineage, combined with cluster 2 of *A. koschevnikovi* from Lamibir in West Kalimantan (HT2) and Sadap in Sarawak (HT10), formed a monophyletic clade with a 100 bootstrap value, and the genetic distances of two clusters ranged from 0.15 to 0.023.

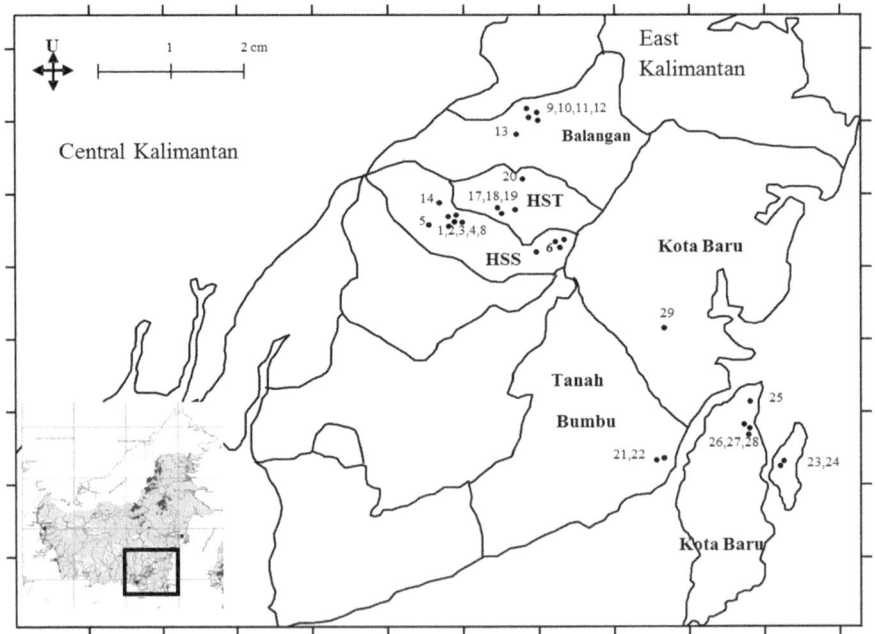

Figure 2.4: *Apis koschevnikovi* sampling sites in South Kalimantan (current study).

Cluster 3 contains two distinctive haplotypes from Crocker Range Park, types 1 and 2 (Fig. 2.5) (Tanaka et al. 2001b, Suka and Tanaka 2005) that have high genetic distances, ranging from 0.067 to 0.075 (Table 2.2). The high genetic distance between *A. koschevnikovi* populations in Borneo may be a result of their restricted habitats in the limited humid tropical forest area in the Sundaland, and the speciation in the Cenozoic Era. Although there were three haplotypes from Crocker Range Park that grouped in cluster 1, evidence that the Crocker Range Park has two additional haplotypes indicates that this forest was isolated for a long period at the geological time scale and that it produced an important habitat for *A. koschevnikovi* diversity (Suka and Tanaka 2005).

Among the Bornean cavity-nesting honey bees, Arias et al. (1996) found that *A. koschevnikovi* is the most distant. Based on the phylogenetic tree of Tanaka

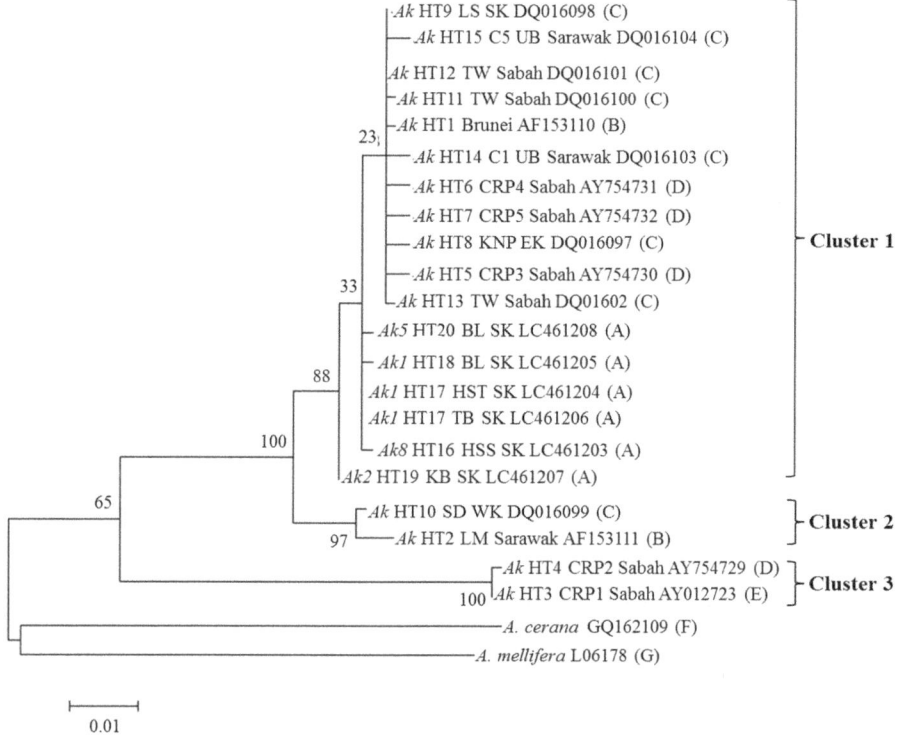

Figure 2.5: Phylogenetic tree of *Apis koschevnikovi* (Ak) based on the *COX1* gene showing the relationship of current study samples from several locations in South Kalimantan (SK) with other samples from Borneo generated using the Maximum-likelihood approach with 1,000 bootstraps. Information following the species names are as follows: colony and individual bee number, HT: haplotype number, Location: South Kalimantan (SK): BL = Balangan, KB = Kota Baru, LS = Loksad, HSS = Hulu Sungai Selatan, and HST = Hulu Sungai Tengah; East Kalimantan (EK): KNP = Kutai National Park; West Kalimantan (WK): SD = Sadap; Sabah: CRP = Crocker Range Park, TW = Tawau; Sarawak: LM = Lamibir, UB = Upper Baram and Brunei, followed by GenBank accession numbers. Letters in brackets indicate samples from: A = current study, B = Tanaka et al. (2001a), C = Tanaka et al. (2003), D = Suka and Tanaka (2005), E = Tanaka et al. (2001b), F = Tan et al. (2011), and G = Crozier and Crozier (1993).

Table 2.2: Genetic distances based on the $COX2$ gene of *Apis koschevnikovi* (Ak) from Borneo Island. Letter in bracket indicates sample from: A = current study of South Kalimantan, B = Tanaka et al. (2001a), C = Tanaka et al. (2003), D = Suka and Tanaka (2005), E = Tanaka et al. (2001b) samples. Sampling locations refer to Figure 2.5.

No.	1*	2	3	4	5	6	7	8	9	10	11	12	13	14	15	16	17	18	19	20
1																				
2	0.001																			
3	0.071	0.070																		
4	0.073	0.071	0.001																	
5	0.070	0.068	0.001	0.003																
6	0.070	0.068	0.001	0.003	0.003															
7	0.072	0.070	0.003	0.004	0.004	0.004														
8	0.072	0.070	0.003	0.004	0.004	0.004	0.006													
9	0.073	0.072	0.004	0.006	0.006	0.006	0.007	0.001												
10	0.068	0.067	0.006	0.007	0.007	0.007	0.009	0.003	0.004											
11	0.071	0.070	0.006	0.007	0.007	0.007	0.009	0.003	0.004	0.006										
12	0.070	0.068	0.004	0.006	0.006	0.003	0.007	0.001	0.003	0.004	0.004									
13	0.073	0.072	0.004	0.006	0.006	0.006	0.007	0.001	0.003	0.004	0.004	0.003								
14	0.072	0.070	0.003	0.004	0.004	0.004	0.006	0.000	0.001	0.003	0.003	0.001	0.001							
15	0.071	0.070	0.006	0.007	0.007	0.007	0.009	0.003	0.004	0.006	0.006	0.004	0.004	0.003						
16	0.075	0.073	0.006	0.007	0.007	0.007	0.009	0.003	0.004	0.006	0.006	0.004	0.004	0.003	0.006					
17	0.075	0.073	0.006	0.007	0.007	0.007	0.009	0.003	0.004	0.006	0.006	0.004	0.004	0.003	0.006	0.006				

18	0.073	0.071	0.006	0.007	0.007	0.009	0.003	0.004	0.006	0.004	0.004	0.004	0.003	0.006	0.006	0.006
19	0.070	0.068	0.018	0.019	0.015	0.015	0.016	0.018	0.015	0.016	0.016	0.015	0.015	0.018	0.018	0.018
20	0.073	0.072	0.021	0.023	0.018	0.018	0.019	0.021	0.018	0.020	0.019	0.018	0.018	0.021	0.021	0.006
21	0.104	0.106	0.091	0.093	0.090	0.088	0.088	0.089	0.091	0.086	0.088	0.088	0.086	0.089	0.088	0.091

* 1. Ak HT4 CRP2 Sabah AY754729 (D), 2. Ak HT3 CRP1 Sabah AY012723 (E), 3. Ak1 HT17 HST SK LC461204 (A), 4. Ak8 HT16 HSS SK LC461203 (A), 5. Ak1 HT18 BL SK LC461205 (A), 6. Ak5 HT20 BL SK LC461208 (A), 7. Ak2 HT19 KB SK LC461207 (A), 8. Ak HT12 TW Sabah DQ016101 (C), 9. Ak HT11 TW Sabah DQ016100 (C), 10. Ak HT5 CRP3 Sabah AY754730 (D), 11. Ak HT6 CRP4 Sabah AY754731 (D), 12. Ak HT13 TW Sabah DQ016102 (C), 13. Ak HT1 Brunei AF153110 (B), 14. Ak HT9 LS SK DQ016098 (C), 15. Ak HT8 KNP EK DQ016097 (C), 16. Ak HT15 UBC5 Sarawak DQ016104 (C), 17. Ak HT14 TW Sabah DQ016103 (C), 18. Ak HT7 CRP5 Sabah AY754732 (D), 19. Ak HT10 SD WK DQ016099 (C), 20. Ak HT2 LM Sarawak AF153111 (B), 21. *A. cerana* GQ162109 (F).

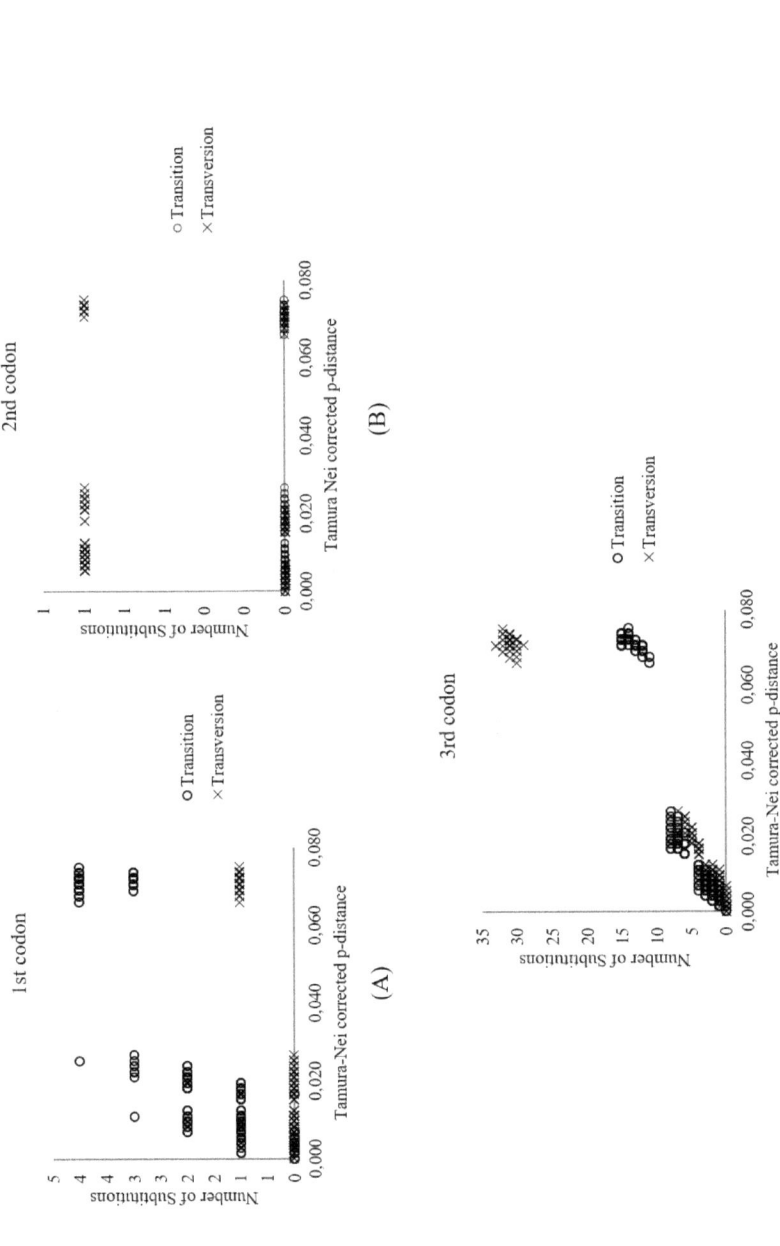

Figure 2.6: Numbers of transition and transversion events in the *COXI* gene of *Apis koschevnikovi* at each codon position, with respect to the p-distance, corrected using Tamura-Nei methods. (A) 1st codon, (B) 2nd codon, and (C) 3rd codon.

et al. (2001b), *A. koschevnikovi* was considered the basal species among Asian cavity-nesting bees, and this species was predicted to have occurred in Southeast Asia during the warmer Cenozoic Era's Tertiary period. As seen in Fig. 2.1 and other analyses of phylogenetic relationships among intraspecies in honey bees using the *ND2* mitochondrial gene and EF1-α intron (Arias and Sheppard 2005), *A. koschevnikovi* is consistently the basal species of the cavity-nesting honey bees *A. cerana*, *A. nigrocincta*, and *A. nuluensis*. The recently completed mitochondrial genome of *A. koschevnikovi* confirmed that *A. koschevnikovi* is the most basal of the Asian cavity honey bees (Wakamiya et al. 2017), and this confirmed the clustering proposed by Raffiudin and Crozier (2007).

Figure 2.6 illustrates that the numbers of transitions and transversions of the *COX1* gene result in different patterns at each codon position. Transition events are more common than transversions in the first codon position (Fig. 2.6), which contrasts the situation at the second codon (Fig. 2.6 B). The third codon shows almost the same rates for both mutational events (Fig. 2.6 C), and had the greatest number of mutations among all the codons, which supported the topology of the *A. koschevnikovi* tree (Fig. 2.5). The numbers of transitions and transversions in each codon positions are consistent with the mutational rate of the *COX1* gene in several species of honey bees (Tanaka et al. 2001a).

An analysis of the *COX1* and *CYTB* genes revealed that they are both rich in A-T sequences (75.5%) compared with G-C sequences (24.5%). The *CYTB* gene of *A. koschevnikovi* has six haplotypes from South Kalimantan, East Kalimantan, and Sarawak (Table 2.3). Fitriya et al. (2012) found that the first haplotype was the most common sequence in the samples from Hulu Sungai Selatan, Balangan, and Hulu Sungai Tengah. The second haplotype occurs in Berau, and that and another sample from Tanah Bumbu both have C to T transition at nucleotide numbers 61 and 160,

Table 2.3: Nucleotide variations and haplotype numbers of *CYTB* in *Apis koschevnikovi*.

Species	Nucleotide site								
	4	36	57	60	111	126	135	156	159
*AK1 KB *CYTB* HT4	T	C	C	T	T	C	T	A	T
AK3 KB *CYTB* HT5	.	T	.	.	.	.	.	.	.
AK1 BR *CYTB* HT2	.	T	T	.	.	T	.	T	C
AK2 TB *CYTB* HT2	.	T	T	.	.	T	.	T	C
AK1 BL *CYTB* HT1	.	T	T	C	.	T	.	T	.
AK5 HSS *CYTB* HT1	.	T	T	C	.	T	.	T	.
AK7 HSS *CYTB* HT1	.	T	T	C	.	T	.	T	.
AK5 BL *CYTB* HT1	.	T	T	C	.	T	.	T	.
AK1 HST *CYTB* HT1	.	T	T	C	.	T	.	T	.
AK2 BR *CYTB* HT3	.	T	T	.	.	T	.	T	.
AK2 SRW *CYTB* HT6	C	.	T	.	C	T	C	T	.

* AK- samples number, sample location source refer to Figure 2.4, HT- haplotype number.

compared with haplotype 1. Another variation occurs in the third haplotype from Berau, in which the C to T transition only occurs at nucleotide 61. The fourth and the fifth haplotypes occurred in two samples from Kota Baru. Haplotype 6 from Sarawak has mostly different point mutations compared with samples from South and East Kalimantan.

Sulawesi and Adjacent Islands: *A. nigrocincta* and *A. d. binghami*

A. nigrocincta. *A. nigrocincta* was first described by Smith (1861) from the collection of Alfred Russel Wallace, and later Maa (1953) mentioned the same name for this Sulawesi honey bee. Hadisoesilo et al. (1995), in the 1989 exploration in South Sulawesi, found two sympatric honey bee morphs that had distinctive characteristics, *A. cerana* (smaller and darker morph) and *A. nigrocincta* (larger with a yellowish clypeus and legs. In current 2019 exploration in Central Sulawesi, we found *A. nigrocincta* in Parigi Moutong (Figs. 2.7 A,B). Discriminant analysis results confirmed that the elliptical cluster of the yellow morph of *A. nigrocincta* is separate from *A. cerana* using a 95% confidence interval (Hadisoesilo et al. 1995). *Apis nigrocincta* is native to Sulawesi, Sangihe, Mindanao, and probably in adjacent islands (Selayar and Buton) (Otis 1996, Damus and Otis 1997). The different drone flight times of *A. cerana* and *A. nigrocincta* serve as pre-mating behavior barriers between the two species and confirm the species states (Hadisoesilo and Otis 1996). *Apis nigrocincta* mostly nests in forest areas, whereas *A. cerana* is predominant in an urban area (Otis 1996). Currently, *A. nigrocincta* and *A. cerana* are reared through traditional beekeeping in several villages in the Parigi Mountong District (Central Sulawesi) (R. Raffiudin, personal observation).

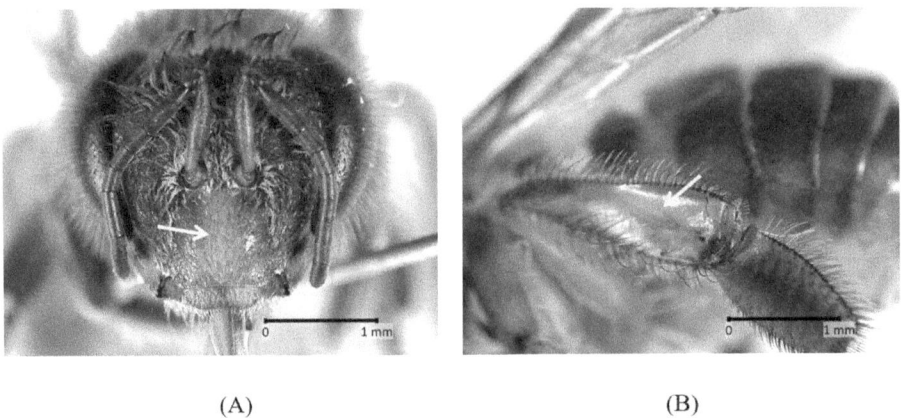

(A)   (B)

Figure 2.7: The morphological characteristics of *Apis nigrocincta* from Parigi Moutong, Central Sulawesi: yellowish (A) clypeus and (B) hind legs.

Color version at the end of the book

Smith and Hagen (1996) found three mitochondrial haplotypes of the non-coding region of *COX1/COX2* in Sulawesi. The "black femur" bees have the same haplotype as bees from Sundaland, and the "yellow femur" bees have the same haplotypes as those from Sangihe. The third group ("short") lacks most of the intergenic region. The "black femur" group is *A. cerana*, and the "yellow" and "short" groups are *A. nigrocincta* (Smith et al. 2000). The sister species relationship of *A. cerana* and *A. nigrocincta* was inferred with a Bayesian method (Fig. 2.1) using *COX2*, *16S rRNA* and *itpr* genes with 30.4% bootstrap values (Raffiudin and Crozier 2007). When the complete mitogenome was used, a more robust 100% bootstrap value was revealed, confirming the relationship between *A. cerana* and *A. nigrocincta* (Eimanifar et al. 2017). Furthermore, Takahashi et al. (2017), analyzed two individuals of *A. nigrocincta* from Sangihe Island using the whole mitogenome, and found a large genetic distance (0.0067), which suggested a high level of diversity among Sulawesi yellow honey bees.

Apis dorsata binghami: genetic variations in the endemic Sulawesi giant honey bees

The formation of Sulawesi Island from three different regions around 50 Mya (Hall 1998) is the plausible source of a large number of endemic species on this island, including the existence of the giant honey bee *A. d. binghami* (Maa 1953). This Sulawesi giant worker honey bee has a black abdomen with a white stripe (Fig. 2.8 A), which is different from the black abdomen with a yellow stripe of the sister subspecies *A. dorsata* (Hadisoesilo 2001, Nagir et al. 2016, R. Raffiudin personal observation). The nesting sites of *A. d. binghami* vary from 1 m to 32 m

(A)   (B)   (C)

Figure 2.8: *Apis dorsata binghami* in South Sulawesi: (A) worker honey bee, (B) nesting in a *Litsea mappacea* (Lauraceae) tree approximately 1 m above the ground, (C) nesting in a *Artocarpus sericocarpus* (Fam. Moraceae) tree at 32.6 m above the ground (Photograph: Muh. Teguh Nagir).

above the ground (Figs. 2.8 B,C) (Nagir et al. 2016). In one tree, generally, only 2–3 colony nests are found, with a maximum of 10 *A. d. binghami* colonies, while hundreds of *A. dorsata* colonies can be found in a single tree (Hadisoesilo 2001). Based on information from local people in Southeast Sulawesi, the annual honey harvest time for *A. d. binghami* is during February–April (Daniel: Head of Tanjung Batikolo and Tanjung Amolengo Wildlife Reserve, Southeast Sulawesi, personal communication, February 2019).

Using the *COX2* gene, the current study revealed four haplotypes from four colonies of *A. d. binghami* collected from two villages in Maros, South Sulawesi. The phylogenetic tree separated this Sulawesi honey bee into two clusters (Fig. 2.9). A low level of genetic variation was observed in group 1 of *A. d. binghami* (colonies 1, 2, and 3 had genetic distances between 0.000 and 0.001) (Table 2.4); however, a high level of variation occurred between *A. d. binghami* groups 1 and 2 (colony 4), with genetic distance between 0.038–0.043. It is surprising that a high level of genetic variation occurred in this Sulawesi giant honey bee, but it is plausible to be a result of local migration, compared with the long migration of *A. dorsata* across the continent (Robinson 2012).

The level of genetic diversity of *A. d. binghami* in Southeast Sulawesi also needs to be determined. This region of Sulawesi had a different historical formation-emerging from the bed of the South Banda Sea (Hall 2013), and thus, potential new haplotypes of *A. d. binghami* could exist.

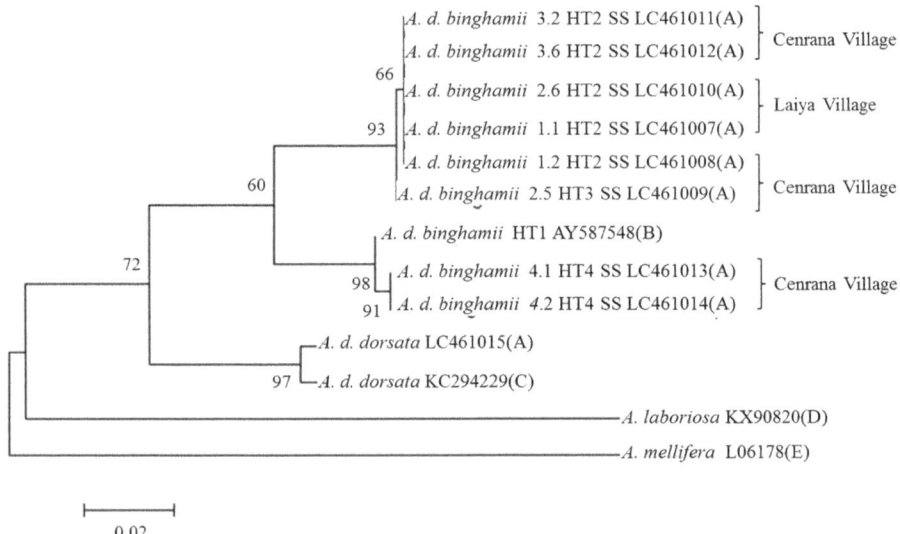

0.02

Figure 2.9: Nucleotide sequence-based phylogenetic tree of the *COX2* gene of *Apis dorsata binghami* collected from Cenrana and Laiya villages in Maros, South Sulawesi generated using the Maximum-likelihood method with 1,000 bootstrap replicates. Information following the species name are: colony and individual number, HT = haplotype number, SS = South Sulawesi, and GenBank Accession Number. Letters in brackets indicate samples from: A = current study, B = Raffiudin and Crozier (2007), C = Yang et al. (2019, unpublished GenBank sequences), D = Wang et al. (2018, unpublished GenBank sequences), and E = Crozier and Crozier (1993).

Table 2.4: Genetic distances based on the *COX2* gene of the Sulawesi giant honey bee *Apis dorsata binghami* with other giant honey bees. Letter in bracket indicates sample from: A = current study, B = Raffiudin and Crozier (2007), C = Yang et al. (2019, unpublished GenBank sequences), D = Wang et al. (2018, unpublished GenBank sequences), and E = Crozier and Crozier (1993).

No	Species	1	2	3	4	5	6	7	8	9	10	11	12
1	A. d dorsata1.2 LC461015 (A)												
2	A. d dorsata KC294229 (C)	0.006											
3	A. d binghami1.1 HT2 SS LC461007 (A)	0.064	0.064										
4	A. d binghami 1.2 HT2 SS LC461008 (A)	0.064	0.064	0.000									
5	A. d binghami 2.6 HT2 SS LC461010 (A)	0.064	0.064	0.000	0.000								
6	A. d binghami 3.6 HT2 SS LC461012 (A)	0.064	0.064	0.000	0.000	0.000							
7	A. d binghami 3.2 HT2 SS LC461011 (A)	0.064	0.064	0.000	0.000	0.000	0.000						
8	A. d binghami 2.5 HT3 SS LC461009 (A)	0.066	0.066	0.001	0.001	0.001	0.001	0.001					
9	A. d binghami 4.1 HT4 SS LC461013 (A)	0.056	0.056	0.043	0.043	0.043	0.043	0.043	0.041				
10	A. d binghami 4.2 HT4 SS LC461014 (A)	0.056	0.056	0.043	0.043	0.043	0.043	0.043	0.041	0.000			
11	A. d binghami HT1 AY587548 (B)	0.056	0.056	0.040	0.040	0.040	0.040	0.040	0.038	0.003	0.003		
12	A. mellifera L06178 (E)	0.092	0.090	0.096	0.096	0.096	0.096	0.096	0.096	0.106	0.106	0.106	
13	A. laboriosa KX908208 (D)	0.103	0.103	0.109	0.109	0.109	0.109	0.109	0.110	0.115	0.115	0.112	0.120

Separate analyses of the *itpr*, *16S* RNA, and *COX2* sequences revealed that the giant honey bee *A. d. binghami* is a sister species of *A. dorsata* (Raffiudin and Crozier 2007), and *A. laboriosa* is at the base of the three species of giant honey bees. This result is in agreement with the phylogenetic tree of Cao et al. (2012). However, when using the same genes, the phylogenetic tree of Lo et al. (2010) showed that *A. dorsata* has a close relationship with *A. laboriosa*.

The Philippine giant honey bee *A. d. breviligula* was determined by Lo et al. (2010) to be a distinct species that is the basal species of all the giant honey bees. Cao et al. (2012), using the *COX2* gene, found that *A. d. binghami* (sequences derived from Raffiudin and Crozier (2007)) is in a sister clade with *A. d. breviligula*.

Apis andreniformis the dwarf honey bee

Apis andreniformis (Figs. 2.10 A–C) is distributed throughout Indo-China and Sundaland, and throughout Thailand and the Malay Peninsula in the mainland (Otis 1996). *Apis andreniformis* is also found in the Southeast Asian archipelago: Sumatra (Salmah et al. 1990), Borneo (R. Raffiudin, personal observation, Koeniger et al. 2000), West and Central Java (Otis 1996), and Palawan Island (de Guzman et al. 1992, Otis 1996). It occurs in both lowland areas with 0–500 m elevations (Salmah et al. 1990), and in highland areas of up to 1,600 m in Thailand (Wongsiri et al. 1997, Wang et al. 2015). A similar distribution, ranging from the eastern foothills of the Himalayas eastward to Indo-China, Sundaland, and the Philippines, was described by Hepburn and Radloff (2011a) for the dwarf honey bee *A. andreniformis*.

(A)   (B)   (C)

Figure 2.10: Asian honey bee *Apis andreniformis*. (A) nest encircled with twigs and worker bees, (B) queen bee, and (C) drone bee (Photograph: Rika Raffiudin).

Color version at the end of the book

Having a vast distribution, the molecular phylogenetic tree of *A. andreniformis* in Thailand (Thailand Peninsula, Phuket Island, and Chiangmai in northern Thailand), based on the *CYTB* gene, revealed two clusters, A and B (Rattanawannee et al. 2007). Most samples are in Cluster A, from the Thailand Peninsula and Tenom (Malaysia) *A. andreniformis*, while the Phuket and mostly Chiangmai samples are closely related and in the B group. Interestingly, the latter has more variations compared with the A group. A low level of genetic variation was also discovered among colonies of *A. andreniformis* in West Sumatra (Indonesia) using the *COX1* gene (GenBank Acc. No. LC427576 - 80, Raffiudin et al. 2019), while the dwarf bees shared a 99% identity with *A. andreniformis* from Nepal (GenBank Acc. No. KF736157, Wang et al. 2015, Raffiudin et al. 2019).

Phylogeny of honey bees inferred from wing venations using a geometric morphometric approach and the *Pyruvate Kinase* (*PK*) gene

Asian honey bees: geometric landmark wing approach. Geometric morphometrics is based on the analysis of a representational form constructed with Cartesian coordinates using landmarks, including distances, angles, and dot coordinates (Slice 2007). Wing geometric morphometrics can be used to identify several species of bees and diverging venation variations between the species, such as stingless bees (Vijayakumar and Jayaraj 2013) and bumble bees (Aytekin et al. 2007), and on the subspecies of *A. mellifera* (Miguel et al. 2011). A study of wing variations in four honey bees in Thailand can be used as a database for bee biodiversity or for comparisons with the fossil records (Rattanawannee et al. 2010). A total of 19 homologous anatomical points of venation were used for the geometric morphometric analysis.

Santoso et al. (2018) found that among five species of honey bees, the morphology of *A. mellifera*, the Asian honey bee *A. cerana*, and *A. koschevnikovi* were highly similar when assessed using a Relative Warp (73% variations) ordination plot (Fig. 2.11). High levels of variation occurred in the wing venations of *A. andreniformis* and *A. dorsata*. The distribution among the five honey bee species revealed clustering of *A. cerana*, *A. koschevnikovi*, and *A. mellifera*. Thus, there is a morphological resemblance in the wing venation among these three species that forms a group of cavity-nesting or medium-sized honey bees. However, *A. andreniformis* and *A. dorsata* form a separate cluster from that of these three species.

Grid deformations clearly show the different honey bee wing venation patterns among the five species. In general, the points of wing venation that were highly variable are at the meeting points of four following landmarks: (1) Landmark no. 1: Radial sector vein (Rs) and pre-*stigma*, (2) landmark no. 10: Rs + medial vein (m) and 2nd abscissa of Rs, (3) landmark no. 14: Rs and 2r–m, and (4) landmark no. 15: at the end of the Rs point (Figs. 12 A–E) (Santoso et al. 2018). Moreover, the geometric morphometric method of studying wing venation can also separate the species of honey bee based on a phylogenetic tree constructed using the Neighbor-Joining approach; therefore, this method is able to distinguish among honey bee

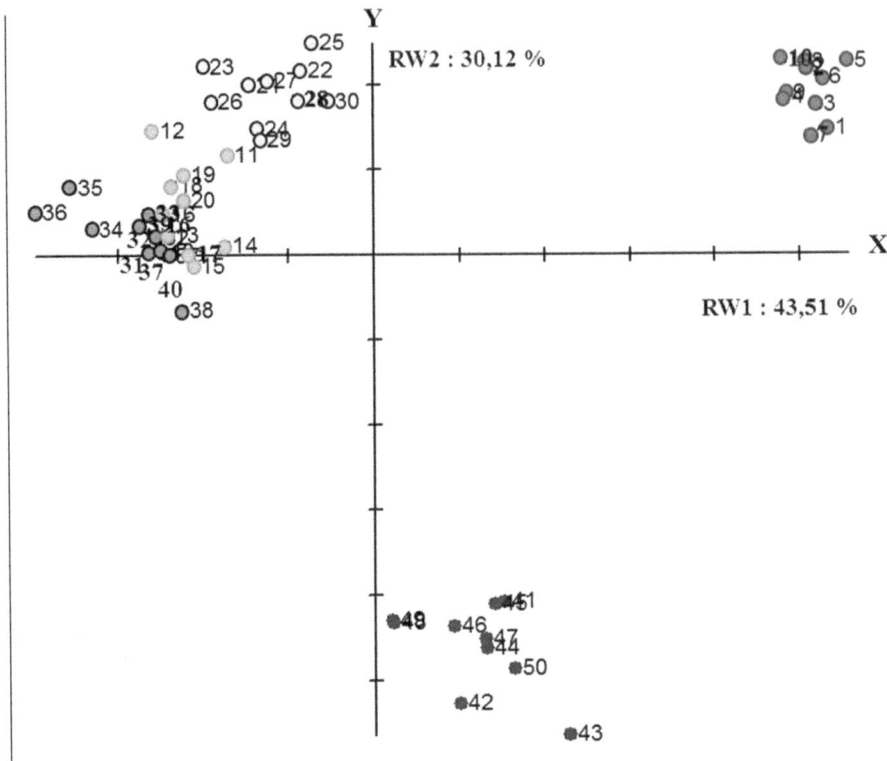

Figure 2.11: The Relative Warp (RW) ordination plot of five honey bee species: geometric morphometric wing venations. 1–10, red = *A. andreniformis*; 11–20, orange = *A. cerana*; 21–30; yellow = *A. koschevnikovi*; 31–40, green = *A. mellifera* and 41–50, blue = *A. dorsata* (Santoso et al. 2018).

> **Color version at the end of the book**

species (Figs. 2.13 A,B). The result of the relative warp was consistent with the topology of the phylogenetic tree constructed using 100% variations in the Neighbor-Joining method. *Apis andreniformis* and *A. dorsata* are in different clades.

Another conformity is the proximity of three cavity-nesting species, *A. mellifera, A. cerana,* and *A. koschevnikovi.* Among these three species, *A. cerana* is closer to *A. mellifera* than *A. koschevnikovi.* The close relationship between *A. cerana* and *A. mellifera* is supported by only two points of venation. This result is different from the bee phylogenetic analyses based on morphology (Engel 1998) and molecular techniques (Raffiudin and Crozier 2007), in which *A. cerana* and *A. koschevnikovi* are sister species.

Wings are important because, during evolution, they were preserved as fossils. The extant Apini wing has highly resembled the fossil Electrapini (genus *Electrapis*) from Eocene Baltic amber because of the jugal lobe in the hind wing (Cockerell 1908). Honey bee fossils have been found in Baltic Amber (Zeuner and Manning 1976) and in China (Zhang 1990). An examination of wing venation in three fossils of genus *Electrapis*, a fossil of *A. armbursteri*, a fossil of *A. henshawi*, and an extant

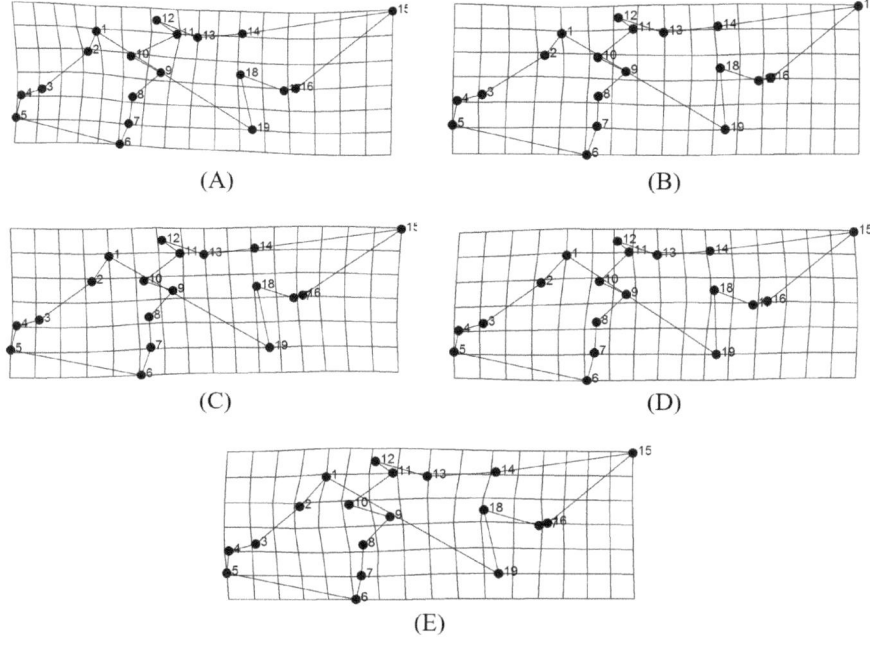

Figure 2.12: Deformation grid of honey bee species: geometric morphometric wing venations in (A) *Apis andreniformis*, (B) *A. cerana*, (C) *A. koschevnikovi*, (D) *A. mellifera*, and (E) *A. dorsata*. The highly variable landmarks of wing venation are: landmark no. 1: Radial sector vein (Rs) and pre-*stigma*, landmark no. 10: Rs + medial vein (m) and 2nd abscissa of Rs, landmark no. 14: Rs and 2r-m, and landmark no. 15: the end of the Rs point (Figs. 2.12 a–e) (Santoso et al. 2018). For the anatomical venation of honey bee, refer to Michener (2000).

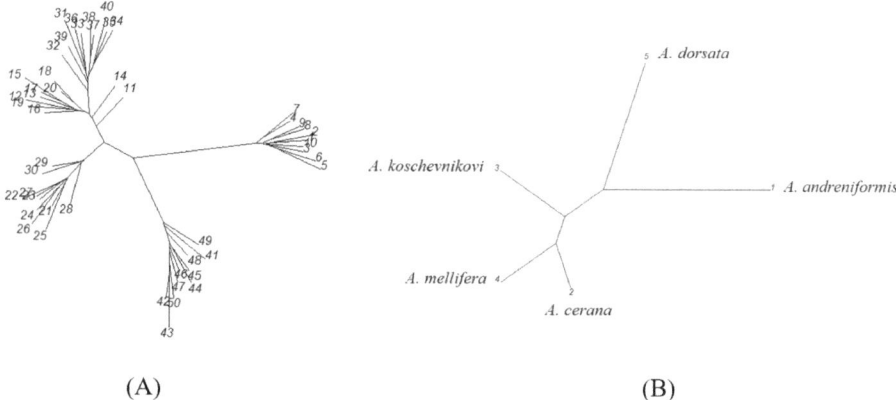

Figure 2.13: Unrooted Neighbor-Joining tree for honey bees based on the geometric morphometric wing venations using (A) all the samples and (B) average values for each honey bee species (Santoso et al. 2018).

A. mellifera revealed no significant differences (Ruttner 1988, Engel 1998). The only difference was that the venation wing length affected the wing cell size.

Phylogenetic analysis using the *PK-like* gene

The carbohydrate-rich diet of honey bees might lead to the rapid evolution of carbohydrate metabolism-related genes in honey bees. Molecular data indicated that carbohydrate-related genes, such as the PK, are among the most rapidly evolving genes in highly eusocial bee lineages relative to non-eusocial and eusocial bee lineages (Woodard et al. 2011). A phylogenetic analysis of the interspecies of honey bees used *PK-like* genes from a variety of species, including those of *A. andreniformis* from Padang Pariaman, West Sumatra, and *A. cerana* from Bogor (GenBank Acc. No. LC318759 - 63, Shullia et al. 2019).

The partial *PK-like* genes of these species consisted of seven exons (Exons 2–8) and six introns (Exons 2–7) (Shullia et al. 2019). A BLAST-n analysis showed that the nucleotide sequence of the *A. andreniformis PK-like* gene is closely related to that of *A. florea*, sharing a 95% identity (GenBank Acc. No. XM_012487945.1, Lowe and Eddy 1997). The nucleotide sequences of the *A. cerana PK-like* gene from Bogor, Indonesia also showed a 99% identity with *A. cerana* from Korea (GenBank Acc. No. XM_017058664.1, Park et al. 2015).

The exon-intron organization of *Apis PK-like* genes revealed that almost all the exon regions have the same nucleotide length. The variations in the deduced *PK-like* exon lengths only occur in Exons 7 and 8 of *A. andreniformis* and *A. florea*, and in Exon 2 of *A. c. javana* (erratum for *A. c. indica* in Shullia et al. 2019) and *A. c. cerana*. Although the intronic regions have more variation in length, the exon-intron boundaries in the *PK-like* genes are consistent, following GT-AG rules and homologous introns have the same phase. The differences in intron lengths were caused by base insertions and deletions (Shullia et al. 2019).

The rapid evolution of the *PK-like* gene in the honey bee can be discerned by the high number of nucleotide variations. The nucleotide variations of *PK-like* exonic regions at each codon position are described in Figs. 2.14 A–C. The greatest number of substitutions occurs at the third codon position (Fig. 2.14 C). The number of transitions is greater than transversions in the first and third codon positions, while in second codon position, the numbers of transversions and transitions are almost equal. These sets of data supported the absolute numbers of transitions and transversions plotted with respect to evolutionary distances in the first and third codon positions of the *COX1* and *COX2* genes (Tanaka et al. 2001a). At the interspecies level of *Apis*, transitions were more common than transversions (Tanaka et al. 2001a). The greatest genetic distance among *Apis* species of 0.061 occurs between *A. mellifera* and *A. florea*, while the genetic distance between *A. andreniformis* and its sister species, *A. florea*, is 0.031. The shortest genetic distance at the intraspecies level occurs between *A. c. cerana* and *A. c. javana*. This suggested that the *PK-like* gene was highly evolved at the interspecies level, and even more so at the intraspecies level of the genus *Apis*. Furthermore, the number of transition events is greater than transversion events in the honey bee *PK-like* gene.

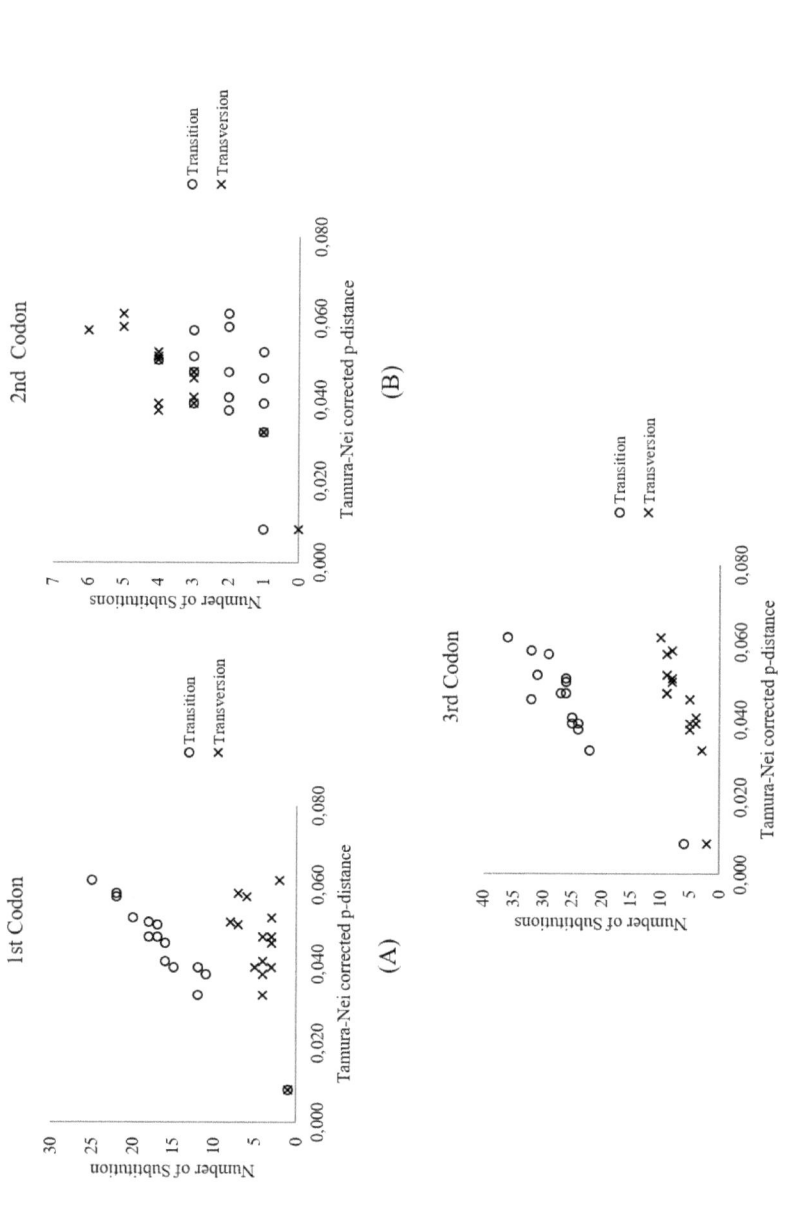

Figure 2.14: Substitution rate at each codon position of the *PK-like* gene in Asian honey bees *Apis andreniformis* and *A. cerana*. (A) 1st codon, (B) 2nd codon, and (C) 3rd codon positions.

The highly evolved nature of this gene compared with other carbohydrate-related genes has also been reported between eusocial insects, such as the honey bee, and solitary insects, such as *Drosophila*. Between the western honey bee *A. mellifera* and *Drosophila*, the phosphofructokinase gene had a 1:1 orthology, while the PK gene had a 2:6 orthology, respectively. The greater diversity level of the *PK-like* gene may result from its position at the end of the glycolytic pathway before pyruvate enters other pathways (Kunieda et al. 2006).

Using either nucleotide sequences or deduced amino acid sequences of the *PK-like* gene, the evolution of the honey bee is presented as Maximum-likelihood phylogenetic trees (Figs. 2.15 A,B). They revealed that the dwarf honey bees (*A. andreniformis* and *A. florea*) were at the basal position. The phylogenetic trees also showed that the medium-sized honey bees (*A. cerana* and *A. mellifera*) and giant honey bee (*A. dorsata*) were grouped into a monophyletic clade. However, the positions of *A. mellifera* and *A. dorsata* resulted in two topologies. The phylogenetic tree based on *PK-like* nucleotide sequences formed the first topology, having the medium-sized honey bee group form a monophyletic clade with *A. dorsata* as its sister species. This tree was congruent with a phylogenetic tree based on the behavioral states and a total of five nuclear and mitochondrial genes (Raffiudin and Crozier 2007). The molecular-based honey bee phylogenetic trees were also congruent with the morphology-based phylogenetic tree (Alexander 1991). This indicated that

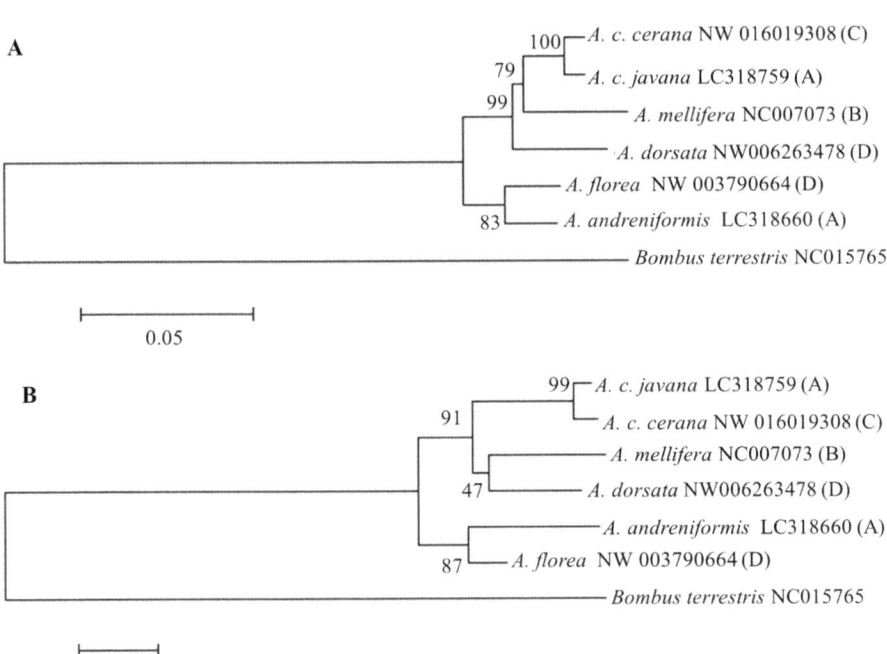

Figure 2.15: Different topologies of Asian honey bee phylogenetic trees constructed using the *PK-like* genes (A) nucleotide and (B) deduced amino acid sequences. Information following the species names are GenBank Accession Numbers, and the letters in the brackets indicate samples from: A = Shullia et al. (2019), B = The Honey bee Genome Sequencing Consortium (2006), C = Park et al. (2015), and D = Lowe and Eddy (1997.

several genes, such as *PK-like*, have the same evolutionary pattern as that derived from the morphological characteristics of the honey bee. The other topology was presented by the phylogenetic tree, based on deduced amino acid sequences, in which *A. mellifera* and *A. dorsata* formed a separate clade from the *A. c. javana* and *A. c. cerana* groups.

Future view

With a wide range of landscapes, several of which are unique, Asia is the home of diverse honey bee species and subspecies. Using mitochondrial sequences, Smith and Hagen (1996), Smith et al. (2000), Sittipraneed et al. (2001), Songram et al. (2006), and Tan et al. (2007) identified a large *A. cerana* cluster on the mainland. Smith et al. (2004) had Southeast Burmese *A. cerana* joining the Sundaland group, while the northern and southwestern *A. cerana* joined the mainland cluster. Using the mitogenome, a new subspecies, *A. c. koreana* that differed from *A. c. cerana* and *A. c. japonica* by 2.57% and 2.58%, respectively, was described by Ilyasov et al. (2018). Thus, this promising tool will be useful in determining the number of *A. cerana* subspecies. Numerous geographical barriers, including mountains, deserts, and islands, in the archipelagoes have the potential to form undiscovered subspecies, such as in Northern Russia, mainland Asia, and Wallacea regions. It will be of interest to re-explore *A. cerana johni* (Skorikov) in Sumatra and other morphs in Lesser Sunda.

Intriguing questions arise regarding other native honey bees in Asia, such as whether the migration patterns of *A. dorsata binghami* and *A. breviligula* help structure the genetic divergence among these adjacent populations of giant honey bees. This phenomenon is also implicated in the high level of variation among Sangihe populations of *A. nigrocincta* (Takahashi et al. 2018).

Acknowledgments

We thank Prof. Gard W. Otis for his valuable scientific comments on the manuscript. We also express our gratitude to the Ministry of Research, Technology and Higher Education of the Republic of Indonesia for the research grant under the Fundamental Research (No: 86/I3.24.4/SPK/BG-PD/2009) and Competitive Grant (No: 026/SPPP/PP-PM/DP3M/IV/2005) schemes, and for the exploration and molecular analyses of *Apis cerana* in Java and *A. koschevnikovi* in South Kalimantan, Indonesia, respectively.

Chapter 3

The Origin of the European Bees and their Intraspecific Biodiversity

Brandorf, A.Z.[1,]* and *Rodrigues, M.*[2]

Introduction

The Western honey bee (*Apis mellifera*), is an insect of great ecological and economic importance that, due to its role in pollination and honey production, is now globally widespread. Its native range is large and diverse, and considerable variation can be observed within many populations and several can be further subdivided into a diversity of "ecotypes". Supported by morphometric and genetic studies, in addition to analyses of ecophysiological, and behavioral traits, Ruttner (1988) developed and introduced the concept of "branches", or evolutionary lineages within *Apis mellifera*. There are now five recognized genetically distinct evolutionary branches distributed throughout Africa, Europe, and Asia: lineage A (African—*lamarckii, adansonii, scutellata, monticola, litorea, capensis, unicolor*, and the newly described *simensis*) (Meixner et al. 2011); lineage C (subspecies east and south of the Alps including those along the northern Mediterranean, including *ligustica, carnica, macedonica, cecropia, cypria*, and *adami*); lineage M (a west Mediterranean and northwest European—*mellifera* and *iberiensis*, but originally also including *intermissa, sahariensis, siciliana, ruttneri* (Sheppard et al. 1997), which were considered as links between the tropical African and the west Mediterranean subspecies); lineage O (Oriental—in the Near East and western Asia, including *caucasica, anatoliaca, syriaca*, meda, *armeniaca, jemenitica*, and the later described *pomonella* (Sheppard

[1] Federal Agricultural Research Center of the North-East, Russia, Kirov city, Lenina street, 166A, 610007, Russia; the head of the "Federal beekeeping research centre", Pochtovaya street, 22, Rybnoe, 391110.
[2] Flora, Bee Research Lab at São Paulo State University, Jaboticabal, Brazil; marisacrodrigues2@gmail.com
* Corresponding author: gordenchuk@mail.ru

and Meixner 2003); and lineage Y ("*yemenitica*" from Ethiopia) (Ruttner 1988, Sheppard et al. 1997, Franck et al. 2000b, Franck et al. 2001, Sheppard and Meixner, 2003, Whitfield et al. 2006a).

Conventionally, intraspecific taxonomy of the honey bee has been based on morphology, and currently at least 27 subspecies of *A. mellifera* are known on the basis of morphometric traits (Ruttner 1988, Sheppard et al. 1997, Sheppard and Meixner 2003). Most recently, the increase in the availability of individual *Apis mellifera* genomes has resulted in significant progress towards a better understanding of its evolution (Kocher and Paxton 2014) and adaptation, therefore allowing for monitoring of the dispersal and admixture of honey bee populations. Moreover, molecular studies have been crucial to reveal that several subspecies and ecotypes can be considered as endangered (De la Rúa et al. 2009, Dogantzis and Zayed 2019), since several factors can lead to a loss of both genetic diversity and specific adaptations to local conditions (Meixner et al. 2013). In one of the first single nucleotide polymorphism (SNP) population genetics studies of *A. mellifera*, Whitfield et al. (2006a) suggested that this species originated in Africa, colonized Europe and Asia, via two or three independent expansions. Notably, the study suggested that M and C lineages colonized Europe via separate migration routes from North Africa and Asia, respectively. Analyses from Han et al. (2012) support a high divergence between western and eastern European populations of *A. mellifera*, suggesting they are likely derived from two distinct colonization routes. Recently, Chapman et al. (2016) developed a test to assign the proportion of ancestry of honey bees to ancestral lineages of West European (M) and East European (C) based on 95 SNPs. During the last decades, morphometrical and molecular efforts have contributed to identifying new subspecies, evolutionary lineages, and a significant number of genes involved with adaptations and colony-level quantitative traits.

With this chapter, we hope to provide a brief description of the origin of European bees and their intraspecific biodiversity. We aim to contribute to conservation and technical programs that seek to improve the economic value of colonies or preserve locally adapted populations and subspecies in Europe, since the original geographic distribution pattern of European subspecies is being dissolved EU-wide (De la Rúa et al. 2009). Some authors defend that mass imports and increasing practice of queen trade and colony movements can endanger regional ecotypes by promoting hybridization (Meixner et al. 2010). Nonetheless, another dimension is added by the deliberate replacement of native subspecies in some regions by non-native bees with more desirable characters and a greater commercial interest (e.g., the replacement of *A. m. mellifera* in northern and central Europe by *A. m. carnica* or *A. m. ligustica*) (Bouga et al. 2011). The downside of these economically-driven processes is an increasing trend towards uniformity of honey bee populations across Europe, and can lead to a loss of both genetic diversity and specific adaptations to local conditions (Meixner et al. 2010). Thus, there is a widely recognized need to encourage regional breeding efforts to preserve local adaptation and to maintain local strains in isolated conservation apiaries. To attain this goal, it is necessary to have a reference base to identify strains to be used for breeding. To provide a stable baseline, it is important that this reference reflects the natural variation of honey bees, since beekeepers and breeders are known to often work with non-native stock.

Here, we place a stronger focus on describing the most common subspecies and a few newly breeds of honey bees in Europe: *Apis mellifera carnica* (Pollman 1879), *Apis mellifera cypria* (Pollman 1879), *Apis mellifera iberiensis* (Goetze 1964), *Apis mellifera ligustica* (Spinola 1806), *Apis mellifera mellifera* (Linnaeus 1758), *Apis mellifera carpathica* (Avetisyan 1969), *Apis mellifera acervorum* (Skorikov 1929), *Apis mellifera caucasica* (Gorbachev 1916), *Apis mellifera remipes* (Gerstäcker 1862), far-eastern honey bees (Linnaeus 2018). It is important to highlight that the scientific description and recognition of honey bee diversity in Europe cannot be regarded as complete, since vast areas, predominantly in the eastern part of the continent, have not yet been studied systematically (Meixner et al. 2013).

***Apis mellifera mellifera* L.** Originated from dark North African (*Apis mellifera intermissa*) (Krivtsov et al. 1999). Their natural habitat is practically the whole of Europe apart from the Balkans, Italy, and the South of the European part of Russia. This breed is endemic for France, Switzerland, Northern and Western slopes of Austrian Alps, England, Denmark, Holland, Belgium, Scandinavian countries, Poland (Brandorf and Ivoilova 2014a).

This breed occupies a wide habitat, which is why it has a lot of population. About 8,000 years ago, bees of this breed started inhabiting growing post-glacial forests, and widely spread in the western, central, and northern parts of Europe. Natural resettlement of *mellifera* reached from the Balkans and the Pyrenees to the Urals, and in the 18th century, they were brought to Siberia. Then the breed spread in Siberia to the Baikal, Middle Asia, to the North of the European part of Russia. In northern European countries following the retreating glacier, it spread up to latitude 62 N. The border of the breed's expansion coincides with the northern border of growth of hazelnut (*Corylus*). The southern border of its habitat is not clearly marked. In Western Europe, this border is the Alps. In America, dark forest bees were brought by settlers in 1622, where bee colonies started inhabiting forests, and although for 100 years they have been breeding exclusively Italian bees in the US, wild bee populations still keep the genes of *mellifera*. From England, these bees were brought into Australia and New Zealand in 1835. These bees demonstrated themselves especially well in cool and damp Tasmania. On the British Isles, for instance, the Isle of Man, humans breed only dark forest bee in purity. In the UK, there are established organizations for the protection of *mellifera*—the Bee Improvement and Bee Breeders Association (BIBBA). There is also the International Association for Conservation of *Apis m. mellifera* (SICAMM), which is actively performing. In Russia, the Russian Association for Conservation of *Apis m. mellifera* (RACAMM) has been acting since 2012 (Brandorf and Ivoilova 2015). In Germany, like in all of Western Europe, beekeepers breed mainly *Carnica* bees, but the supporters of *mellifera* still presented, and in the City of Kiel, established the Society for Breeding Dark European Bees.

A basic feature of this breed is the chitin coloring in worker bees, which is pure dark gray. The length of the proboscis varies between 5.9 and 6.4 mm. The width of the third tergite in workers is on average 5.0 mm (4.8–5.2 mm), cubital index is 60–65% (according to the European system, it is 1.5–1.9, with the average value being 1.5), tarsal index is 54.0–55.5%, discoidal shift is negative (its value being

−10 to 0 degrees). Currently in measurement of the front wings there is hantel index, which comprises 0.60–0.93 in *mellifera*. Another feature of this breed is a short proboscis and the biggest body (Avetisyan and Cherevko 2001). The weight of a young worker is about 100–110 mg, of a virgin queen around 200 mg and more.

The advantage of *mellifera* is that it has the best winter hardiness. Bees are adapted to dramatic changes in weather and temperature, which allow the breed to have wide habitat (Avetisyan 1982, Krivtsov and Lebedev 1995). During wintering, bees of this breed maintain quite a high level of carbon dioxide in overwintering colonies (up to 4%), which has an effect on the colony keeping its stability and bees having a low response to changes in micro- and macroclimate. Only *mellifera* are evolutionarily fit to long-term, within 6–7 months, staying inside the nest in the form of a club at the temperature of outside air as low as minus 40°C. The weight of rectum gets to 40 mg, and the maximum is 70 mg (Avetisyan 1982, Krivtsov and Lebedev 1995). In winter season one colony uses 9–12 kg within 5 months. The queen starts laying eggs (diapausing in the queen's laying eggs stops) at the end of February.

Mellifera efficiently use abundant mid-term and late nectar flow. They use monofloral nectar flow better than polyflorous, which results from the breed's restricted flower specialization and weak flower migration. Bees efficiently use the honey flow from forest raspberry, blooming rosebay willowherb, different types of linden, from buckwheat and other crops, which abundantly produce nectar. They have no rivals with the heather honey flow. The weight of honey sac can reach 80 mg, with an average value of 40 mg. Summer activity starts at temperatures over 14°C (Brandorf 2016).

A number of researchers claim that as worker bees have a short proboscis they are not capable of efficiently pollinating bean plants (Thomas and Sirivat 1992, Boitsenyuk 2008). According to Gubin (1947), all breeds of honey bees are capable of working on meadow clover, which is proved by Ruttner (2006). The efficiency of *mellifera* with difficult to pollinate crops increases with training. Due to their breed's qualities, the bees prefer visiting no more than 2–3 species of crops during a working day, and the existence of blooming competing flora does not reduce attendance of the crops (Brandorf and Ivoilova 2016).

They have high wax and honey productivity, and high egg-laying capacity of the queen (Grankin 1998, Krivtsov et al. 1999). According to Grankin (1981) and Krivtsov and Lebedev (1995), the queen's fertility can get as high as 3,000 eggs in a day. This breed's particular feature is that they seal honeycombs with white (dry) wax cappings, and build combs of high quality. Another positive feature they have is that there are no brace combs. In nests during swarming, bees build no more than 5–8 queen cups, but they have the maximal amount of royal jelly in the cup, it can get as much as 400–600 mg per queen cup, with 10-HDA content being 1.8–3.0% (Brandorf and Ivoilova 2014b).

During abundant honey flow, the amount of commercial honey can be 70–120 kg per colony, with a hoarding of up to 18 kg in a day. *A. m. mellifera* is resistant (compared to other breeds) to such diseases as varroosis, toxicosis caused by honeydew, nosemosis. This breed differs from the others by their increased hygienic behavior (Grankin 2006). Pritchard (2014) claims that *mellifera* chew off mite *Varroa destructor* from their bodies.

The disadvantages of this breed include high aggressiveness and high swarming ability, and inability to protect their nests from robber-bees. However, high aggressiveness in these bees can manifest itself when the rules of inspection of nests are violated. It is necessary to consider that high aggressiveness is characteristic for hybrid bees. These bees have high negative phototaxis (during inspection bees run to the bottom part of the comb). Up to 50% of the colonies in one apiary can get into swarming conditions (Kokorev and Chernov 2005).

On the territory of Russia, *mellifera* is represented by several populations: Bashkirskaya, Uralsko-Gornotaezhnaya, Altayskaya, Tatarskaya, Krasnoyarskaya, Permskaya *A. mellifera* were the base for breed types Priokskiy, Orlovskiy, Tatarskiy, Burzyanskaya Bortevaya, and of the new breed Bashkirskaya.

Breed type Priokskiy was bred in All Russian Institute of Apiculture on the basis of crossbreeding of *Apis mellifera caucasica* and *Apis mellifera mellifera* (copyright certificate # 5818 from 21.10.1992). These bees' exterior features differ from the original basic breeds: the length of the proboscis (6.7–6.8 mm) and the width of the third tergite (4.65–4.85 mm). Their coloring is predominantly gray, 20–30% of bees have yellow spots on the first two tergites. The weight of a one-day-old bee is 105 mg, of a virgin queen is 185 mg, of a mated queen is 205 mg, of a drone is 230 mg. These bees are peaceful, and unlike *mellifera*, they keep calm working on combs which have been removed from the nest for inspection, but, nonetheless, they are not as peaceful as *Apis mellifera carpathica*, *Apis mellifera carnica*, and *Apis mellifera caucasica*. They are better than *Apis mellifera mellifera* at orienting themselves, they get less lost, and less drift to other bee colonies. Capping is predominantly light "dry". These bees are characterized by wide flower specialization (within a day they visit 5 to 7 species of crops), their resourcefulness in finding sources for feeding, and their speed of switching from one source to another exceed those in *Apis mellifera mellifera*. "Quiet" replacement and cohabitation of queens are rare in this breed type. They are comparatively good at protecting nests from robber bees, but they have a tendency to rob themselves. They moderately propolize nests. Hoarded nectar is firstly stored in the super, and then in the brood combs.

For breeding Priokskiy bees, it is preferable to use hives with a vertical arrangement. Winter hardiness of these bees is the same as in *Apis mellifera mellifera*. Comparative testing showed that the incidence of nosemosis in both Priokskiy bees and *Apis mellifera mellifera* was practically the same, but it was twice as big in *Apis mellifera caucasica*. Their spring development starts a little earlier than in *Apis mellifera mellifera* and in this period they grow 15.4% more brood. Queen's laying ability in the period of intensive development in Priokskiy breed type before the honey flow is 1,600–2,000 eggs a day. Swarming ability in these bees is 2 times weaker than in *Apis mellifera mellifera*. They efficiently use weak long-term unstable honey flow, bean plants included, and strong honey flow from linden and buckwheat. According to findings of comparative tests, honey productivity of Priokskiy bee colonies in central European Russia is 30–50 kg. Bees of this origin are marked by high wax productivity (Borodachev and Krivtsov 2000, Krivtsov and Sokolskiy 2002, Borodachev and Savushkina 2012).

Orlovskiy type of *Apis mellifera mellifera* was bred using 11 populations of *Apis mellifera mellifera*, which were brought from their natural habitat (patent #

4110 from 23.06.2008). The bees are dark gray, their weight is 104 mg, the weight of a virgin queen is 218 mg, of a mated queen during intensive egg-laying is up to 300 mg, of a drone is—253 mg. Unlike *Apis mellifera mellifera* these bees are less aggressive, and during the inspection they run to the bottom bar of the frame and hang there in bunches, honey capping is light "dry". They do not tend to rob. Hoarded nectar is first stored in the super, then in the nest. There is no cohabitation or "quiet" replacement of queens.

These bees are remarkable for their excellent winter hardiness and high resistance to disease. Their spring development is later, but it is more intensive. Queen's laying ability during quick population increase before the honey flow is 1,800–2,000 eggs a day, with a maximal value of 3,000. During an intensive honey flow, these bees have the insignificant swarming ability (no more than 5% of bee colonies).

These bees are most productive in strong late honey flow from linden (*Tilia cordata*), blooming rosebay willowherb (*Chamaenerion angustifolium*), buckwheat (*Fagopyrum esculentum*). Honey productivity is 40–70 kg per bee colony. They are characterized by high wax productivity, building during one season no less than 10 new combs per colony. They moderately propolize nests. They stock up a big amount of pollen and bee-bread, which can be collected as marketable products (Grankin 2008).

The Tatar type of *Apis mellifera mellifera* was bred in collaboration of the Center of Bee Selection "Tatarskiy" and the Research Institute of Apiculture, and based on the Tatarskiy population of *Apis mellifera mellifera* (patent # 5476 from 28.01.2010). These bees are dark colored, their weight is 111 mg, the weight of a virgin queen is 202 mg, of a mated queen is 230 mg, of a drone is 267 mg. They are less aggressive than *Apis mellifera mellifera*, and during the inspection they run from the comb, honey capping is light "dry". They do not tend to rob. These bees do not experience "quiet" replacement or cohabitation of queens. They accept manufactured supers faster than the original population.

These bees are remarkable due to their exclusive winter hardiness. They are more resistant to disease compared with the original population. Spring development is comparatively late but intensive. Queen's laying ability during quick population increase before the honey flow is 2,000 eggs a day. They have a weaker tendency to swarming (up to 18% of bee colonies). These bees are most productive in strong late honey flow from linden (*Tilia cordata*), blooming rosebay willowherb (*Chamaenerion angustifolium*), buckwheat (*Fagopyrum esculentum*). They are good for pollinating entomophilic agricultural crops.

Their honey productivity is 40–60 kg per bee colony. This type is characterized by high wax productivity, building during one season no less than 10 new combs per colony. They moderately propolize nests. They stock up a big amount of pollen and bee-bread, which can be collected as marketable products. Burzyanskaya tree trunk hollow bee was created during selection work with the unique genetic material of Burzyanskiy tree trunk hollow bees in severe conditions of wildlife reserve "Shulgan-Tash" in Bashkortostan (Sharipov 2019).

These bees are dark colored, the weight of a virgin queen is 202 mg, of a mated queen is 230 mg, of a drone is 267 mg. These bees are characterized by high aggressiveness, and during the inspection, they flee to the bottom bar of the frame and hang there in bunches, honey capping is light "dry". They do not tend to rob. These bees do not experience "quiet" replacement or cohabitation of queens. During a honey flow, they accept domesticate manufactured supers fast.

They have high winter hardiness, and are resistant to nosemosis, European foulbrood, honeydew toxicosis. Spring development starts late, but it has an intensive character. Average queen's laying ability before the main honey flow is 2.000 eggs in a day. They tend to swarm, which allows to form up to 50% of new bee colonies. These bees effectively use the strong honey flow from linden, Umbelliferae, buckwheat. Honey productivity in tree trunk hollow is 20–30 kg, and in hives is 50–60 kg per bee colony. This type is characterized by high wax productivity, building during one season no less than 10 new combs per colony, and by the high stock of propolis and pollen (Kosarev et al. 2001).

Bashkirskaya breed type was created by Bashkir Apiculture and Apitherapy Research Centre based on the Bashkir population of *Apis mellifera mellifera* L. (Ishemgulov 2008). These bees are characterized by dark gray coloring without yellowishness. Their weight is 113 mg, the weight of a virgin queen is 197 mg, of a mated queen is 215 mg, of a drone is 243 mg. Unlike *Apis mellifera mellifera*, these bees are less aggressive, are characterized by marked phototaxis (during an inspection of the frame they tend to run from the lit part of the comb and hang in bunches on the bottom bar of the frame), honey capping is light "dry". They do not tend to rob, but they are bad at protecting their nest from robber bees. They prefer to store nectar in the super, thus not limiting the queen in laying eggs. They are very hard working (summer activity in honey flow lasts no less than 16 hours a day). They propolize nests moderately. These bees are characterized by high winter hardiness, and they are resistant to foulbrood, nosemosis, honeydew toxicosis.

Their spring development starts a little later, but it has a much more intensive nature. Queen's laying ability during quick population increase before honey flow is 2,000 and more eggs in a day. Swarming ability is 20–25% weaker than in *mellifera*. These bees build up to 15 queen cups, and no more than 3 swarms leave a colony. Kept in big-sized hives and with well-timed enlargement of the nest, no more than 30% of bee colonies swarm. With the honey flow of 3 kg in a day, swarming stops.

These bees most effectively use the strong honey flow from linden, buckwheat, melilot. Their honey productivity is 40–70 kg per bee colony. They are characterized by higher (compared to *Apis mellifera mellifera*) wax production. They are capable of stocking up a big amount of pollen and bee-bread, which can be collected as marketable products.

Apis mellifera carpathica. *Apis mellifera carpathica*'s natural habitat is mountain regions of western Ukraine, south-west of European Russia, as these bees are massively brought into central Russia. According to current statistics, bee colonies of this breed compose 17% of an overall number of bee colonies in Russia. A number of features make this breed close to *Apis mellifera carnica*, hence this breed is considered its population. Their coloring is pure gray without yellowishness. The length of the proboscis is 6.3–6.7 mm, the cubital index is 45–50% (according to European system 2.3–3.0 with an average value of 2.7), the width of the third tergite is 4.8 mm, discoidal shift is positive, hantel index is 0.92. The weight of a one-day-old bee is 110 mg, of a virgin queen is 185 mg, of a mated queen is 205 mg.

This breed is peaceful, and they keep calm working on removed combs. It allows working without a smoker during the whole season. They seal honeycombs with light and dark capping, but mainly closer to the light one. These bees have a weak swarming ability. During swarming, they build up to 50–80 queen cups in the nest, the amount of royal jelly in a queen cup is on average 200–250 mg. These colonies experience "quiet" replacement of queens. They are capable of effectively using weak long honey flow when they are able to stock up to 50 kg of marketable honey. With the onset of the honey flow, they store honey in the manufactured super. Flights start at a temperature of 6 to 8°C, and these bees visit a lot of species of plants. They have high wax production.

These bees are characterized by good winter hardiness, low feed consumption in the winter period in northern parts of Russia it is 15 kg per bee colony (within 5 months). They are resistant to honeydew toxicosis, nosemosis, European foulbrood. The weight of rectum gets to 30 mg. Queen's diapausing ends in January. This breed's spring development starts early and vigorously. Queen's laying ability is 1,100–1,170 eggs a day. Disadvantages of this breed include a tendency to rob, low production of propolis, and indifferent attitude to wax moth (Alpatov 1948, Cherevko 2006, Gaidar and Pilipenko 1989). In Russia, there is a Maikopskaya population of this breed.

Apis mellifera acervorum. *Apis mellifera acervorum* is spread around Ukraine. It represents a southern branch of *Apis mellifera mellifera*, and its formation was influenced by *Apis mellifera carpathica*. They are well adapted to steppe conditions. Their coloring is light gray, sometimes there is the yellow coloring of the first three tergites. The length of the proboscis is 6.3–6.7 mm. Cubital index is 55–60% (2.2–2.5). Cappings are predominantly white. The bees are moderately aggressive. They keep calm on removed combs, and go on working. They propolize their nests weakly, but they build combs of high quality without brace combs.

Using honey flow they manifest features of *Apis mellifera mellifera*. They are poor at finding new sources and they slowly transfer from certain species of crops to others. They are good at using the abundant honey flow from linden and buckwheat. Honey production is 30–40 kg, and during abundant honey flow the amount of marketable honey can get to 100 kg per colony.

They have good winter hardiness, and they are resistant to honeydew toxicosis, European foulbrood, and nosemosis. Queen's egg laying ability is 1,950–2,300 eggs in a day. They intensively increase the population for honey flow from locust tree and buckwheat. Wax production does not differ from that of *Apis mellifera mellifera*. Up to 30% of bee colonies get into swarming condition, and during abundant honey flow they stop swarming. The number of queen cups during swarming is 9–18. They collect little propolis. Their main disadvantage is bad resistance to wax moth.

Apis mellifera caucasica. *Apis mellifera caucasica* is widely spread in western countries and the USA. They are the second most popular breed in the world. They can be seen in the north-east of Anatolia (Turkey), along the eastern coast and in the south of Russia, along the north-eastern coast of the Black Sea, and also in foothills of the Caucasus. They are spread in 40 countries. Some researchers consider this breed as the opposite of *Apis mellifera mellifera*. This means that positive features of *Apis mellifera mellifera* become negative in *Apis mellifera caucasica* and negative features become positive accordingly.

The color of the chitin is pure gray. Bees of this breed are characterized by a maximal length of the proboscis (6.7–7.2 mm). Thanks to this, *Apis mellifera caucasica* are considered the best pollinator for bean plants (Fabaceae), particularly red clover (Gorbachev 1916). Although, Gubin (1947) claimed that all breeds of bees (with long proboscis and short proboscis) can work on red clover with no exception. According to current data, the length of the proboscis is only one factor determining the ability of bees to collect pollen and nectar from flowers with a complicated structure. According to Ruth et al. (1993) and Ruttner (2006), *Apis mellifera caucasica* did not prove the best pollinizers of red clover. The width of the third tergite is 4.7 mm, the weight of a young bee is 80–90 mg, of a virgin queen is 180 mg, of a mated queen is 200 mg.

Bees of this breed are characterized by good flower migration, high resourcefulness in finding new sources of nectar, and consequently, they are very efficient in the polyflorous honey flow. With the beginning of honey flow, nectar is first stored in the brood nest, thus limiting the queen in laying eggs. They use weak honey flow well, thus they are convenient to breed in an area with weak honey flow.

They are peaceful. During an inspection of the nest, they work calmly on removed combs without showing any aggression. These bees protect their nests from robber-bees well, and they have a tendency to rob. They seal honey with dark capping. This breed has low wax production. Bee colonies have a weak swarming ability, as within a season no more than 3% of colonies swarm, and the number of queen cups in the nest is 2–5. The amount of royal jelly in a queen cup never exceeds 150 mg, with the content of 10-HDA 1.8 percent. Queens' fertility is not high (1,100–1,500 eggs in a day). Colonies experience cohabitation of two queens.

Their disadvantages are bad winter hardiness, high sensitivity to nosemosis and European foulbrood. The weight of rectum reaches 20 mg. In overwintering colonies they do not keep more than 1–2% of carbon dioxide, which is why they are quite active in the winter period. Queens practically never stop laying eggs, and diapausing lasts 2–3 weeks. In the natural habitat of these bees, there are several populations: Megrelskaya, Abkhazskaya, Kartalinskaya, Kakhetinskaya, Imeretinskaya, Guriyskaya, and others (on the territory of the former USSR). On the territory of Russia, about 14% of all bee colonies belong to this breed.

Apis mellifera remipes. *Apis mellifera remipes* is spread in lowland areas of Armenia, Georgia, and Azerbaijan. Chitin of these bees is a gray color with significant yellowishness. In worker bees, the length of the proboscis is 6.5–6.9 mm, the width of the third tergite is 4.7–4.75 mm. The cubital index is 50–55%. The weight of a worker is 80–90 mg, of a virgin queen is 180 mg, of a mated queen is 200 mg. Queens' egg laying ability is low—up to 1,700 eggs a day. They seal honey with dark "wet" capping. These bees are peaceful, and they are characterized by a tendency to rob. They are fastidious for climate conditions, as they are adapted to the southern climate and not fit for long cold winters (poor winter hardiness). They are not resistant to a lot of diseases. These bees have high swarming ability. Up to 80% of bee colonies get to swarming, the number of queen cups in a colony can be 300–350, the number of swarms can be up to 12. The cohabitation of two queens is noted in their nests. This breed is joined by Kubanskiye bees, living in North Caucasus of Russia.

Far-eastern honey bee. The Far-eastern honey bee is certified as a breed in 2018 (Patent # 9428 from 11.01.2018). It is adapted to abundant honey flow from linden. These bees were bred by the haphazard crossing of *Apis mellifera mellifera*, *Apis mellifera acervorum*, *Apis mellifera caucasica*, *Apis mellifera remipes*, and *Apis mellifera ligustica*.

Their chitin is gray with yellowishness. The length of the proboscis is 6.2–6.75 mm. The width of the third tergite is 5.1 mm. The cubital index is 42.1–45.4%, and the discoidal shift is positive. The tarsal index is 56.0–57.7%. The weight of a worker is 105 mg, of a virgin queen is 205 mg, of a mated queen is 230 mg. Spring development starts early and intensively, queen's laying ability is 1,300–1,700 eggs a day. They are marked by high swarming—up to 50% of colonies can get to swarming. The number of queen cups during swarming can be 4–150.

These bees are comparatively peaceful, and they are calm during nest inspection. Honey capping is mixed, predominantly light. Their tendency to rob is moderate. They are good at protecting their nests from the wax moth. This breed is characterized by good winter hardiness, and bees are resistant to nosemosis, honeydew toxicosis, and European foulbrood, but less than *Apis mellifera mellifera*. Weakness during wintering is 11.2–25.3%, and feed consumption per frame during wintering is 0.82–1.52 kg. These bees are adapted to sudden changes in temperature in the winter period. Far-Eastern honey bee efficiently uses rapid honey flow. In some years they bring up to 30 kg of linden nectar in a day, and the amount of marketable honey is 70–150 kg from one bee colony- it is honey flow from 3 linden trees (Sharov 2018).

Apis mellifera iberiensis. Classified by Engel 1999, and found in Iberian peninsula (Spain and Portugal), this subspecies is considered well studied in most of its current range. It belongs to the morphological A lineage and to the mitochondrial A/MS lineages group of *Apis mellifera*. No distinct ecotype variants were reported at present for the mainland, but ecotypes described for island populations were found (Meixner et al. 2013). *Apis mellifera iberiensis* show a marked gradient in genetic background, as northern populations are more similar to the black European *A. m. mellifera*, whereas those from the south are more similar to African *A. m. intermissa* (Cánovas et al. 2008). De la Rúa et al. (2002a, 2005) reported an absence of cytonuclear disequilibrium in honey bees from eastern Spain, and postulated that *A. m. iberiensis* may derive from a complex of events, including multiple hybridizations, selection, and human manipulation. In the study of Arias et al. (2006), allozyme, mtDNA, and morphological variation in honey bee populations sampled from southern France to northern Morocco were investigated. The results were discussed in view of current hypotheses concerning the Iberian Peninsula as a glacial refuge and secondary contact between lineages. While the basis of subspecific classification derives from the accumulation of genetic changes in isolated populations, hybridization can occur between subspecies of the same or even different lineages through secondary contact during interglacial periods (Badino et al. 1984, Comparini and Biasiolo 1991, Meixner et al. 1993, Franck et al. 2000a). *A. m. iberiensis* has been considered of particular interest, since it has been shown to have a morphology somewhat intermediate between the subspecies *A. m. mellifera* that occurs in France and *A. m. intermissa*, that occurs in Morocco (Ruttner et al. 1978, Ruttner 1988, Cornuet and Fresnaye 1989, Hepburn and Radloff 1996). Allozyme data (*MDH - 1*) has also been reported to be consistent with a south-north cline. The allele *Mdh100* appears to be fixed or nearly fixed in *A. m. intermissa* (Cornuet 1982). and shows a high frequency in southern Spanish honey bee populations. The frequency of *Mdh100* decreases in northern Spain, where allele *Mdh80* predominates (Cornuet 1982, Smith and Glenn 1995). The allele *Mdh80* occurs in high frequency in samples of *A. m. mellifera* throughout France and northwestern Europe (Sheppard and Berlocher 1984, Cornuet et al. 1986). Mitochondrial genetic markers originating from both African (A) and western European (M) honey bee lineages can be found in Iberia. Initial studies revealed two different haplotypes, one predominant in the south (*A. m. intermissa*-like) and the other in the north (*A. m. mellifera*-like) (Smith et al. 1991, Garnery et al. 1995). Such results reinforced the hypothesis that Spain constituted a hybrid zone of secondary contact between *A. m. intermissa* and *A. m. mellifera* (Cornuet and Garnery 1991, Smith et al. 1991, Smith and Glenn 1995). However, recent authors have suggested that the introgression originated through human import of African lineage honey bees into Iberia, based on fine-scale analysis of mtDNA haplotypes (Garnery et al. 1998b) and assessment of microsatellite variation (Franck et al. 1998a,b).

Apis mellifera ligustica. Classified as Spinola, 1806, it is known as the Italian bee. Considered as the most commonly-kept race in North America, South America, and southern Europe. It belongs to the morphological C lineage and to the mitochondrial C (M) lineages group of *Apis mellifera*. Comparatively well studied, although ecotype variation is not known, it is likely to be diffused by colony movements (Meixner et al. 2013). This subspecies is the most widely exported because of its adaptability to a wide range of climatic conditions, its ability to store large quantities of honey without swarming, and its docility if disturbed. Consequently, *A. m. ligustica* is certainly the honey bee subspecies that is most implicated in human introductions. However, a few undesirable characteristics were also reported. Colonies tend to maintain larger populations through winter, so they require more winter reserves than other temperate zone subspecies. Phenotypically, Italian bees tend to be light colored and mostly leather colored, but some strains are golden. Ruttner (1988) classified both Italian subspecies, *A. m. ligustica* in continental Italy, and *A. m. sicula* in Sicily, within the north Mediterranean branch C, pointing that *Apis mellifera ligustica* bees across the entire Italian Peninsula were considered to be quite similar to *A. m. carnica* from which they differ essentially by the yellow color of their abdominal tergum. While *A. m. sicula* differs greatly from continental

A. m. ligustica, it is separated geographically only by the narrow channel of Messina, and shows morphometric measurements close to those of the Greek subspecies *A. m. cecropia* and *A. m. macedonica* (Ruttner 1988). Efforts to replace native populations with *A. m. ligustica* were made in several countries, such as Israel and Norway (Bar-Cohen et al. 1978). Italy is an important country to investigate honey bee phylogeography: it was potentially a refuge region for European populations during the Pleistocene ice period (Ruttner 1988); it is at the southwestern end of the distribution area of the Mediterranean evolutionary branch (branch C); and in contact with the western European branch (branch M) along the Alpine arc and the African branch (branch A) in Sicily (Ruttner et al. 1978, Franck et al. 2000a). These contacts between evolutionary branches affect the genetic structure of adjacent populations in which nuclear and mitochondrial compartments can be affected differently (Franck et al. 1998). In the phylogeographical study of honey bee populations, and especially in the Mediterranean area, it is necessary to take into account the possible human influences during historic times. For instance, Sicily was a major trading post under the successive domination of Phoenicians, Greeks, Carthaginians, and Italians. Franck et al. (2000) studied the genetic variability in honey bee populations from Italy and Sicily, for the better understanding of the recolonization of northern Europe after Quaternary ice periods, as well as to assess the role of humans on honey bee biodiversity during the last three millennia. There is a hypothesis that hybridization can occur between subspecies of the same or even different lineages through secondary contact during interglacial periods (Badino et al. 1984). For example, the subspecies *A. mellifera ligustica* and *A. mellifera mellifera* are separated by the Alps. However, in an area of northwestern Italy where the mountains are less than 2,000 m in elevation, a hybrid zone has been described (Badino et al. 1983, Manino and Marletto 1984). Similarly, in north-eastern Italy, introgression of morphology and mitochondrial DNA markers has been reported between the Italian subspecies *A. m. ligustica* and neighboring *A. m. carnica* (Meixner et al. 1993). In *A. m. ligustica* populations, most of the genetic studies concerned enzymatic polymorphism at the Malate Dehydrogenase-1 (*MDH-1*) locus in northern Italy (Badino et al. 1982, 1983, Manino and Marletto 1984, Marletto et al. 1984) showed hybridization between *A. m. ligustica* and *A. m. mellifera* along the Alpine arc and the Ligurian coast. Franck et al. (2000a) hypothesized that the distribution of mtDNA polymorphisms found in *A. m. ligustica* is best explained by isolation of honey bee populations containing multiple mtDNA haplotypes into glacial refugia in southern Liguria. The authors have also shown that both subspecies, *Apis mellifera ligustica* and *Apis mellifera sicula*, have a "hybrid" origin, in the sense that their genomes include elements from at least two different evolutionary branches. These results provided new insight into the evolutionary history of European honey bees and helped to better evaluate the impact of Italian queen exportations around the world.

Apis mellifera cypria. Classified as Pollmann, 1879, *A. m. cypria* was shown by Ruttner (1988) to belong to the morphological O lineage and to the C/O considering mitochondrial analysis (Franck et al. 1998). The island of Cyprus is situated at the eastern end of the Mediterranean Sea, south of Turkey (~ 75 km), west of Syria and

Lebanon (~ 105 km), and north of Egypt (~ 380 km). While other island populations and subspecies of honey bees in the Mediterranean Islands have received more scientific interest (Sinacori et al. 1998, Franck et al. 2001, Sheppard et al. 1997, De la Rúa et al. 2001a, 2003), very little is known about the honey bee of Cyprus. The geographic location of Cyprus positions *A. m. cypria*, is in close proximity to subspecies to both the mitochondrial C and O lineages and with the geographic region of transition between them. In the study of Bouga et al. (2005), using ten gene enzymic systems on five Greek populations, allozymic data support for *A. m. cypria* as a distinct subspecies was found, but there was no allozymic support for the distinction of the other subspecies existing in Greece. These authors, using a Wagner tree based on the discrete character parsimony method, supported the hypothesis that the population from Cyprus is the most distant population (Fig. 3.1).

Bouga et al. (2005, 2011) also reported the occurrence of α-GPDH[*]80 and ACPH[*]80 alleles in the Cyprus population, showing lower frequencies in comparison with Greek populations, again supporting a strong genetic differentiation of that population from Greek ones. Besides, Kandemir et al. (2006) presented allozymic data showing that of the six loci of the northern Cyprus honey bees studied, four were polymorphic. These results are strong evidence supporting the existence of a distinct subspecies in Cyprus, a situation reflecting the isolated status of the Cypriot population (Bouga et al. 2005). Papachristoforou et al. (2013), using a combination of mtDNA and microsatellite analyses, verified a double pattern of both C and O lineages in the *A. m. cypria* genome. The authors also concluded that this analysis could provide an accurate criterion to discriminate the domestic honey bee

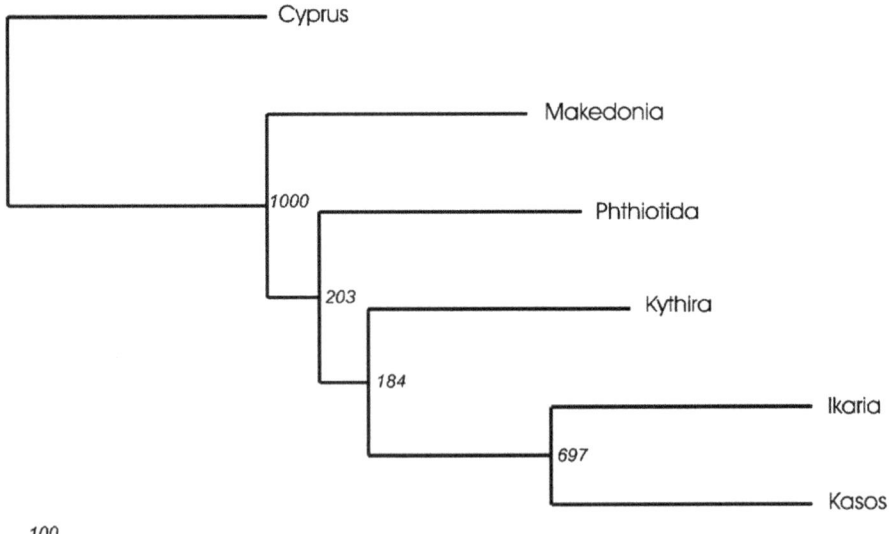

Figure 3.1: Adapted from Bouga et al. (2005), showing the relationships of populations studied in Greece (Macedonia, Phthiotida, Kithira, Ikaria, and Kasos) and Cyprus, using Wagner parsimony dendrogram. Numbers indicate the percentage of 1,000 bootstrap replicates at each node in the consensus tree.

subspecies of the island, which remains homogenous in most of Cyprus. In terms of behavior, this subspecies has the reputation of being more defensive when compared with other European subspecies, especially from Italian subspecies, from which it is isolated by the Mediterranean Sea. Cyprian honey bees have developed effective defensive behaviors to reduce the impact of hornets; these defenses, such as an efficient asphyxia-balling behavior, have been described by Butler (1954), Morse (1978), and Papachristoforou et al. (2008).

—————————— CHAPTER 4 ——————————

The Classic Taxonomy of Asian and European Honey Bees

Dar, S.A.,[1] *Dukku, U.H.,*[2] *Ilyasov, R.A.,*[3,*]
Kandemir, I.,[4,*] *Kwon H.W.,*[3,*] *Lee, M.L.*[5] *and*
Özkan Koca, A.[6]

Diversity of honey bees in the genus *Apis*

Honey bees (genus *Apis*) belong to the family Apidae (social bees) and superfamily Apoidea (all bees) in the insect order Hymenoptera (bees, wasps, ants, etc.). Based on an analysis of the extant and extinct species, Kotthoff et al. (2013) hypothesized that *Apis* originated in Europe in Oligocene and expanded to Asia, through a southeastern migration, and diversified on both continents during Miocene. The bees then invaded North America, from Asia, via Miocene connections across Beringia. At the end of Miocene, *Apis* invaded Africa from Europe, via the Iberian Peninsula, and, subsequently, evolved into the modern *A. mellifera*. Whereas the ancient climatic conditions in Asia favored further diversification that saw the evolution of all the

[1] Entomology in Sher-e-Kashmir University of Agricultural Sciences and Technology of Kashmir, Jammu and Kashmir, 192401, India; sadar@skuastkashmir.ac.in
[2] Department of Biological Sciences, Abubakar Tafawa Balewa University, Bauchi, 740004, Nigeria; udukku@yahoo.com
[3] Division of Life Sciences, Major of Biological Sciences, and Convergence Research Center for Insect Vectors, Incheon National University, Academy-ro 119, Yeonsu-gu, Songdo-dong, Incheon, 22012, Korea; Ufa Scientific Center, Institute of Biochemistry and Genetics Russian Academy of Sciences, Prospect Oktyabrya 71, Ufa, 450054, Russia.
[4] Department of Biology Faculty of Science Ankara University, Beşevler/Ankara, 06100, Turkey.
[5] National Institute of Agricultural Science, 166 Nongsaengmyeong-ro, Iseo-myeon, Wanju-gun, Jeollabuk-do, 55365, Korea; Convergence Research Center for Insect Vectors, Incheon National University, Academy-ro 119, Yeonsu-gu, Songdo-dong, Incheon, 22012, Korea; mllee6@korea.kr
[6] Department of Gastronomy and Culinary Arts Faculty of Fine Arts Maltepe University, Marmara Education Village, 34857, Maltepe/Istanbul Turkey; ayca.queenbee@gmail.com
* Corresponding authors: apismell@hotmail.com; ikandemir@gmail.com; hwkwon@inu.ac.kr

modern species of *Apis*, except *A. mellifera*, the conditions in Europe and North America resulted in the extinction of the genus. Thereafter, *A. mellifera* expanded its territory, from Africa, into Europe and western Asia during Holocene. Meanwhile, North America remained devoid of honey bees until their reintroduction by humans in the 17th century A.D. Examples of fossil honey bees include: *A. armbrusteri* Zeuner 1931; *A. dalica* Engel et al. 2018, and *A. nearctica* Engel et al. 2009 from Miocene Europe, Asia, and North America, respectively. In the case of Africa, the only fossil records so far are those of modern *A. mellifera* from late Quaternary of East Africa. For details on the diversity and distribution of fossil honey bees, see Engel et al. (2018), Kotthoff et al. (2013), Engel et al. (2011), Engel et al. (2009), and Engel (2006).

The number of extant species of *Apis*, and their respective diagnoses has been a matter of debate, over the last couple of decades, and varies from twenty-four (Maa 1953) to ten (Lo et al. 2010) or six (Engel 1999). Most of the controversy surrounds the status of some Southeast Asian populations (Koeniger et al. 2010, Radloff et al. 2011). While several analyses have examined the phylogeny of *Apis*, the following species were recognized in the combined analysis of Engel and Schultz (1997): *A. mellifera* Linnaeus 1758, *A. cerana* Fabricius 1793, *A. koschevnikovi* Enderlein 1906, *A. nuluensis* Tingek, Koeniger and Koeniger 1996, *A. florea* Fabricius 1787, *A. andreniformis* Smith 1858, *A. dorsata* Fabricius 1793, and *A. laboriosa* Smith 1871 (at that time authors did not consider *Apis nigrocincta* specifically distinct from *Apis cerana*). Subsequently, *A. nigrocincta* Smith 1861 was added to this list of honey bee diversity (Fig. 4.1) (Hadisoesilo et al. 1995, Hadisoesilo and Otis 1996, 1998, Engel 1999, Smith et al. 2000, 2003). *Apis* species are divided into three lineages: (1) the cavity-nesting bees with multiple combs (*A. mellifera*, *A. cerana*, *A. koschevnikovi*, *A. nigrocincta*, and *A. nuluensis*); (2) open-nesting dwarf bees (*A. florea*, and *A. andreniformis*), and (3) open-nesting giant bees (*A. dorsata*, and *A. laboriosa*). All the open-nesting species build a single comb. Phylogenetic analyses based on nuclear DNA (nDNA) and mitochondrial (mtDNA) markers also strongly support clustering these species into three distinct groups (Willis et al. 1992, Tanaka et al. 2001a, Arias and Sheppard 2005, Raffiudin and Crozier 2007).

While *A. mellifera* is native to western Asia (Middle and Near East), Africa, and Europe, all the other species are restricted to Asia. Before interference by humans, *A. mellifera* was allopatric with the other species of *Apis* in Asia (Ruttner 1988). Now it is sympatric with *A. florea* in Oman, Jordan, and Sudan, following the introduction of *A. mellifera* in Oman (Dutton et al. 1981) and *A. florea* in the other countries (Lord and Nagi 1987, Mogga and Ruttner 1988, Haddad et al. 2009b). Moreover, *A. mellifera* has been introduced, by man, to all other continents, except Antarctica (Meixner et al. 2013) and is used intensively in pollination and honey production all over the world. Of the nine species of *Apis*, mentioned above, only *A. mellifera* (the European or Western honey bee) and *A. cerana* (the Asian or Eastern honey bee) have been "domesticated" for a long time (Koeniger 1976a), and are of major commercial importance. Thus, this chapter is devoted to the taxonomy of these two species.

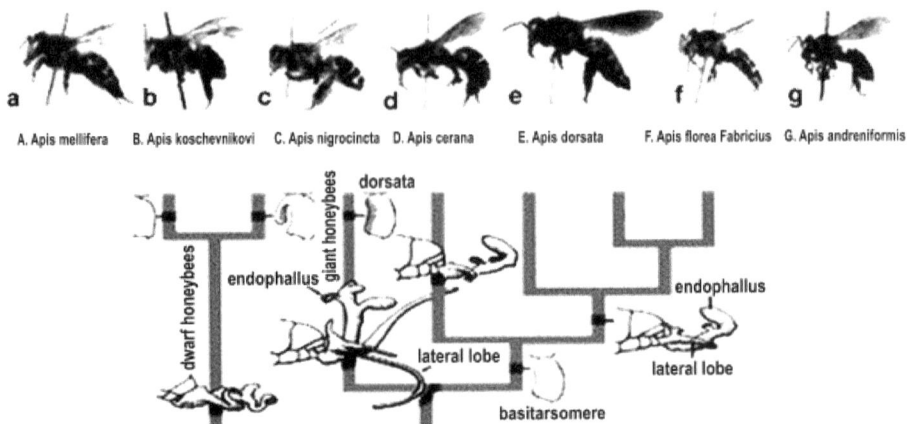

Figure 4.1: Honey bee diversity. *A. Apis mellifera* Linnaeus; B. *Apis koschevnikovi* Enderlein; C. *Apis nigrocincta* Smith; D. *Apis cerana* Fabricius; E. *Apis dorsata* Fabricius; F. *Apis florea* Fabricius; G. *Apis andreniformis* Smith. (Gupta 2014).

Classic taxonomy of the Asian honey bee, *Apis cerana*

Speciation of *Apis cerana* in Asia. *A. cerana*, or the Asiatic honey bee (or the Eastern honey bee), are small honey bees of India and South-eastern Asian countries, such as Malaysia, Indonesia, Philippines, Singapore, Thailand, Vietnam, Bangladesh, and Sri Lanka, and North-eastern Asian countries, such as China, Mongolia, Taiwan, Korea (North and South), Japan, and Far Eastern Russia.

This species is the sister species of *A. koschevnikovi* and both are in the same subgenus as the European honey bee, *A. mellifera*. There are about 20,000 species of bees belonging to the Superfamily Apoidea. Honey bees of genus *Apis* belong to a small sub-group of this superfamily comprising nine species and *A. cerana* is one of the five cavity-nesting species (Arias and Sheppard 2005, Raffiudin and Crozier 2007, Koeniger et al. 2010). The critical importance of this review is the recognition of the genetic diversity present within *A. cerana*. Initial studies on the species refer to "races", "strains", and "subspecies" (Ruttner 1988). Recent studies note that this species may be subject to "cryptic speciation" and that its taxonomy is by no means resolved (Oldroyd and Wongsiri 2006). Three variants, once thought to be members of *A. cerana*, are now recognized as distinct species, e.g., *A. nigrocincta*, *A. koschevnikovi*, and *A. nuluensis* (Lo et al. 2010). Currently recognized variants of *A. cerana* (e.g., *indica, japonica, javana*), to avoid confusion, should be referred to as "genotypes" in particular locations.

Differentiation of *A. cerana* among sympatric medium-sized bees

The sympatric occurrence of *A. cerana* with other medium-sized bees, *A. koschevnikovi, A. nigrocincta,* and *A. nuluensis,* in southeastern Asia, unfortunately, means that previous "*A. cerana*" literature may inadvertently include data derived from other species (Hepburn et al. 2001). To assist in overcoming this problem, Radloff et al. (2011) list biometric characters that, in combination, can discriminate these four species. Firstly the cubital indexes of the forewings, which are 3.9 for *A. cerana*, 7.2 for *A. koschevnikovi*, 3.7 for *A. nigrocincta*, and 2.4 for *A. nuluensis* can quickly separate paired comparisons for all, with the exception of an *A. cerana* and *A. nigrocincta* option. Three characters may be used to separate *A. cerana* from *A. nigrocincta*: (1) the length of the basal part of the radial cell of the forewing, which is 1.2 mm in *A. cerana* and 1.8 mm in *A. nigrocincta*; (2) the length of the apical part of the radial cell, which is 1.8 mm in *A. cerana* and 1.1 mm in *A. nigrocincta*; (3) the length of the labial palp, which is 1.8 mm in *A. cerana* and 3.7 mm in *A. nigrocincta*.

Distribution of *A. cerana*

A. cerana is very widespread across temperate and tropical Asia, reaching from Afghanistan to Korea and Japan, north into the foothills of the Himalayas and eastern Russia, and south through Indonesia (Fig. 4.2) (Ruttner 1988, Crane 1999, Hepburn and Radloff 2011a, Koetz 2013a). *A. cerana* range covers many climatic zones, from tropical rainforest and tropical savannah to mid-latitude grasslands, moist continental deciduous forests to taiga (Hepburn and Radloff 2011b). The current area of *A. cerana* has expanded across the world due to human interference (Koetz 2013b).

A. cerana is also called the oriental honey bee, because they are widely distributed throughout Asia, from Iran in the east to Pakistan in the west and from Japan in the north to the Philippines in the south (Ruttner 1988). Thus, *A. cerana* does not live only in tropical and subtropical areas of Asia, but also in colder areas, such as Siberia and Manchuria, Northern China, and the high mountain area of the Himalayan region (Koeniger 1976a).

The south-eastern Asian range of *A. cerana* on the west of the Wallace Line in Indonesia (Flores Island) later has been extended since a Java genotype of *A. cerana* was introduced during the 1970s to New Guinea (Annand 2009, Anderson 2010). Now *A. cerana* is found throughout New Guinea (including Papua New Guinea), its offshore islands, and on the Solomon Islands (Anderson 2010, Anderson et al. 2012). *A. cerana* is found only in remote mountainous areas, almost like a relic, which could soon become "an endangered species" (Ruttner 1988). However, just the opposite occurred in parts of the Solomon Islands where the introduction of *A. cerana* led to the total extinction of the exotic *A. mellifera* populations on some islands (Anderson 2010).

A. cerana is the third smallest of the nine species of honey bee (Koeniger et al. 2010). Some genotypes of *A. mellifera* are smaller than medium-sized *A. cerana* genotypes, thus there is no significant difference between the smallest genotypes of both species (Ruttner 1988). *A. cerana* was intentionally introduced from Java into the Indonesian province of Papua New Guinea in the late 1970s. It was then established throughout New Guinea (Anderson 1994). In 1993, swarms of *A. cerana* were detected on Boigu, Saibai, and Dauan islands in the Torres Strait (Dunn 1992). *A. cerana* has been intercepted and destroyed on vessels at Australian seaports since

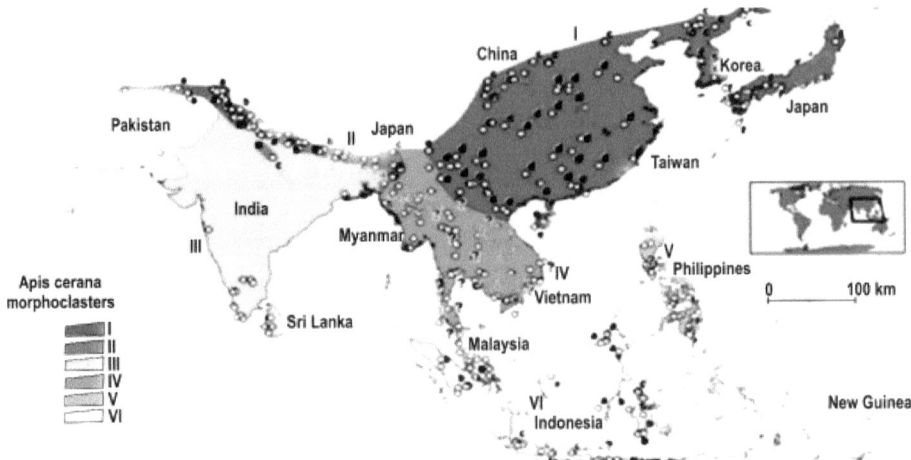

Figure 4.2: The distribution of *A. cerana*. Different colors depict different *A. cerana* morphoclusters based on a complex of multivariate morphometric analysis) (Radloff et al. 2010).

1995, namely Cairns, Brisbane, Melbourne, and South Australia (Barry et al. 2010). A nest was found in Darwin in 1998—it was destroyed and an eradication and surveillance program established (Anderson 2010). In 2003, *A. cerana* was detected over 1,000 kilometers further east on the Solomon Islands (Anderson et al. 2012).

Importance of *A. cerana* in Asia

Eastern honey bee *A. cerana* is an important native pollinator of crops in Asia, and essential producer of honey, wax, etc. (Dietz 1992). Beekeeping has been promoted as an economic concept for honey production in each country. However, its roles in enhancing agricultural productivity and maintaining biodiversity are more important. Until today, most governments have emphasized beekeeping itself to increase honey production, which resulted in an extensive introduction of *A. mellifera* to Asian countries.

A number of resources, time, and many research efforts have been invested to improve the productivity of *A. mellifera* (Partap 2011). Even though *A. mellifera* produces large quantities of honey per colony, the indigenous *A. cerana* has distinct advantages over *A. mellifera*. To produce large quantities of honey, *A. mellifera* requires intensive management, standardized equipment, and larger foraging areas. Furthermore, the race of *A. mellifera* introduced to Asia region is not very suitable for cold, mountainous areas. *A. cerana* exhibits high tolerance of seasonal low temperatures and as a result, it is the first active honey bee on cool mornings in the northern tropics, and the only species to extend significantly north of the tropics (Corlett 2011).

On the other hand, *A. cerana* is highly suitable for small-scale stationary beekeeping. This bee is very well adapted to cold mountain conditions and colonies can be kept at the same place throughout the year without much input. *A. cerana* has adapted to local diseases, parasites, and enemies, and does not need any medicines or chemicals to treat them. Moreover, *A. cerana* bees are better pollinators of early blooming crops and flora (Partap 2011).

It is reported that *A. cerana* are more efficient pollinators of fruit and vegetable crops than *A. mellifera*. The field experiments conducted by Kathmandu valley of Nepal, Partap and Verma (1992, 1994) and Verma and Partap (1993, 1994) showed that the foragers of *A. cerana* started working on cauliflower and cabbage bloom more early in the morning and ceased later in the evening in comparison with *A. mellifera*. Further, in *A. cerana*, pollen collectors outnumbered nectar collectors, while in *A. mellifera*, nectar collectors outnumbered pollen collectors.

A. cerana matches *A. mellifera* in commercial use value and has high potentials for further genetic improvement by selective breeding based on molecular markers. Now *A. cerana* is threatened across its native area due to the spreading of Korean Sacbrood Virus (kSBV) and importing of *A. mellifera* (Choi et al. 2010, Koetz 2013a, Vung et al. 2017, Ilyasov et al. 2018).

Morphological characters for classification of *A. cerana*

The honey bee classification has slowly moved away from the fixed abstractions of the Linnaean system to the analysis of population dynamics in multivariate probability terms (Ruttner 1988, Hepburn and Radloff 1998, Hepburn et al. 2001). Modern classification of honey bees originated from multivariate methods of analysis originally advanced by DuPraw (1964, 1965a), and substantially developed by Ruttner et al. (1978), Ruttner (1988), and Daly (1991, 1992). Ruttner (1988) completed the pioneering task of providing the first multivariate analytical attempt at the comprehensive macroscale honey bee classification for the genus Apis, also including the biogeography of the honey bees in Asia. It provided a motive in many successive regional honey bee studies in Asia (Verma 1990, 1992).

Instead of a plain description of characters in individual honey bees, the multivariate analytical method uses numeric data resulting from exact measurements of colony characters for statistical analyses. A set of morphological characters of body size, color, wing veins, and hairiness is measured. However, there is no universally accepted standard set of characters for use in classical morphometry of honey bees. Ruttner et al. (1978) listed a total of 41 morphological characters of worker honey bees selected from several sources (Alpatov 1948, DuPraw 1964, Goetze 1964) for the statistical analysis of the geographic variability of *A. mellifera*. These are composed of seventeen size characters, thirteen wing vein lengths and angles, seven color characters, three hair characters, and a number of hamuli (Table 4.1 and Fig. 4.3).

These appear to be most favored in most researchers of *A. cerana* as well as *A. mellifera*. Verma et al. (1994) listed total 55 characters for the morphometric study of *A. cerana* in India, which were composed of nineteen characters on the forewing, twelve on the abdomen, ten on the hind wing, six on the tongue, four on the antenna, and four hind legs.

Table 4.1: List of the characters used for the morphometrical analysis of honey bees (Ruttner et al. 1978).

No.	Character	Author
1	Length of hairs on tergite 5	Goetze
3	Width of the tomentum band on the side of tergite 4	Goetze
4	Width of the dark stripe between the tomentum and the posterior rim of the tergite	Goetze
5	Length of the stretched proboscis (glossa + mentum + submentum)	Alpatov
6 – 8	Length of the hind leg (femur No. 6, tibia No. 7, metatatarsus No. 8)	Alpatov
9	Width of metatarsus 3	Alpatov
10 - 12	Pigmentation of tergite 2-1, evaluated according to a scale of 10 grades between the darkest (0) and the brightest (9)	Goetze
13, 14	The diameter of tergite 3 and 4, longitudinal	Alpatov
15	Sternite 3, longitudinal	Alpatov
16, 17	Wax mirror, sternite 3, longitudinal and transversal	Alpatov
18	Distance between wax mirror, tergite 3	Ruttner
19, 20	Sternite 5, longitudinal and transversal	Ruttner
21, 22	Fore wing, length, and width	Alpatov
23, 24	Pigmentation of the scutellum	Ruttner
25, 26	Pigmentation of the labrum	Ruttner
27 - 30	Segment a and b cubital cell 3, right and left	Alpatov
31 - 41	11 angles between lines connecting cross points of the venation on the fore wing (No 31= angle A 4, 32= B 4, 33= D 7, 34= E 9, 35= G 19, 36= J 10, 37= J 16, 38= K 19, 39= L 13, 40= N 23, 41= O 26)	Goetze
42	Numbers of hooks on the hind wing	DuPraw

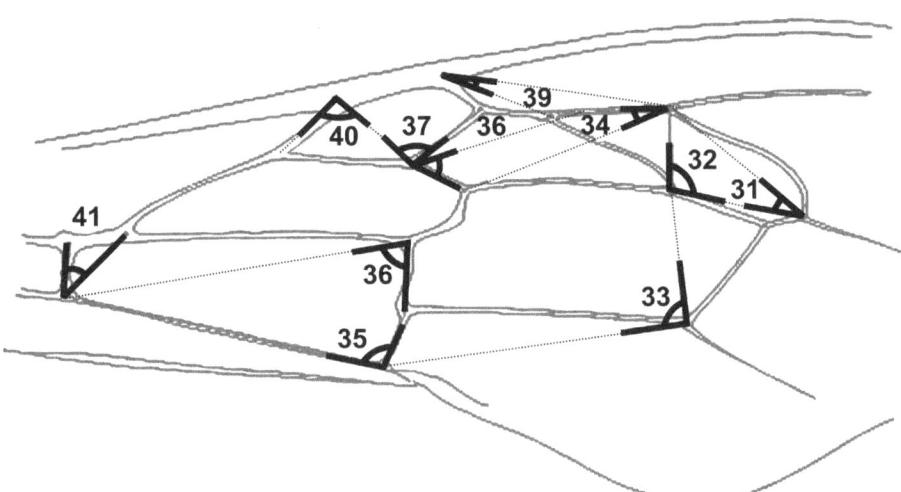

Figure 4.3: Angles of wing venation of *A. cerana* (No. 31–41) (Ruttner 1988).

Consideration of samples for morphometrics of *A. cerana*

For the gene pool of honey bee colonies formed by one queen and many drones (Moritz and Southwick 1992), the intra-colonial genetic variation of quantitative traits, particularly morphological characters, may be higher than intercolonial (Adams et al. 1977, Oldroyd and Moran 1983, Rinderer et al. 1990, Oldroyd et al. 1991). Consequently, in the morphology analysis at the level of population, the deficiency of honey bee samples from each colony can be compensated by the analysis of the greater numbers of colonies. It is proved that the samples of five colonies per locality with 10 bees per colony have shown to be adequate in the analysis of morphological data in studies of honey bee populations (Radloff et al. 2003). It was shown, in the morphology analysis, that the number of sampled colonies affects the variation in the mean character values more than the number of sampled bees per colony. However, the morphology characters related to the size and wing venation are considerably less affected than pigmentation (Alpatov 1929, Falconer 1989, Radloff et al. 2003). It was demonstrated that the seasonal variations affect a large number of morphological characters of *A. cerana* (Mattu and Verma 1984) and *A. mellifera* (Mikhailov 1927b, Gromisz 1962, Antontseva 1975, Mizis 1976, Dianov 1977). Thus, it is recommended that the samples should be collected in summer or autumn, and treated by immersion in the hot water or by ether anesthesia for the extension of the proboscis. The differentiation and identification of honey bees should be based on Principal Component Analysis (PCA), Factor Analysis (FA), k-means clustering, and Linear Discriminant Analysis (LDA) clustering methods in Stat graphics plus, Statistica, SPSS, and JMP computer programs (Meixner et al. 2013).

Morphogenetic divergence of *A. cerana*

Considerable genetic and morphological variation has been shown within *A. cerana* covering its wide range of many climatic zones (Ruttner 1988, Smith et al. 2000, Radloff et al. 2010). Extensive debates have appeared after Fabricius in 1793 first published reclassification and renaming of *A. cerana* species. Recent use of more sophisticated morphological and genetic techniques have started to shed light into the taxonomy and sub-groupings of *A. cerana*.

Morphological diversity of *A. cerana*

Morphologically and genetically, *A. cerana* is subdivided across its range. Most recent studies found that there are six "morphoclusters" (Fig. 4.2), i.e., groupings within *A. cerana* based on complex statistical multivariate morphometrical analyses of 12 morphological characters (Radloff et al. 2010). The genetic strain of *A. cerana* found in New Guinea and the Solomon Islands (as determined by D. Anderson, personal communication) falls within morphocluster VI, which is distributed across southern Thailand, Malaysia, and Indonesia (Fig. 4.4) (Radloff et al. 2010). Morphoclusters V and VI (Philippine and Indo-Malayan clusters, respectively) also occur in the tropical wet climate. All other morphoclusters occur outside wet tropical climates, although some subclusters may fall within wet/dry tropical or subtropical climates (within morphocluster I: Indus, central and eastern China, and *japonica* subclusters,

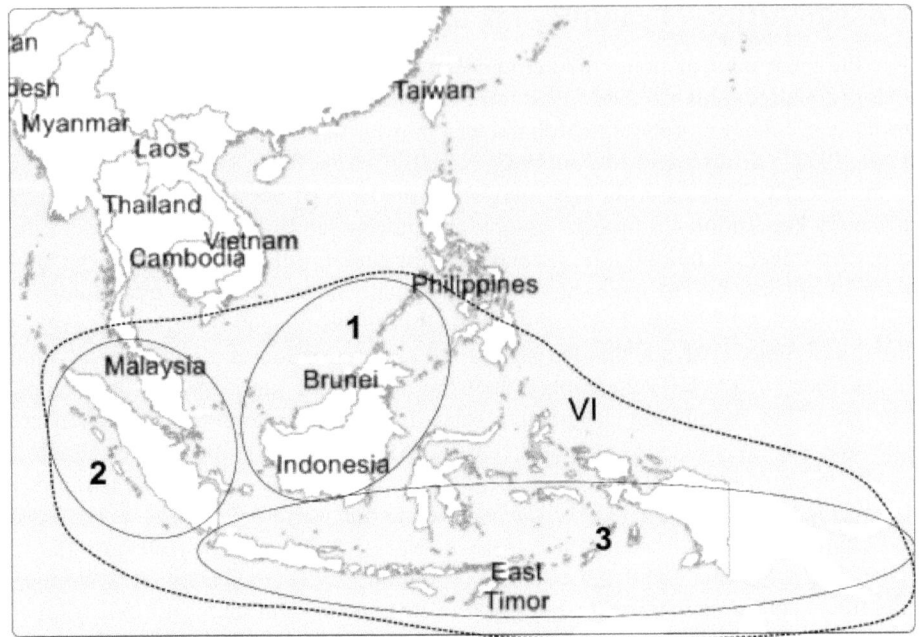

Figure 4.4: The subgroupings found within morphocluster VI, the Indo-Malayan *A. cerana* according to Radloff et al. (2010). (1) Palawan (Philippines), North Borneo (Malaysia) and Kalimantan (Indonesia); (2) Malay Peninsula, Sumatra, and some Sulawesi; (3) Indonesia (Java, Bali, Irian Jaya, some Sulawesi, and Sumatera).

within morphocluster IV: Thailand subcluster). Subtle morphological differentiation has been detected within some of the morphoclusters, which is generally linked to biogeographical and climatic boundaries (Radloff et al. 2010).

Within the Indo-Malayan morphocluster VI (containing *A. cerana* Java genotype), three main subgroups were found (Radloff et al. 2005b, Radloff et al. 2010): (1) Palawan (Philippines), North Borneo (Malaysia), and Kalimantan (Indonesia); (2) Malay Peninsula, Sumatra, and some Sulawesi; (3) Indonesia (Java, Bali, Irian Jaya, some Sulawesi, and Sumatera). It must be noted here that morphological subdivision, particularly based on extremely subtle changes as found in *A. cerana*, does not imply division into strains, nor changes in behavior or genetics.

Genetic diversity of *A. cerana*

Most recent genetic studies generally agree with these morphological studies. They divide the species into four main genetic groups (Fig. 4.4) (Smith 2011, Smith et al. 2000). One of these groups (the Sundaland group) corresponds with morphocluster VI (= Indo-Malayan *A. cerana*) that contains *A. cerana* Java genotype (Fig. 4.5). This genetically and morphologically distinct subgroup is confined to the Asian tropics south of 10°N latitude (Smith et al. 2000, Rueppell et al. 2011, Smith 2011, Songram et al. 2006). Further genetic subdivision can be found within the Sundaland/Indo-Malayan group. Relevant here is the fact that *A. cerana* samples from Java,

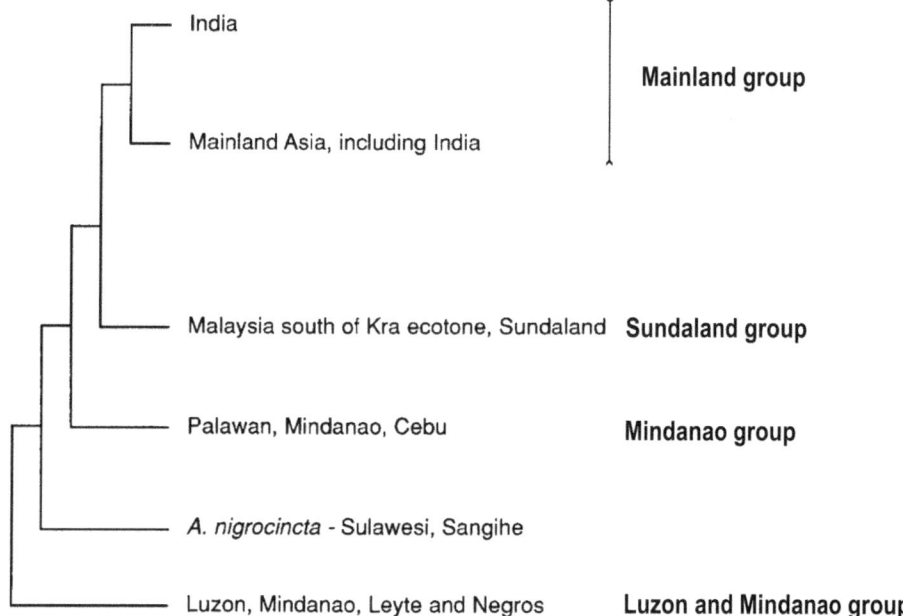

Figure 4.5: Phylogenetic tree of the main *A. cerana* haplotypes (mitochondrial DNA subgroupings) and their corresponding geographic regions (Smith et al. 2000). Also shown is the placement of *A. nigrocincta* within the tree.

Bali, Flores, Timor, and Sulawesi cluster together, as do samples from Bali and Lombok (Smith et al. 2000, Smith 2011). Genetic clustering within the Sundaland/Indo-Malayan group seems to be linked to location upon the Sunda continental shelf and sea level fluctuations during Pleistocene glaciations. Islands on the Sunda Shelf (Sumatra, Java, Bali, Lombok, Timor, and Flores) would have been connected by dry land during glaciations periods, whereas Borneo and Sulawesi remained separated by deep channels (Smith 2011). It needs to be noted here that sharp genetic boundaries between populations (e.g., between the Mainland Asia group and the Sundaland/Indo-Malayan group) are linked specifically to the genetic marker used. Mitochondrial DNA is maternally inherited, and so any gene flow and admixture between populations represents female gene flow (migration), whereas drone gene flow is "invisible" using this marker. The nuclear genome differences evolved much more slowly and were inherited both maternally and paternally (Smith 2011). In addition, mitochondrial DNA gives a good picture of past population genetic events, while it gives little or no information about adaptation to local environments (Smith 2011). This means that differentiation within the Sundaland/Indo-Malayan group (as, indeed, between, and within other morphoclusters) is very slight, and broad habitat differences rather than genetic differences may explain differences in behavior.

A. cerana Subspecies/Races

Although *A. cerana* bees must have shared a common ancestor with *A. mellifera*, they have evolved into separate species. It is not possible to cross *A. cerana* with

A. mellifera even using instrumental insemination, because the two species are now genetically incompatible and viable eggs do not result from the cross-fertilization. Other differences include their differing reactions to diseases, infestations, and predators. *A. cerana* can tolerate *Varroa* and has developed an effective defense strategy against the Giant Hornet, against which *mellifera* bees have no defense. *A. cerana* is, however, highly susceptible to the Acarine mite, which arrived with the introduction of *mellifera* bees into *cerana* territory. It is also highly susceptible to sac brood and foulbrood, but not markedly so to *Nosema*. A high degree of variation in size and coloration probably may reflect the ecological diversity of *A. cerana*. The influence of latitude and altitude on the size of worker bees was also found for *A. cerana* in Vietnam. This wide climatic range has led to substantial variations among geographical races of the honey bees, particularly, tropical and temperate races differ in size of the body, nest, and colony, as well as in swarming and absconding behavior. The temperate and sub-tropical races store greater quantities of food than the tropical races, which are characterized by more migrating, swarming, and absconding behavior than the former. The intraspecific classification of the Asiatic honey bee species, *A. cerana* is in a state of flux and uncertainty (Hepburn et al. 2001). Studies carried out by the International Centre for Integrated Mountain Development (ICIMOD) reveal that *A. cerana* populations can be divided into three subspecies, namely *A. c. cerana*, *A. c. himalaya*, and *A. c. indica*. Of these, *A. cerana* is distributed over Northwest Himalayas in India, Northwest Frontier Province of Pakistan, and Jumla region of Nepal. *A. c. himalaya* is found in hills of Nepal, Uttar Pradesh, North-east Himalayas, and Bhutan. *A. c. indica* is found in plain areas and foothills of the region.

Similar studies carried out in China reveal the presence of five sub-species of *A. cerana*. These include *A. c. cerana*, *A. c. skorikovi*, *A. c. abaensis*, *A. c. hainanensis*, and *A. c. indica* (Zhen-Ming et al. 1992, Partap 1999). The morphometric analysis of *A. cerana* F. in China showed that the "Chinese Eastern race" belonging to "*A. c. cerana*" and the "South Yunnan race" being "*A. c. indica*", the "South Yunnan race", and the "Aba race" (Aba *cerana*) could be discriminated. However, this analysis failed to discriminate among "South Yunnan race" (*A. c. indica*), "Hainan race" (Hainan *cerana*), and "Tibet race" (Tibet *cerana*). Molecular analyses revealed that the mitochondrial genotypes of *A. cerana* were the same as that of all samples originated from India without variation and belonging to the "Mainland Asia" group of *A. cerana*. It was approved that there was abundance for mitochondrial genotypes of *A. cerana* in Southern Gansu and Northern Aba area. The description of the following species is available in the literature.

1. *A. c. cerana*—This subspecies with the biggest body size of *A. cerana* occurs in northern parts of China, in the northwest of India, in the north of Pakistan and Afghanistan, and in the north of Vietnam. On average, the proboscis and forewing length measure 5.25 and 8.63 mm respectively, and they are found in Afghanistan, Pakistan, North India, China, and North Vietnam.

2. *A. c. indica*—This subspecies has the smallest body size, lives in the south of India, in the south of Thailand, Cambodia, and Vietnam. The length of proboscis and forewing is 4.58–4.78 mm and 7.42–7.78 mm, respectively (Ruttner 1988).

It is also distributed in Sri Lanka, Bangladesh, Burma, Malaysia, Indonesia, and the Philippines.

3. *A. cerana japonica*—This subspecies is endemic in Japanese temperate climates except for the island of Hokkaido. This subspecies is divided into two separate ecotypes: Honshu and Tsushima. The body size of *A. c. japonica* is relatively big, with an average proboscis length of 5.18 mm and an average forewing length of 8.69 mm. *A. c. japonica* gradually has been replaced by introduced *A. mellifera* (Okada 1986).

4. *A. c. skorikovi* or *A. c. himalaya*—The body size of this subspecies is intermediate between *A. c. cerana* and *A. c. indica*. It occurs in the east of the Himalayas from Nepal to northern Thailand. On average, the proboscis and forewing length measure 5.14 and 8.03 mm, respectively. It is native to Asia between Afghanistan and Japan, and from Russia and China in the north to southern Indonesia. They were recently introduced to Papua New Guinea and found in Central and east Himalayan mountains (Ruttner 1988). *A. cerana* builds a nest consisting of a series of parallel combs, similar to *A. mellifera*, and builds its nest within a cavity.

5. *A. c. nuluensis*—It is a subspecies of honey bee described in 1996 by Tingek, Koeniger, and Koeniger (Tingek et al. 1996). The geographic distribution of the subspecies is the Southeast Asian island of Borneo, politically divided between Indonesia, Malaysia, and Brunei. *A. c. nuluensis* is one of a number of Asiatic honey bees, including the more obscure *A. koschevnikovi* and *A. nigrocincta* (the latter of which has habitat on nearby Sulawesi and Mindanao islands). While this was originally described as a species, it has since been determined to represent a geographic race (subspecies) of the widespread *A. cerana* (Engel 1999). Like many honey bees, *A. c. nuluensis* is liable to infestation by the parasitic *Varroa* mite, although in this case the particular species is *Varroa* underwood. (In this aspect, *A. c. nuluensis* is similar to *A. nigrocincta*).

A. cerana morphoclusters

Very recently, Radloff et al. (2010) have classified this bee into six clusters cited hereunder:

1. Morphocluster I: named "Northern *cerana*". The bee extends from northern Afghanistan and Pakistan through northwest India, across southern Tibet, northern Myanmar, China, and then north-easterly into Korea, far eastern Russia, and Japan. These bees were previously named as follows: *A. c. skorikovi, A. c. abansis, A. c. abanensis, A. c. bijjieca, A. c. cathayca, A. c. cerana, A. c. fantsun, A. c. hainana, A. c. hainanensis, A. c. heimifeng, A. c. indica, A. c. japonica, A. c. javana, A. c. kweiyanga, A. c. maerkang, A. c. pekinga, A. c. peroni, A. c. skorikovi, A. c. shankianga,* and *A. c. twolareca.*

   Six sub-clusters or populations are morphometrically discernible within this morphocluster:

   (a) An "Indus" group in Afghanistan, Pakistan, and Kashmir.

(b) A "Himachali" group in Himachal Pradesh, India.

(c) An "Aba" group in southern Ganshu, and central and northern Sichuan provinces in China, northern China, and Russia (larger bees).

(d) A subcluster in central and eastern China.

(e) A "southern" *cerana* sub-cluster in southern Yunnan, Guangdong, Guangxi, and Hainan in China.

(f) A "*Japonica*" group in Japan and Korea.

2. Morphocluster II: named "Himalayan *cerana*". This includes the bees of northern India: (a) northwest, (b) north-east; and some of southern (c) Tibet and Nepal. These bees have previously been named *A. c. skorikovi*, *A. c. indica*, *A. c. himalayana*, and *A. c. himalaya*. Two subclusters are discernible within this morphocluster: (a) the bees of the northwest the "Hills" group, and (b) those of the north-east, the "Ganges" group.

3. Morphocluster III: named "Indian Plains *cerana*" occurs across the plains of central and southern India and Sri Lanka as a fairly uniform population, long known as "plains *cerana*" for this subcontinent. These bees have only previously been termed *A. c. indica*. The variety of "Plains" within Morphocluster III (previously *A. cerana indica*) has recently been genetically split into a new species (*A. indica*) (Lo et al. 2010).

4. Morphocluster IV: named "Indo-Chinese *cerana*" form a compact group in Myanmar, northern Thailand, Laos, Cambodia, and more southern Vietnam. Morphocluster IV bees have previously been named *A. c. indica* and *A. c. javana*.

5. Morphocluster V: named "Philippine *cerana*" is restricted to the Philippines, but excluding most of Palawan Island. The bees of this cluster have previously been named as *A. philippina*, *A. c. philippina*, and *A. c. samarensis*. Within these islands, there are subclusters, and we term these bees respectively after the major island groups there: "Luzon" bees, "Mindanao" bees, and "Visayas" bees.

6. Morphocluster VI: named "Indo-Malayan *cerana*", extend from southern Thailand, through Malaysia, and Indonesia. This large area consists of a morphometrically rather uniform bee below the South China Sea. These bees have been previously termed as *A. cerana*, *A. indica*, *A. javana*, *A. c. johni*, *A. lieftincki*, *A. peroni*, *A. Vechti linda*, and *A. v. vechti*. Three subclusters are discernible within this morphocluster:

(a) Philippines (Palawan), Malaysia (North Borneo), Indonesia (Kalimantan) bees.

(b) Malay Peninsula, Sumatera, and some Sulawesi bees.

(c) Indonesia (Java, Bali, Irian Jaya, some Sulawesi, and Sumatera).

A. cerana ecotypes in India

Comprehensive studies on the biometry and taxonomy of *A. cerana* in India revealed intra-specific variation in Indian *A. cerana* into seven ecotypes indicated

Table 4.2: Ecotypes of *A. cerana* F. in India.

Geographic region	Latitude	Altitude	Location of sample collection	Remarks
Kashmir Valley	34°05'	1.586	Srinagar, Jammu and Kashmir	Largest ecotype in the country
Western Himalayas	31°43'	761	Mandi, Himachal Pradesh	Possibly includes the next two variants
Western Sub-Himalayas	30°05'	700	Kangra, Himachal Pradesh	Possibly variant of Western Himalayas
Western Sub- Himalayan Foot Hills	30°10'	630	Ranipokhari, Uttar Pradesh	Possibly variant of Western Himalayas, and not ecotype
Eastern Himalayas	26°53'	1.500	Kurseong, West Bengal	Verma (1992) proposes 3 races in this region
Indo-Gangetic Plains and Aravalli Hills	29°13'	440	Haldwani, Uttar Pradesh	Mahabaleshwar included due to its high altitude
	26°06'	53	Muzaffarpur, Bihar	
	26°05'	54	Guahati, Assam	
	24°36'	1.195	Mount Abu, Rajasthan	
	17°56'	1.382	Mahabaleshwar, Maharashtra	
Central Peninsula	20°48'	27	Cuttack, Orissa	
	17°50'	767	Lammasingi, Andhra Pradesh	
	17°00'	670	Petlond, Maharashtra	
Western and Eastern Ghats	15°20'	700	Castle Rock, Karnataka	Kodaikanal included due to its high altitude
	14°57'	700	Yellapur, Karnataka	
	12°57'	650	Sakleshpur, Karnataka	
	10°14'	2.343	Kodaikanal, Tamilnadu	
Western and Eastern Peninsular Coastal strips	14°25'	0	Kumtha, Karnataka	Smallest ecotype in the country
	11°55'	0	Pondicherry, Pondicherry	
	10°46'	97	Palghat, Kerala	
	08°44'	51	Tirunelveli, Tamil Nadu	
	08°05'	37	Kanyakumari, Tamil Nadu	

by Kshirsagar (1983) and redefined here (Table 4.2). It is possible that by further detailed investigations, additional ecotypes and races can be found.

Notes on *A. cerana* in North-east Asian countries

Two North-east Asian *A. cerana* subspecies are described recently using mtDNA (complete genome), nDNA (two genes), and morphology (six parameters): *A. cerana koreana* and *A. cerana ussuriensis* (Ilyasov et al. 2018, Ilyasov et al. 2019). The subspecies *A. c. koreana* and *A. c. ussuriensis* have been separated from Chinese

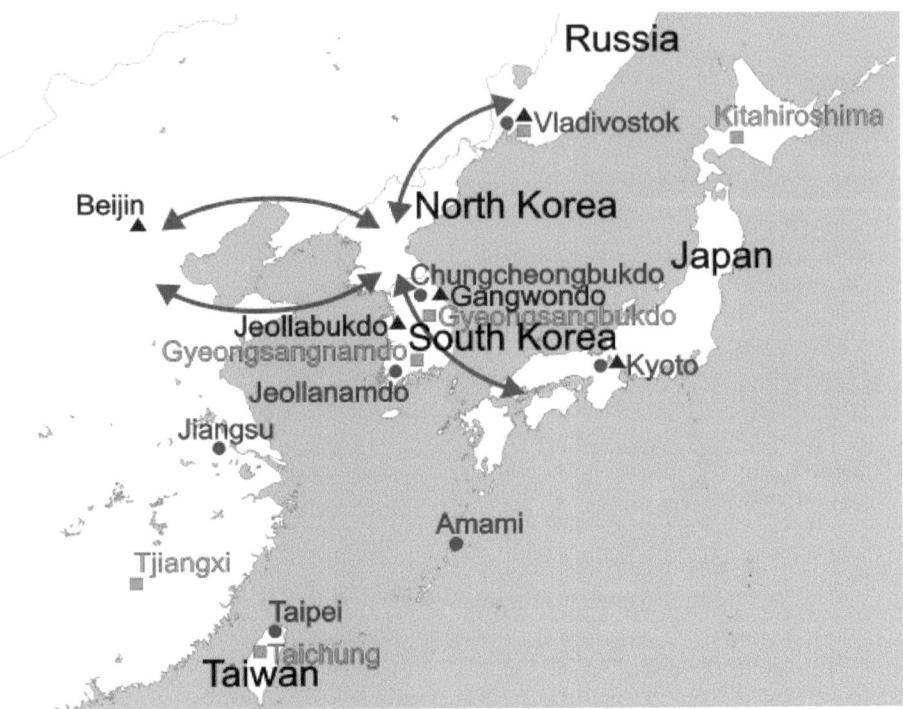

Figure 4.6: Geographical distribution of *A. cerana* samples used in a comparative analysis of complete mtDNA sequences and two possible ways of their migration in northern Asia (Ilyasov et al. 2019).

A. c. cerana and Japanese *A. c. japonica* subspecies by 0.8% in nDNA, by 2.6% in mtDNA, and by 2.2% in morphology, on average.

The most of all historical migration from South to North Asia, it has been assumed, took place across the Korean Peninsula in two directions: to Far East of Russia and Japan archipelago during glaciation period 30 thousand year ago (Ilyasov et al. 2018, Ilyasov et al. 2019) (Fig. 4.6). The phylogenetic trees based on mtDNA, nDNA, and morphology totally matched each other, which show the separate location of all four northern subspecies *A. c. cerana*, *A. c. japonica*, *A. c. koreana*, and *A. c. ussuriensis*. The average phylogenetic tree constructed using a Neighbor-joining algorithm and Euclidian distances based on the all mtDNA, nDNA, and morphology data of *A. cerana* samples and rooted with the *A. mellifera* out-group sample shows the difference between analyzed subspecies (Ilyasov et al. 2018, Ilyasov et al. 2019) (Fig. 4.7).

Classic taxonomy of the European honey bee, *A. mellifera*

Evolution and distribution of European honey bee *A. mellifera*. This section includes classical morphometric and geometric morphometric studies on the lineages and subspecies of the European honey bee, *A. mellifera*. Firstly, *A. mellifera* species and subspecies are reviewed briefly. Then, the methodologies used to distinguish the European bee lineages and subspecies are mentioned and major studies in this area

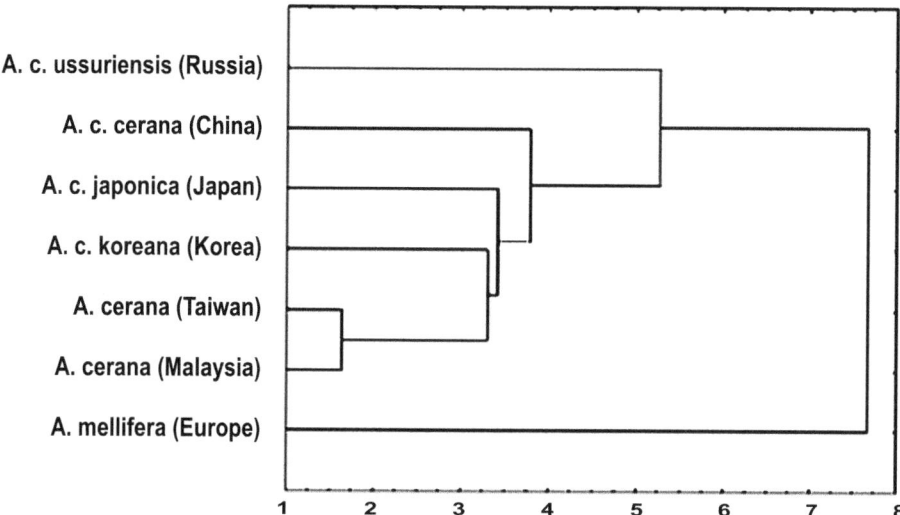

Figure 4.7: The consensus phylogenetic tree constructed using a Neighbor-joining algorithm and Euclidian distances based on the mtDNA, nDNA, and morphology data of *A. cerana* samples and rooted with the *A. mellifera* as an out-group sample.

are discussed. The geographical origin of *A. mellifera* is not certain. Whereas some authors (Ruttner et al. 1978, Ruttner 1988, Garnery et al. 1993, Han et al. 2012, Wallberg et al. 2014) favor an Asian origin, others (Wilson 1971, Whitfield et al. 2006b, Kotthoff et al. 2013) favor an African one.

A. mellifera is of great importance to agriculture, environment, and the global economy. This species has a wide geographical range and subspecies occur naturally throughout almost all of Europe, Africa, Near East and Central Asia (Ruttner 1988, Sheppard and Meixner 2003, Chen, C. et al. 2016), and these subspecies are distinguishable based on behavior, physiology, and morphology (Ruttner 1988). In addition, subspecies of the western honey bee have been spread extensively beyond their natural range due to economic benefits related to pollination and honey production. Across this large area, the species occupies quite varied ecological regions, from deserts to tropical rainforests and from mountainous regions to swamps. The combined effect of ecological changes, cooling, and vegetation changeover has led to widespread morphological and behavioral diversity in the species. Within the wide distribution area, numerous subspecies have been described (last reviewed by Engel 1999, Ruttner 1988, Sheppard et al. 1997). Using classical morphometry, Ruttner (1988) classified *A. mellifera* into 24 geographic subspecies. In a recent review of the intraspecific classification of *A. mellifera*, Engel (1999) added four subspecies and upheld the synonym, *A. m. remipes*, Gerstäcker as the valid name for Ruttner's *A. m. armeniaca* and *A. m. jemenitica* Ruttner as the valid name for Ruttner's *A. m. yemenitica*. In addition, three new subspecies have been described, namely, *A. m. pomonella* (Sheppard and Meixner 2003), *A. m. simensis* (Meixner et al. 2011), and *A. m. sinisxinyuan* (Chen, C. et al. 2016). Details of *A. mellifera* subspecies are given in Table 4.3 and Fig. 4.8.

Table 4.3: *Apis mellifera* Linnaeus, subspecies based on Ruttner (1988, 1992), Engel (1999), Sheppard and Meixner (2003), Meixner et al. (2011), Chen, C. et al. (2016).

Adapted Regions	Subspecies-Scientific Name	Common Name	Morphometric lineage	Distribution
Near East -Oriental (Eastern Mediterranean and Iran)	*Apis mellifera adami* Ruttner 1975	Crete Honey bee	O	Mediterranean island of Crete
	Apis mellifera pomonella Sheppard and Meixner 2003	The Tien Shan Honey bee	O	Tien Shan Mountains of Central Asia
	Apis mellifera cypria Pollman 1879	The Cyprian Honey bee	O	Cyprus only
	Apis mellifera syriaca Skorikov 1929	The Syrian Honey bee	O	Along the eastern shores of the Mediterranean Sea, from Syria in the north, to the Negev Desert in the south.
	Apis mellifera meda Skorikov 1929	The Median Honey bee	O	Iran, Iraq, northern Syria, and southern Turkey
	Apis mellifera caucasica Gorbachev 1916 Reinstated name: *Apis mellifera caucasia* Engel 1999	The Caucasian Honey bee	O	Caucasus Mountains
	Apis mellifera armeniaca Skorikov 1929 Reinstated name: *Apis mellifera remipes* Gerstäcker 1862	Armenian Honey bee	O	Armenia
	Apis mellifera anatoliaca Maa 1953	The Anatolian Honey bee	O	Turkey

Table 4.3 contd. ...

...Table 4.3 contd.

Adapted Regions	Subspecies-Scientific Name	Common Name	Morphometric lineage	Distribution
Tropical Africa	*Apis mellifera lamarkii* Cockerell 1906	The Egyptian Honey bee	A	Nile Valley of Egypt
	Apis mellifera yemenitica Ruttner 1988 Reinstated name: *Apis mellifera jemenitica* Engel 1999 [1]	The Arabian or Nubian Honey bee	A	Arid zones of eastern Africa and the Arabian Peninsula. Found in Chad, Saudi Arabia, Somalia, Sudan, and Yemen
	Apis mellifera litorea Smith 1961	The East African Honey bee	A	Eastern coast of tropical Africa
	Apis mellifera adansonii Latreille 1804	The West African Honey bee	A	From *Niger* in the north to Zambia in the south; and from Senegal in the west to Sudan in the east
	Apis mellifera scutellata Lepeletier 1835	The African Honey bee	A	Southeastern Africa, from South Africa to Somalia
	Apis mellifera monticola Smith 1961	The East African Mountain Honey bee	A	Mountains of eastern Africa
	Apis mellifera capensis Escholtz 1821	The Cape Honey bee	A	Cape region of South Africa
	Apis mellifera unicolor Latreille 1804	The Malagasy Honey bee	A	Madagascar
	Apis mellifera simensis Meixner et al. 2011 [2]	The Ethiopian Mountain Honey bee	A	Mountain systems of Ethiopia
Eastern Mediterranean and Southeastern European	*Apis mellifera macedonica* Ruttner 1988	The Macedonian Honey bee	C	From southern Romania in the north to northern Greece in the south
	Apis mellifera ligustica Spinola 1806	The Italian Honey bee	C	Italian Peninsula
	Apis mellifera carnica [3] Pollman 1879	The Carnolian Honey bee	C	South of the Alps, west into northern Italy, and east into Serbia and Romania
	Apis mellifera cecropia Kiesenweiter 1860	The Greek Honey bee	C	Greece and surrounding Aegean islands
	Apis mellifera sicula Montagano 1911 Reinstated name: *Apis mellifera siciliana* Grassi 1881	The Sicilian Honey bee	C/A	Mediterranean island of Sicily

Table 4.3 contd. ...

...Table 4.3 contd.

Adapted Regions	Subspecies-Scientific Name	Common Name	Morphometric lineage	Distribution
Western Mediterranean and Northwestern European	*Apis mellifera mellifera* Linnaeus 1758	The Western or European Honey bee	M	Western and central Europe
	Apis mellifera iberica Goetze 1964 New name: *Apis mellifera iberiensis* Engel 1999	The Iberian Honey bee	M	Iberian Peninsula
	Apis mellifera sahariensis Baldensperger 1924	The Saharan Honey bee	M	Northwestern Africa, along the southern side of Atlas mountains
	Apis mellifera intermissa Maa 1953 [4]	The Tellian Honey bee	M	North of the Atlas Mountains, from Morocco in the west to Tunisia in the East
	Apis mellifera ruttneri Sheppard et al. 1997	The Maltese Honey bee	M/A	Mediterranean island of Malta only
Lesser known honey bee subspecies	*Apis mellifera artemisia* Engel 1999 [5]	The Russian Steppe Honey bee	Not available	Central Russian steppes
	Apis mellifera sinisxinyuan Chen et al. 2016	Xinyuan Honey bee	Not available	Xinyuan, China
	Apis mellifera sossimai Engel 1999	The Ukrainian Honey bee	Not available	Mostly in Ukraine
	Apis mellifera taurica Alpatov 1938 [6]	The Crimean Honey bee	Not available	Along the north-central shores of the Black Sea, in the Crimea

[1] *A. m. nubica* Ruttner 1976 is a synonym of *A. m. jemenitica* while *A. m. sudanensis* and *A. m. bandasii*, nomina *nuda*, are used to describe populations of this subspecies (Engel 1999).

[2] A new subspecific name, for the honey bees of Ethiopia, duly proposed by the authors to replace several nomina *nuda*, earlier reported in the literature (Meixner et al. 2011).

[3] *A. m. carpathica* Foti et al. 1965, according to Ruttner (1988), is a sub-population of *A. m. carnica*.

[4] *Apis mellifera intermissa* Maa and *A. m. major* Ruttner 1978 are synonyms (Engel 1999).

[5] Replacement names proposed by Engel (1999) for *A. m. acervorum* Skorikov 1929 and *A. m. cerifera* Gestäcker 1862, respectively, for preoccupied subspecific taxa.

[6] This subspecies was dropped by Ruttner (1988), due to insufficient information, but upheld by Engel (1999).

Classical morphometry of *A. mellifera*

Classical morphometry (also referred to as traditional morphometry), based on measurement of morphometric characters, has been widely used to first distinguish the subspecies and then group them into evolutionary lineages (Alpatov 1929, DuPraw 1964, 1965a, Daly et al. 1982, Ruttner 1988). The earlier methods of classifying honey bees were based mainly on qualitative descriptions of morphology.

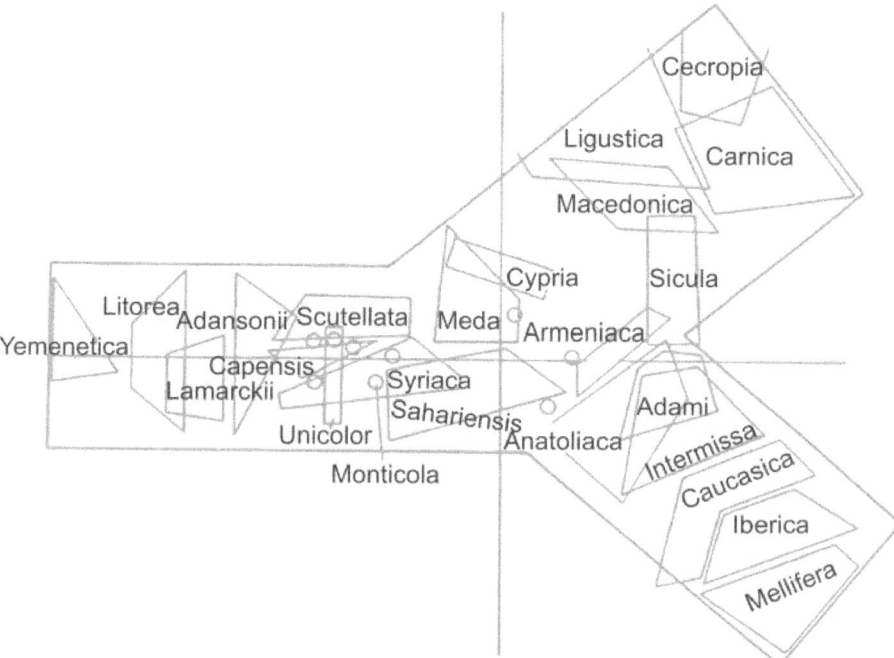

Figure 4.8: Graphical presentation of the principal component analysis of all 24 subspecies of *A. mellifera* (Ruttner 1988).

Though adequate in discriminating higher taxa, these methods proved insufficient at discriminating subspecies of honey bees. This necessitated the evolution of classical morphometry. Instead of mere description of characters of individual honey bees, this method uses numeric data resulting from exact measurements, from which means of colony characters are obtained for statistical analyses (Ruttner 1988). The concept of numerical taxonomy was introduced into honey bee taxonomy by DuPraw (1964, 1965b), and further elaborated by Ruttner et al. (1978). A set of morphological characters of body size, color, and pilosity are measured. Although there is not yet a universally accepted standard suite of characters for use in classical morphometry, the standard 36 characters compiled, from several sources, and used by Ruttner (1988) in his monograph or a subset of them, appear to be most favored (Table 4.4).

Indexes calculated between the 36 morphometric characters are also utilized in various studies. Besides, there are additional characters that are rarely measured (Meixner et al. 2013). Meixner et al. (2013) reported that character sets given in Table 4.4 were used in many studies, but there was considerable variation in the number and selection of characters, used in many studies starting with Ruttner's study. In particular, the wing, which is the most commonly used part for distinguishing species—its shape has been characterized using angles and vein length measurements (Ruttner 1988). The usefulness of wing angles in phylogenetic studies was well shown by Diniz-Filho et al. (2000).

Table 4.4: Standard morphometric characters. For characters described in Ruttner (1988), characters are given, with abbreviations according to Ruttner et al. (1978). The abbreviations used in the last column are taken from Meixner et al. (2013).

Character	Ruttner	Abbreviations	Character	Ruttner	Abbreviations
HAIR			COLOR		
Length of cover hair on tergite 5	H	HLT5	Pigmentation of scutellum, Cupolla	Sc	PSC1
Width of tomentum on tergite 4	A	TOM A	Pigmentation of scutellum, B and K	B, K	PSC2
Width of stripe posterior of tomentum	B	TI			
			WING		
SIZE			Forewing		
Proboscis	-	PROBL	Fore wing length	F_L	FWL
Femur	Fe	FEM	Fore wing width	F_W	FWW
Tibia	Ti	TIB	Cubital vein, distance a	a	CUBA
Basitarsus length	M_L	TAL	Cubital vein, distance b	b	CUBB
Basitarsus width	M_T	TAW	Forewing-angles		
Tergite 3, longitudinal	T_3	T3	Wing angle A1	A1	A1
Tergite 4, longitudinal	T_4	T4	Wing angle A4	A4	A4
Sternite 3, longitudinal	S_3	LS3	Wing angle B3	B3	B3
Wax mirror of sternite 3 longitudinal	W_L	WML	Wing angle B4	B4	B4
Wax mirror of sternite 3, transversal	W_T	WMT	Wing angle D7	D7	D7
Distance between wax mirrors st. 3	W_D	WD	Wing angle E9	E9	E9
Sternite 6, longitudinal	L_6	S6L	Wing angle G7	G7	G7
Sternite 6, transversal	T_6	S6T	Wing angle G18	G18	G18
Length of the hind leg	$Fe + Ti + M_L$	LEG	Wing angle J10	J10	J10
			Wing angle J16	J16	J16
COLOR			Wing angle K19	K19	K19
Pigmentation of tergite 2	-	PT2	Wing angle L13	L13	L13
Pigmentation of tergite 3	-	PT3	Wing angle N23	N23	N23
Pigmentation of tergite 4	-	PT4	Wing angle O26	O26	O26

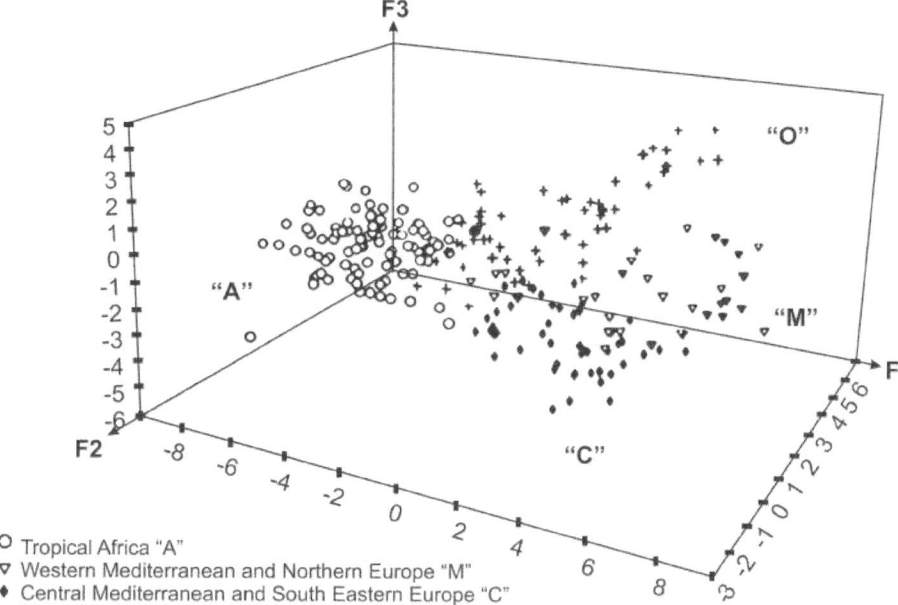

Figure 4.9: Spatial allocation of groups of subspecies resulting from a principal components analysis on morphometric characters (Kauhausen-Keller et al. 1997).

The details of the subspecies of *A. mellifera*, presently recognized on the basis of morphometric characters and grouped into four evolutionary lineages, are shown in Table 4.3 and Fig. 4.8 (Ruttner 1988, 1992, Sheppard et al. 1997, Engel 1999, Sheppard and Meixner 2003, Meixner et al. 2013, Chen, C. et al. 2016). These subspecies are also described as "geographic races" because their distributions correspond to distinct geographic areas.

Kauhausen-Keller et al. (1997) carried out morphometric studies on the micro taxonomy of the *A. mellifera* species. The data consisted of 34 characters taken from the morphometric data bank of honey bees in Oberursel, Germany (Ruttner et al. 1978) were used in principal components analysis (PCA). The analysis was performed on 252 samples representing 21 subspecies of *A. mellifera*. In this study, 3-D scatter-plot of PCA shows four clearly separated branches of four evolutionary lineages: "A" with samples from tropical Africa, "C" with samples from southeastern Europe and the central Mediterranean, "M" with samples from the western Mediterranean and North Europe, and "O" with Caucasian and Near East samples (Fig. 4.9). In addition, in the PCA analysis applied in this study, the distribution of the subspecies in each evolutionary lineage is revealed.

In addition to classical morphometry, biochemical and molecular methods, including analyses of isozymes, mitochondrial DNA polymorphism, nuclear DNA, and microsatellites, have been used to evaluate subspecific diversity of *A. mellifera* (reviewed in Sheppard and Smith 2000). Molecular analyses mostly confirmed the morphological lineages, with a few minor discrepancies (Franck et al. 1998, Whitfield et al. 2006b, Alburaki et al. 2011).

Geometric morphometrics of *A. mellifera*

An emerging variant of classical morphometry is geometric morphometrics (GM), which analyses the shape of an organ (such as the wing) based on the coordinates of selected landmarks, such as the intersection of veins (Adams et al. 2004, Rohlf and Marcus 1993, Francoy et al. 2008, Tofilski 2004, 2008, Kandemir et al. 2011). Presently the coordinates of 19 landmarks, on the forewing, as defined by Smith et al. (1997), are mostly used (Fig. 4.10). According to Bookstein (1991), the use of size-free coordinates instead of distances, rotations, or angles leads to more exhaustive descriptions of geometric forms in biology. While traditional morphometry is restricted to distance and ratios of distances, GM not only includes these measurements indirectly, but also allows for shape analysis by using the landmark approach.

In the measuring process, landmark coordinates are superimposed by translation, scaling, and rotation so that the effect of size is removed (Rohlf and Slice 1990, Bookstein 1991). After superimposition, the landmark configurations differ only in shape. Shape differences can then be analyzed by multivariate statistical methods (Rohlf and Marcus 1993, Zelditch et al. 2004a).

Essentially the same set or a slightly reduced set of landmark coordinates have been employed in most studies of Hymenoptera (Robin 2012), and wing shape variation of various insects at different taxonomic levels (De La Riva et al. 2001, Pretorius and Scholtz 2001, Monteiro et al. 2002, Houle et al. 2003, Schachter-Broide et al. 2004, Pretorius 2005, Aytekin et al. 2007, Sadeghi et al. 2009).

Forewings have been used in all studies, but some have also used data from the hind wings (Klingenberg et al. 2001, Aytekin et al. 2007, Kandemir et al. 2009, Dolati et al. 2013). In addition, wing shape variation based on GM has been used to identify and discriminate honey bee species (Oleksa and Tofilski 2015, Santoso et al. 2018) and subspecies (Francoy et al. 2008, Tofilski 2008), the heritability of wing shape (Monteiro et al. 2002) and the influence of hybridization on fluctuating asymmetry (Schneider et al. 2003). The use of GM to detect differentiation and hybridization in the honey bees of Greece (Bouga and Hatjina 2005) was probably the earliest application of this method to the study of subspecific variation of

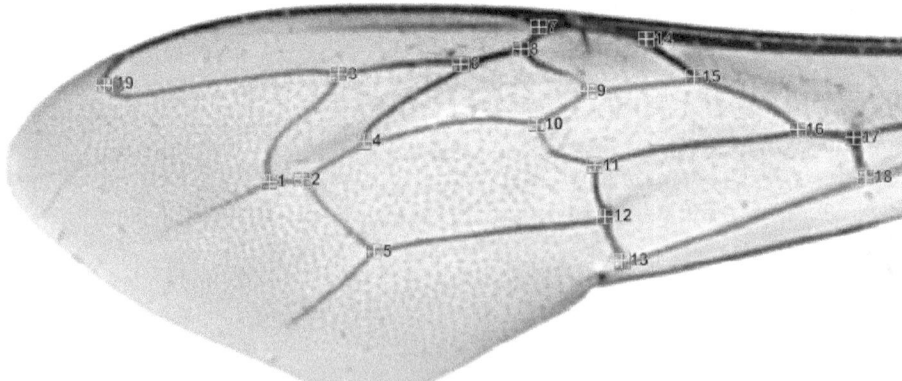

Figure 4.10: Location of the 19 landmarks, on the left forewing of *A. mellifera*, used in the geometric morphometrics analysis.

A. mellifera. Previous studies (Tofilski 2008, Francoy et al. 2008, 2009, Miguel et al. 2011, Kandemir et al. 2011) provide a robust demonstration of the usefulness of landmark-based GM for honey bee micro taxonomy, including the discrimination of honey bee lineages and subspecies. Other studies have shown that this method is easier, faster, more precise, and better at discriminating subspecies and castes of *A. mellifera* than classical morphometrics (Özkan Koca and Kandemir 2013, Rafie et al. 2014, Rašić et al. 2015, Gomeh et al. 2016). For example, in their re-evaluation of the subspecific taxonomy of *A. mellifera*, using landmark-based GM, Kandemir et al. (2011) revealed that GM approach provided consistent discrimination among the four honey bee lineages and 24 subspecies, as in previous studies based on classical morphometry (Fig. 4.11). Recently, Barour and Baylac (2016) used this method to

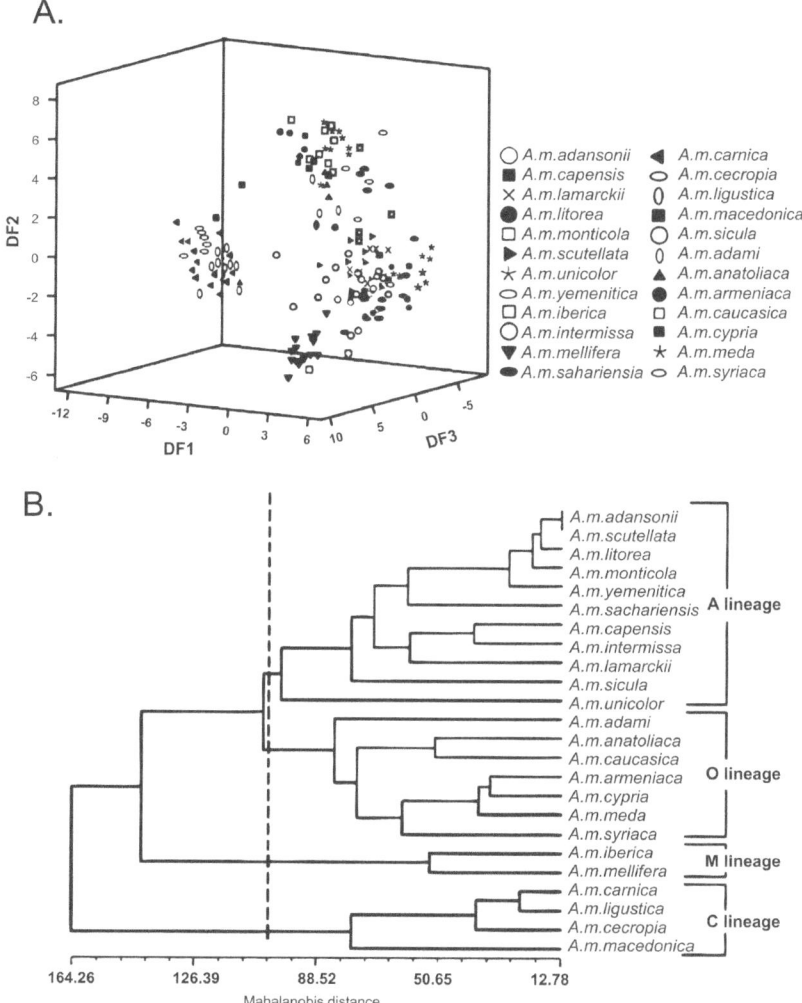

Figure 4.11: Discrimination of honey bee subspecies based on the landmark-based GM method. (A) A scatter plot showing the discrimination of honey bee subspecies, (B) UPGMA phenogram showing the relationship among honey bee subspecies.

discriminate three African subspecies—*A. m. intermissa, A. m. sahariensis,* and *A. m. capensis*—with a very high cross-validation classification rate of 96.7 percent.

Thus, GM analysis of wing shape could be used as a reliable and powerful tool to discriminate among honey bee subspecies, to evaluate and resolve the subspecific taxonomy of honey bees, and may have advantages over classical morphometry. In addition, Bloch et al. (2010) referred outline-based method for a single wing cell to identify *A. mellifera* subspecies in ancient apiculture during biblical times in the Jordan valley. Bloch et al. (2010) also suggested that outline-based GM analysis of a single wing cell might be a useful tool for fossil bees and outline of wing cell could be helpful to identify, for example, honey bee specimens with damaged wings.

Consequently, the morphometric approach has been applied in a number of different methods to problems in the European bee, *A. mellifera,* and has proved to have an important and useful set of data for answering interesting questions. It is particularly useful for distinguishing populations, subspecies, and lineages. The geometric morphometric methods appear to be as sensitive as the traditional methods for many problems. The best approach would be to bring together morphometric (traditional and geometric morphometric methods) and genetic methods to help answer questions of systematics and taxonomy in many insect groups, including *A. mellifera.*

CHAPTER 5

Genetic Diversity of Honey Bee *Apis mellifera* in Siberia

Ostroverkhova, N.V.,[1,]* *Kucher, A.N.,*[1]
Konusova, O.L.,[1] *Kireeva, T.N.,*[1] *Rosseykina, S.A.,*[2]
Yartsev, V.V.[1] and *Pogorelov, Y.L.*[1]

Introduction

The natural distribution area of *Apis mellifera* L. includes Africa and Eurasia. In these geographical areas subspecies exist, which are locally adapted to environmental conditions, including climate, vegetation, pests, pathogens, and other factors (Ruttner 1988, Meixner et al. 2010). Considerable differences in morphological, genetical, behavioral, and biological characters can be identified both between and within the bee subspecies; some of the bee subspecies can be further subdivided into many "ecotypes" (Meixner et al. 2013).

Preservation of specificity and optimal level of genetic diversity of native *A. mellifera* populations is one of the most important conditions of the existence and sustainable development of beekeeping (Meixner et al. 2010, Büchler et al. 2014, Pinto et al. 2014). The loss of the native genetic honey bee diversity is primarily due to human activity. Beekeepers are actively replacing the native bee subspecies by human-selected, more docile and productive bee colonies; there are recurrent introductions of commercial bee colonies (De la Rúa et al. 2009, Pinto et al. 2014, Muñoz et al. 2015). Import of non-native bee subspecies and hybrids, as well as the subsequent difficulty in controlling mating further lead to bee hybridization and uncontrolled gene flow between local and imported (commercial)

---

[1] Institute of Biology, National Research Tomsk State University, av. Lenina, 36, Tomsk, 634050, Russia; kucheraksana@gmail.com; olga.konusova@mail.ru; emilia30@mail.ru; vadim_yartsev@mail.ru; yury_pogorelov@mail.ru
[2] Tomsk State University, av. Lenina, 36, Tomsk, 634050, Russia; rosseykina75@mail.ru
* Corresponding author: nvostrov@mail.ru

bees within a geographic area. Mass bee hybridization leads to the destruction of the evolutionarily established genetic complexes of the honey bee subspecies, the loss of unique gene pools of different bee breeds, and the purebredness of bee colonies. The emerging interbreed hybrids are characterized by unpredictable combinations of genetic material, undesirable phenotypic traits, a decrease in economically valuable indicators and immunity compared to the original breeds, as a result of low disease resistance. Thus, the main causes of mass honey bee colony deaths in Europe are considered the spread of honey bee pests and parasites, such as the mite *Varroa destructor* and the microsporidian *Nosema ceranae* (vanEngelsdorp and Meixner 2010). In connection with this, today an important problem in beekeeping is the preservation of the gene pools of local populations (breeds, subspecies) of honey bees (Meixner et al. 2013, Büchler et al. 2014).

Currently, in Europe, ten *A. mellifera* subspecies among the 30 recognized worldwide are described. These European subspecies have been grouped into two evolutionary lineages: the western European (lineage M) and the eastern European (lineage C) (De la Rúa et al. 2009). In recent years, throughout Europe, mass bee hybridization has been observed both between subspecies of different evolutionary lineages (M and C), and different bee subspecies belonging to the lineage C. Several subspecies and ecotypes can be considered as endangered, especially the *Apis mellifera mellifera* L. subspecies (the dark-colored forest bee) (Jensen and Pedersen 2005, Soland-Reckeweg et al. 2009, Munoz et al. 2015). For example, in Western Europe, where the dark-colored forest bee lived, there is a rapid spread of two subspecies—*A. m. ligustica* and *A. m. carnica* (C lineage) (Bouga et al. 2011). As a result, in Europe, a change in the representation of the honey bee subspecies and their genotypic composition is observed. These sets of data show the importance of studies on changes in the genotypic composition of bee colonies and the constant monitoring of the pure breed of bees.

The distribution of the honey bee subspecies and the process of bee hybridization, primarily the introgression of the lineage C genes (*A. m. ligustica*, *A. m. carnica*, and other subspecies) in the lineage M (subspecies *A. m. mellifera* and *A. m. iberica*), was studied in most European countries (Garnery et al. 1998a, De la Rúa et al. 2001a,b, 2002a,b, 2003, 2006, 2009, Radloff et al. 2001, Jensen et al. 2005, Miguel et al. 2007, Soland-Reckeweg et al. 2009, Muñoz et al. 2009, 2011, 2012, 2013, 2014, 2015, Oleksa et al. 2011, Nedić et al. 2014, Pinto et al. 2014, Uzunov et al. 2014a, Péntek-Zakar et al. 2015). However, the scientific description of honey bee diversity in Europe cannot be regarded as complete, since vast areas, predominantly in the eastern part of the continent, have not yet been studied systematically (Meixner et al. 2013).

A significant limitation to the generalization of gene geographic studies and the data on the genetic diversity of honey bees of different subspecies is lack of information about the vast territory of Russia (for example, Siberia, Altai, and the Far East) (Ostroverkhova et al. 2019b). According to the methodological manual on the honey bee called "BeeBook", published by European researchers (Meixner et al. 2013), data on the honey bees inhabiting Russia is practically absent. Minor information on morphometric characteristics of the honey bees inhabiting the territory of the European part of Russia is given. However, in Russia, research of the honey bees is actively carried out in different directions (biological and genetic diversity, the incidence of bee, breeding work, etc.). The goal of this study is to describe the genetic diversity of honey bee populations and identify characteristics of the bee hybridization in Siberia.

Material and methods

Region. Siberia is a huge region of the north-eastern part of Eurasia and is located from the Urals to the Baikal ridge. The territory of Siberia is bounded in the north by the Arctic Ocean and in the south by Kazakhstan and Mongolia. Siberia occupies almost two-thirds of the territory of Russia and is divided into Western and Eastern Siberia. Western Siberia, which occupies the main part of the West Siberian Plain, is located from the Urals to the Yenisei River. In the south-east of Western Siberia, there are mountainous areas, such as the Altai, the Salair Ridge, the Kuznetsk Alatau, and others. Eastern Siberia is located on the Central Siberian Plateau from the Yenisei River to the Baikal ridge.

Climate. The significant scale of the territory of Siberia determines a wide variety of natural and climatic conditions. The climate of Western Siberia is typical continental, but there is a significant difference in climatic conditions from the north to south due to changes in the amount of solar radiation and circulation of air masses. The average annual temperatures are from –10.5°C in the north to 1–2°C in the south. The warmest month is July; the average July temperature varies from 3.6 to 21–22°C in different areas of Western Siberia. The bulk of the sediments (75–80%) fall in summer and is, on average, 300–600 mm. The winter is cold; the average January temperature varies from –18°C in the south to –30°C in the north-east of Western Siberia. In winter, the territory is covered with snow; snow cover is 260 days in the north and 170–180 days in the south of the region. Snow depth is significant and varies from 20–40 cm in the steppe zone to 50–60 cm in the taiga zone (Gvozdetsky and Mikhailov 1963).

The climate of Eastern Siberia is sharply continental. In winter frosts reach –60°C, and in summer the temperature rises to 35°C. The average annual temperature is –1.5°C; the average temperature in July is 18.1°C and in January is –21.6°C. The frost-free period is 100–110 days. Unlike Western Siberia, in Eastern Siberia, there is less precipitation (200–350 mm); snow cover is usually small, and permafrost is widespread in the north. The terrain is more rugged than in neighboring Western Siberia. Marshes in Eastern Siberia are small, and occur mainly in lowlands on flat, poorly drained interfluves (Bazhenova 2006).

The climate of mountainous areas of Siberia is rather severe for these latitudes. The average annual temperatures are negative almost everywhere due to the long duration of winter and low temperatures of the cold season. The average January temperature ranges from –20–27°C to –15–18°C. The average July temperature in the warmest hollows is 20–22°C. The amount of precipitation varies from 100–200 to 1500–2500 mm in different areas. For example, the greatest amount of precipitation falling in the summer in the form of long heavy rains is characteristic of the western slopes of the Altai and Kuznetsk Alatau.

Natural areas. The soils, vegetation, and fauna of the West Siberian Plain vary from north to south; there is the clear latitudinal zonality: tundra, forest-tundra, forest (taiga and sub-taiga), forest-steppe, and steppe (Shumnyi et al. 2006). Beekeeping is developed only in the forest, forest-steppe, and steppe zones.

In the forest zona, forests occupy 40% of the area and are limited by small hills between the rivers. In the river valleys, large floodplain meadows are located. The zone is characterized by widespread swamps. The amount of precipitation is 400–500 mm per year; the average temperature in July is 18–19°C. Winter lasts from 5.5 to 6 months. The forest zone is very favorable for beekeeping, which is developed mainly in the southern taiga and sub-taiga (Parayeva 1970).

Forest-steppe zone occupies a significant part of the West Siberian plain from the Urals to the Ob River. Summers are hot and windy. The average July temperature is 18–20°C. Rainfall is 300–400 mm. Compared with the European forest-steppe zone, in the forest-steppe of Siberia, winter lasts from 5 to 5.5 months, which is 10–15 degrees colder. The most favorable area for beekeeping is a forest-steppe zone on the right bank of the Ob River. Beekeepers use both wild and cultivated plants.

The steppe zone occupies the extreme southern part of the West Siberian Plain. In the steppe zone, characterized by a diverse landscape, the food base of honey bees is often insufficient.

For the Altai and Kuznetsk Alatau mountains, mountain taiga landscapes are characteristic. Forests cover more than 60% of the territory (Gvozdetsky and Mikhailov 1963). Rugged terrain and rich vegetation of mountain taiga create excellent conditions for beekeeping.

History of beekeeping in Siberia

The honey bee *A. mellifera* was introduced into Siberia at the end of the 18th century. It was the dark-colored forest bee *A. mellifera mellifera*, or the Middle Russian race (a term adopted in Russia), that was cultivated in Siberia as the most adapted to the harsh climatic conditions of the region. The first apiaries were created in the mountainous regions of Western Siberia. During the next century, beekeeping became widespread in the mountainous and lowland parts of the region. In addition, beekeeping began to develop gradually in Eastern Siberia. It is assumed that throughout the 19th century, colonies of the dark-colored forest bee were cultivated in Siberia. The first cases of introduction to Siberia in other honey bee subspecies are known from the beginning of the 20th century. In the second half of the 20th century and at the beginning of the current century, this process becomes widespread and almost uncontrollable, which leads to a high level of bee hybridization. In this regard, the study of bee colonies of spatially isolated apiaries preserved in remote areas of Siberia and existing for a long time without the influence of other bee subspecies is of great interest. For example, in Eastern Siberia, the long isolated apiaries located in the deep taiga in the Old Believers are identified.

Administrative districts of Siberia

The development of beekeeping in administrative regions is largely determined by natural conditions of these territories. Beekeeping is developed only in some administrative districts of Siberia. If in Western Siberia, beekeeping is well developed in all administrative districts (Tomsk, Kemerovo, Omsk, Novosibirsk, Tyumen regions, and the Altai Krai), then in Eastern Siberia, beekeeping is well developed only in the Krasnoyarsk Krai, in contrast to other districts, such as the Irkutsk region, the Republic of Buryatia, and the Republic of Tyva.

The Tomsk region is located in the forest zone. The Omsk and Tyumen regions are located in the forest and forest-steppe zones. The Kemerovo region is located in three different zones: the northern part- in the forest zone, the north-western part- in the forest-steppe zone, whereas for a large part of the region- in mountainous areas. The Central and Eastern parts of the Novosibirsk region are in the forest-steppe zone, while other areas are in the forest zone. A small part of the Altai Krai is located in the forest zone, and the north-eastern and eastern parts in the forest-steppe zone (Gvozdetsky and Mikhailov 1963). The Krasnoyarsk Krai is a predominantly taiga region; forests occupy about 70% of the territory. The Republic of Altai is located in mountainous areas.

We studied honey bee populations in five regions of Siberia: the Tomsk region, the Kemerovo region, the Krasnoyarsk Krai, the Altai Krai, and the Republic of Altai (Fig. 5.1).

Figure 5.1: Map of studied territories of Siberia where honey bee samples have been collected (A–E). Note: A—the Tomsk region; B—Southeast part of the Krasnoyarsk Krai, C—the Kemerovo region; D—the Altai Krai, and E—the Republic of Altai. The Burzyan dark-colored forest bee population located in the reserve "Shulgan-Tash" (the Republic of Bashkortostan, Ural) and used for comparison is indicated by a point F (data obtained from Ilyasov 2016, Ilyasov et al. 2016).

The algorithm on the study of honey bee colonies

At the first stage of the research, we investigated bee colonies inhabiting the different regions of Siberia (northern and southern territory, isolated apiaries, forest areas, and others) to characterize the origin (subspecies) of local bees and identify the *A. mellifera mellifera* populations. At the second stage of the research, we described the process of bee hybridization in Siberia to assess the level of introgression of genes of the M and C lineages.

Each bee colony has been studied using the mtDNA analysis (locus *COI-COII*) and morphological analysis to determine the conformance of the bee colony to the *A. m. mellifera* standard (see details in Ostroverkhova et al. 2015, 2016, Konusova et al. 2016). We use the following research algorithm:

1. mtDNA analysis (variability of the locus *COI-COII*) to determine the origin of the bee colony on the maternal line. If the bees had variants *PQQ* or *PQQQ* of the *COI-COII* mtDNA locus, this colony was of the *A. m. mellifera* origin on the maternal line (evolutionary lineage M). If the bees had a variant *Q* of the *COI-COII* locus, the bee colony maternally originates from the southern subspecies (*A. m. carnica*, *A. m. carpathica*, *A. m. caucasica*, *A. m. ligustica*, and others) (evolutionary lineages C or O).

2. Morphological analysis (parameters of wing including the cubital index, the hantel index, and the discoidal shift, and body painting) to establish the bee colony compliance with the breed standard. If the morphometric parameters of bees derived from bee colony of the maternal *A. m. mellifera* origin (variants *PQQ* or *PQQQ* of the *COI-COII* mtDNA locus) correspond to the dark-colored forest bee's standard, this bee colony is considered potentially "pure" *A. m. mellifera*.

   In connection with the fact that the *A. m. carpathica* subspecies (southern subspecies) is mainly imported into Siberia, we only investigated the compliance of bee colonies originating from the southern subspecies (a variant *Q* of the *COI-COII* mtDNA locus) with the breed standards of this southern subspecies. Other honey bee subspecies of southern origin are more rarely imported into Siberia.

   We considered bee colonies as hybrids in the following cases: (1) morphometric parameters are not consistent with breed standards namely *A. m. mellifera* or *A. m. carpathica* (Cauia et al. 2008); (2) there is a discrepancy between the mtDNA variants and the morphometric parameters of the corresponding bee subspecies.

3. Analysis of microsatellite loci to study genetic diversity of bee subspecies of different evolutionary lineages (M and C) and to determine the level of gene introgression in hybrids of different origin. For this, we initially studied the variability of 11 microsatellite loci in two honey bee subspecies (*A. m. mellifera* and *A. m. carpathica*). To assess the introgression of genes of the evolutionary lineage C (*A. m. carpathica*) into the lineage M (*A. m. mellifera*), a comparative

study of the genetic diversity of purebred bees (*A. m. mellifera* and *A. m. carpathica*) and hybrid bees (both on the basis of *A. m. mellifera* and on the basis of southern subspecies according to analysis of the *COI-COII* mtDNA locus) was carried out using the complex markers of the nuclear genome (11 microsatellite loci).

Materials. We investigated 92 apiaries of 69 populated localities from various regions of Siberia (Ostroverkhova et al. 2015, 2016, 2017, 2018a,c). In total, about 400 bee colonies were studied. Collected honey bee workers were immediately killed by immersion in 96% ethanol and stored at –20°C until morphological examination and DNA extraction. Minimum 20 bees from each colony were studied by the morphometric method; to exclude mixed bee colonies from analysis, a minimum of 5 individuals from each colony were examined by mtDNA analysis. We analyzed 11 microsatellite loci; the minimum number of individuals analyzed for the locus was 269, and the maximum number of bees was 524 (from ten to thirty individuals from each bee colony).

Morphometric method. Morphometric parameters (wing venation), including the cubital index, the hantel index, and the discoidal shift, as well as body painting were studied (Konusova et al. 2016, Ostroverkhova et al. 2016).

Molecular genetic methods. DNA isolation and polymerase chain reaction (PCR) was carried out according to standard techniques with some modifications (Nikonorov et al. 1998, Ostroverkhova et al. 2013). To amplify the *COI-COII* mtDNA locus, the following sequences of primers were used: 3'-CACATTTAGAAATTCCATTA, 5'-ATAAATATGAATCATGTGGA (Nikonorov et al. 1998). Amplification products were fractionated in 1.5% agarose gel and the results were documented with the use of Gel-Doc XR +.

We examined the variability of 11 microsatellite loci (Table 5.1). PCR was performed using specific primers and reaction conditions according to Solignac et al. (Solignac et al. 2003). Amplification products were analyzed with ABI Prism 3730 Genetic Analyser and GeneMapper Software (Applied Biosystems, Inc., Foster City, CA) in the collective center Medical Genomics (Research Institute of Medical Genetics, Tomsk National Research Medical Center, Russian Academy of Sciences). Two microliters of PCR products were mixed with GeneScan500-ROX size standards (Applied Biosystems, Inc.) and deionized formamide. Samples were run according to the manufacturer's recommendations.

Population parameters (expected and observed heterozygosity, number and frequency of alleles) were calculated according to Nei (1975). Principal Components Analysis (PCA) (Peakall and Smouse 2012) was used to investigate the genetic structure and/or differentiation between the Siberian bee populations. STRUCTURE software version 2.3.4 (Pritchard et al. 2000) was used to identify genetically similar groups of individuals in our data set. The results were generated for a given number of K genetic populations (K = 2) using an admixture model based on 5000 Markov Chain Monte Carlo iterations.

Table 5.1: Characteristics of microsatellite loci used for analyses of honey bees in Siberia.

Name of the locus	Chromosome	Size of the sequenced allele (bp)	Motifs, repeats between primers	Annealing temp (°C)	MgCl$_2$ (mM)	The sequence of primers: forward (F) and reverse (R)
A008 (rs267233127)	2	160	(GA)$_{15}$ (GCTCG)$_5$	55	1.2	F: GCGAAGGTAAGGTAAATGAAAC R: GGGCGGTTAAAGTTCTGG
Ap049 (rs267233076)	1	142	(AGG)$_7$	58	1.2	F: CCAATAGCGGCGAGTGTG R: GGGCTTCGTACGTCCACC
AC117 (rs267233481)	12	181	(TTTC)$_5$	50	1.5	F: CGGTTCATCTTCCCTTTATTTC R: CCACGGATTATTATCGTTTATC
A113 (rs267233291)	6	220	(TC)$_5$TT(T C)$_8$TT(TC)$_5$	60	1.0	F: CTCGAATCGTGGCGTCC R: CCTGTATTTTGCAACCTCGC
Ap243 (rs267233098)	1	260	(TCC)$_9$	50	1.5	F: AATGTCCGCGAGCATCTG R: TGTTTACGAGAATTCGACGGG
A024 (rs267234016)	7	100	(CT)$_{11}$	55	1.2	F: CACAAGTTCCAACAATGC R: CACATTGAGGATGAGCG
A043 (rs267233033)	1	140	(CT)$_{12}$	55	1.5	F: CACCGAAACAAGATGCAAG R: CCGCTCATTAAGATATCCG
H110 (rs267233914)	5	160	(ATCC)$_4$(ATCT)$_2$	56	1.5	F: CGCTCGCGGTGGATTTCATTT R: GGCAAAAGTTGGCGGAGAAAGA
SV185 (rs267233900)	5	272	(AAC)$_{12}$	55	1.5	F: AGCTCACGCAGCACATGC R: GACGTTGTTTCCATCACCACTC
K0820 (rs267234092)	8	130	(TG)$_{10}$	55	1.5	F: CATCGATGCGTCGAGGAT R: CCGATCGCGTGATATTACG
MRJP3	11	350–530	fragment length polymorphism	55	1.5	F: ATGTAAITTTTGAAGAAITGAACTTG R: TGTAGAITGACTTAATGAGAAAACAC

Results and discussion

Distribution of honey bee subspecies in Siberia. We previously completed a comprehensive study of some bee populations (about 300 bee colonies from 60 apiaries) in Siberia (the Tomsk region, the Krasnoyarsk Krai, and the Altai Krai) and identified the "pure" *A. m. mellifera* populations in the Krasnoyarsk Krai and the Tomsk region using morphometric and molecular genetic methods (Ostroverkhova et al. 2016, 2018a). In this study, we added previously unexplored bee colonies obtained from new apiaries of already studied regions (the Tomsk region, the Krasnoyarsk Krai), as well as from new territories of Siberia (the Kemerovo region and the Republic of Altai).

Honey bee population of the Tomsk region. It was established that 60% of bee colonies originate from *A. m. mellifera* on the maternal line, 32% of bee colonies were of southern origin, and 8% of samples were mixed colonies according to the analysis of the variability of the *COI-COII* mtDNA locus. Three variants of the mtDNA *COI-COII* locus were registered: *PQQ*, *PQQQ* specific to *A. m. mellifera*, and Q specific to subspecies of southern origin.

According to a comprehensive study including morphometric and mtDNA analysis, the majority of bee colonies of the Tomsk region were mostly hybrids between subspecies of *A. m. mellifera* and *A. m. carpathica* (Fig. 5.2); in some colonies, the mismatch of morphometric and mtDNA data was observed.

The structure of the honey bee populations in the Tomsk region is quite complex and mosaic, especially in the southern areas of the region (for example, in the Tomsk district), where both purebred bee colonies of different origin and hybrids are registered. In contrast to the northern regions, more developed beekeeping and the active import of southern bee subspecies and hybrids are characteristic of the southern areas of the Tomsk region. Populations of the southern districts of the Tomsk region are characterized by a greater genetic diversity of honey bees on mitochondrial and nuclear DNA markers; for example, several variants of the *COI-COII* locus or their combinations and wider allele spectrum of microsatellite loci are recorded (Ostroverkhova et al. 2015, 2017). Genetically, homogeneous bee colonies and apiaries dominate in the northern regions; one or two variants of the *COI-COII* mtDNA locus are registered in 96% of bee colonies and 73% of apiaries.

No large territories where the *A. m. mellifera* bees live are found in the Tomsk region. At the same time, in some districts of the Tomsk region (Teguldetsky, Molchanovsky, Zyryansky, and others), apiaries where the "pure" *A. m. mellifera* subspecies are preserved were identified.

Honey bee population of the Kemerovo region. Most of the studied apiaries (including industrial apiaries) in Promyshlennovsky, Novokuznetsky, and Yashkinsky districts are represented by honey bees of southern origin (*A. m. carpathica*, *A. m. carnica*). All investigated workers of these apiaries had variant Q of the *COI-COII* mtDNA locus and corresponded to the breed standards of *A. m. carpathica* or *A. m. carnica* on morphometric parameters (Fig. 5.3). In the present work, the *A. m. carnica* bees were not studied using microsatellite loci.

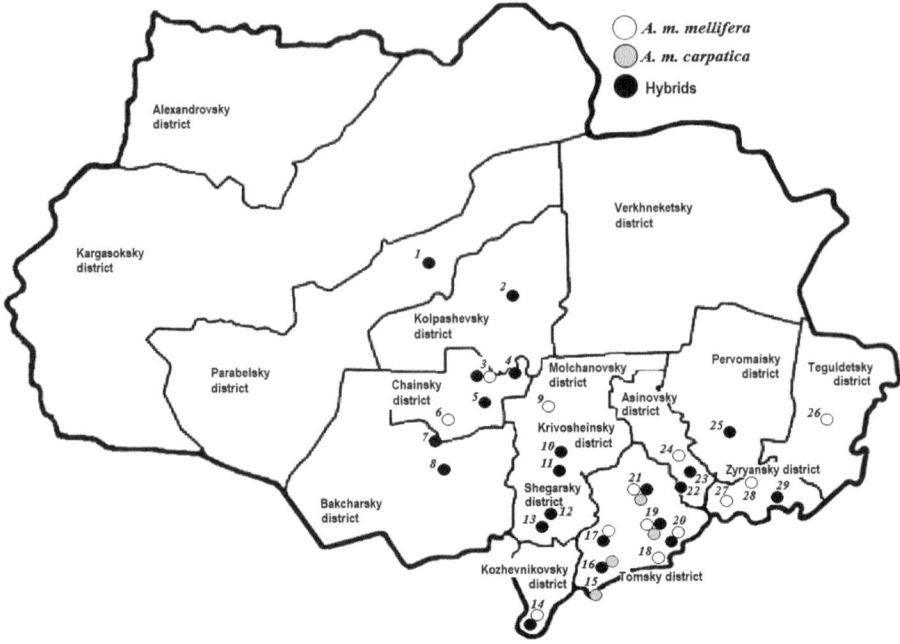

Figure 5.2: Distribution of bee subspecies and hybrids in the apiaries of the Tomsk region on the basis of morphometric and mtDNA analysis. Studied settlements are indicated by numbers: 1: s. Parabel; 2: vicinity of g. Kolpashevo; 3: s. Podgornoe; 4: s. Leboter; 5: d. Strelnikovo; 6: s. Gorelovka; 7: s. Vysoky Yar, d. Krylovka; 8: s. Bakchar; 9: s. Mogochino, s. Volog; 10: s. Krivosheino; 11: s. Volodino; 12: s. Kargala; 13: s. Batkat; 14: d. Elovka; 15: s. Yar; 16: s. Kurlek, d. Kandinka; 17: s. Zorkaltsevo, s. Rybalovo, d. Kudrinsky uchastok, d. Gubino, d. Berezkino, p. Zarechny (Mezheninovskoe rural settlement), s. Nizhne-Sechenovo; 18: s. Mezheninovka; 19: p. Sinii Utes, d. Magadaevo, d. Prosekino, s. Kolarovo, vicinity of Tomsk; 20: d. Bodazhkovo, d. Bolshoe Protopopovo; 21: s. Semiluzhki, p. Zarechny (Malinovskoe rural settlement); 22: d. Tikhomirovka; 23: vicinity of Asino; 24: ur. Kuzherbak; 25: d. Krutolozhnoe; 26: s. Teguldet; 27: s. Dubrovka; 28: s. Zyryanskoe; 29: s. Okuneevo. Apiaries located at a distance less than 15 km from each other are marked as a single point.

One industrial apiary, where the *A. m. mellifera* subspecies was bred and various available techniques are used to preserve the purebred honey bees, was found in the Novokuznetskiy district. Hybrid colonies were found only in one apiary in Yashkino village: different variants of the *COI-COII* mtDNA locus (Q and *PQQ*) were found in individuals of the studied colonies and the deviation of morphometric parameters from the breed standard was shown.

Honey bee population of the Altai Krai and the Republic of Altai. The Altai Krai and the Republic of Altai are territories with well-developed beekeeping and closely located apiaries, where there is a constant genetic exchange between bee colonies. On the territory of the Altai Krai, bee colonies of different origin, including purebred colonies (*A. m. mellifera* and *A. m. carpathica*) and hybrids were identified (Fig. 5.3). For example, we identified the *A. m. carpathica* bees at the apiary in the Loktevsky district and hybrid colonies (hybrids between *A. m. mellifera* and *A. m. carpathica* subspecies) at the apiaries in the Altaisky, Barnaulsky, Charyshsky, Tretyakovsky, and Zmeinogorsky districts according to morphometric and mtDNA analysis. At the

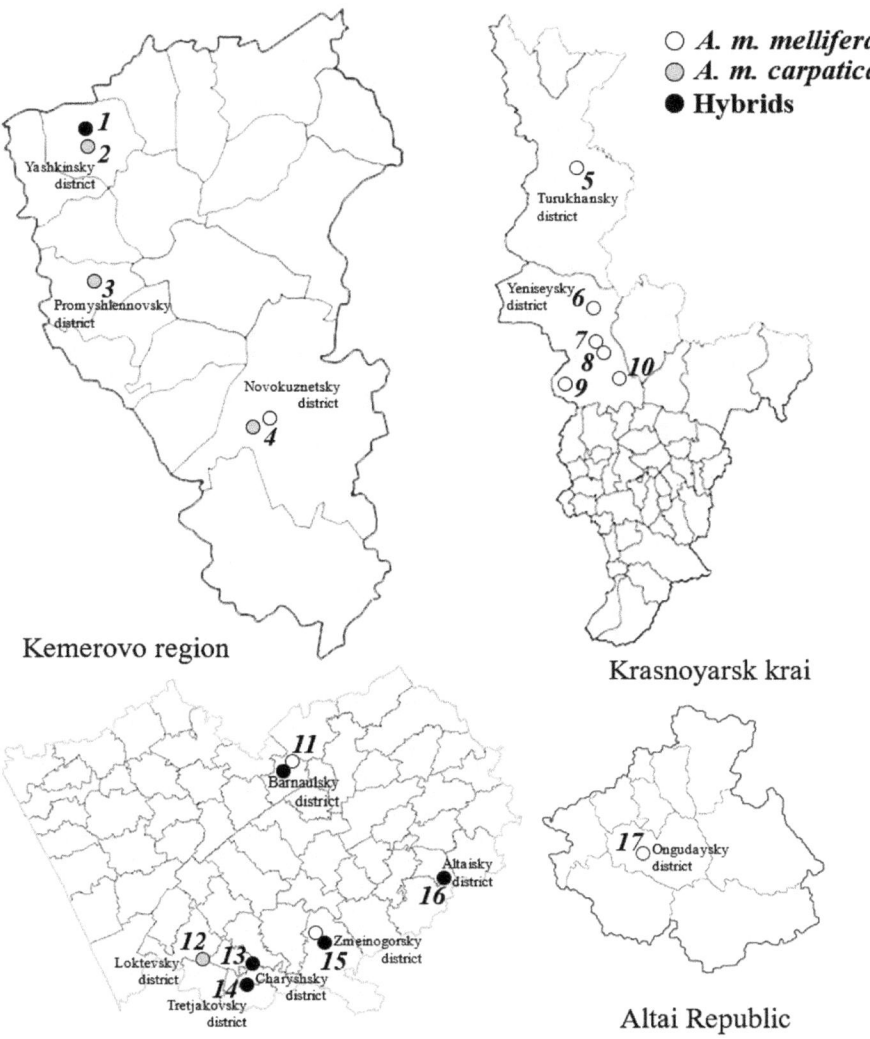

Figure 5.3: Distribution of bee subspecies and hybrids in the apiaries of four regions of Siberia (the Kemerovo region, the Krasnoyarsk Krai, the Altai Krai, and the Republic of Altai) on the basis of morphometric and mtDNA analysis. Studied settlements are indicated by numbers: 1: Yashkino village; 2: s. Nizhneyashkino; 3: Promyshlenny village; 4: vicinity of Novokuznetsk (two apiaries); 5: vicinity of s. Turukhansk; 6: s. Yartsevo; 7: s. Ostyatskoe; 8: s. Kolmogorovo; 9: Yaksha village; 10: s. Ozernoe; 11: vicinity of Barnaul; 12: Masalsky village; 13: vicinity of s. Charyshskoe; 14: s. Staroaleiskoe; 15: Baranovka village; 16: s. Aya; 17: vicinity of s. Ongudai.

same time, an apiary where the *A. m. mellifera* subspecies is cultivated (the *PQQ* variant of the *COI-COII* mtDNA locus was found in bees) and constant monitoring of the pure breeding of bee colonies is carried out was found in the Zmeinogorsky district. In the Republic of Altai, we identified one isolated apiary where the *A. m. mellifera* bees live in the vicinity of Ongudai (the Ongudaysky district).

In the *A. m. mellifera* bees, two variants of the *COI-COII* mtDNA locus (*PQQ* and *PQQQ*) were recorded. The results of the morphometric analysis confirmed the origin of bee colonies from the *A. m. mellifera* subspecies, but for some bee colonies, some parameters studied (wing parameters) deviated from the *A. m. mellifera* standard, which may indicate the influence of subspecies of southern origin. For example, in the colony No. 1 at the apiary in Baranovka village, workers had a positive and zero value for the "discoidal shift" parameter, which is not typical for the *A. m. mellifera* standard.

Honey bee population of the Krasnoyarsk Krai. Honey bees living in apiaries of the Krasnoyarsk Krai (the Yenisei population) are of considerable interest. According to the mtDNA analysis (only the *PQQ* variant of the *COI-COII* locus is registered), all the bee colonies studied were derived from the *A. m. mellifera* subspecies (Fig. 5.3). The variant Q of the *COI-COII* mtDNA locus, specific for subspecies of southern origin (*A. m. carpathica*, *A. m. carnica*), has not been identified. Earlier, we studied the genetic diversity of honey bees of the Yenisei population using a complex of microsatellite loci (see the details in Ostroverkhova et al. 2018a,c).

Unlike the Altai Krai, where there is a high level of beekeeping, the Yenisei population has long-isolated apiaries (s. Ostyatskoe, s. Kolmogorovo, s. Turukhansk, s. Yartsevo, s. Ozernoe, s. Yaksha) located in the taiga, in a sparsely populated district. However, despite the isolation of apiaries in the Krasnoyarsk Krai (the influence of other bee subspecies is excluded), for the bees of the Yenisei population, the deviation of some morphometric parameters from the Russian and European *A. m. mellifera* standards have also been shown. For example, according to the results of the morphometric analysis, for most bee colonies, the deviation of the mean value of the cubital index from the *A. m. mellifera* standard was registered. Since the Yenisei population is characterized by prolonged isolation and lack of influence of other bee subspecies, the observed deviation of some morphometric parameters from accepted standards may be a consequence of either loss of genetic diversity (due to the isolation of apiaries) or the adaptation of colonies to the harsh climatic conditions and features of the flora.

Genetic diversity of the honey bees in Siberia on the microsatellite loci

A study of honey bees of different origin using microsatellite loci was carried out to characterize the genetic diversity of bees of the *A. m. mellifera* and *A. m. carpathica* subspecies, to search for unique or specific DNA markers characteristic of different subspecies (lineages M and C), and to assess the level of gene introgression between lineages M and C. To characterize the genetic diversity of honey bees and identify highly polymorphic DNA markers, we studied worker bees from apiaries of Siberia using 11 microsatellite loci (*A008*, *AC117*, *A043*, *A113*, *A024*, *Ap243*, *Ap049*, *H110*, *SV185*, *K0820*, and *MRJP3*) (Table 5.2).

Four bee groups were formed, including samples of purebred and hybrid bees:

1. *Apis mellifera mellifera* bees (M lineage). Honey bees received from bee colonies corresponding to the *A. m. mellifera* standard according to the analysis

Table 5.2: Allele frequencies of the 11 microsatellite loci in honey bee subspecies (*A. m. mellifera,* *A. m. carpathica*) and hybrids of different origin.

Allele size, bp	*A. m. mellifera* (variants *PQQ* or *PQQQ COI-COII* mtDNA, lineage M)	*A. m. carpathica* (variant Q *COI-COII* mtDNA, lineage C)	Hybrids based on	
			A. m. mellifera (variants *PQQ* or *PQQQ COI-COII* mtDNA)	*A. m. carpathica* (variant Q *COI-COII* mtDNA)
Locus *A008*				
151	0.020±0.006	0	0	0
154	0.011±0.004	0	0.005±0.003	0.037±0.009
162	0.868±0.014	0.026±0.013	0.721±0.018	0.687±0.021
170	0	0.026±0.013	0	0.004±0.003
172	0.051±0.009	0.077±0.021	0.114±0.013	0.029±0.008
174	0.023±0.006	0.455±0.040	0.083±0.011	0.124±0.015
176	0.028±0.007	0.026±0.013	0.036±0.008	0.055±0.013
178	0	0.039±0.015	0.020±0.006	0.039±0.009
180	0	0.180±0.031	0.021±0.006	0.026±0.007
182	0	0.026±0.013	0	0
189	0	0.147±0.028	0	0
Ho	0.226±0.024	0.500±0.057 * * *	0.409±0.028	0.463±0.032
He	0.243±0.023	0.729±0.028	0.459±0.023	0.505±0.026
N	306	78	316	246
Locus *AC117*				
173	0.093±0.012	0	0.084±0.013	0.022±0.007
177	0.098±0.012	0.010±0.010	0.130±0.016	0.165±0.017
181	0.292±0.019	0.031±0.017	0.443±0.024	0.278±0.020
185	0.517±0.020	0.959±0.020	0.343±0.023	0.534±0.022
Ho	0.389±0.028 * * *	0.082±0.039	0.446±0.034	0.391±0.031
He	0.629±0.014	0.079±0.037	0.662±0.012	0.609±0.015
N	301	49	220	248
Locus *A043*				
120	0.002±0.002	0	0.009±0.005	0
128	0.831±0.015	0.076±0.023	0.844±0.017	0.637±0.024
130	0.012±0.004	0	0.005±0.003	0.020±0.007
136	0	0.083±0.024	0	0
138	0	0.197±0.035	0	0.020±0.007
140	0.156±0.015	0.629±0.042	0.138±0.016	0.292±0.023
142	0	0.015±0.011	0.005±0.003	0.032±0.009
Ho	0.279±0.026	0.242±0.053 * * *	0.249±0.029	0.451±0.035
He	0.285±0.021	0.553±0.042	0.269±0.025	0.507±0.020
N	305	66	221	204

Table 5.2 contd. ...

...*Table 5.2 contd.*

Allele size, bp	*A. m. mellifera* (variants *PQQ* or *PQQQ COI-COII* mtDNA, lineage M)	*A. m. carpathica* (variant Q *COI-COII* mtDNA, lineage C)	Hybrids based on	
			A. m. mellifera (variants *PQQ* or *PQQQ COI-COII* mtDNA)	*A. m. carpathica* (variant Q *COI-COII* mtDNA)
Locus *A024*				
92	0.660±0.019	0.066±0.021	0.500±0.025	0.472±0.024
94	0	0	0.094±0.015	0
96	0.007±0.003	0.015±0.010	0.023±0.008	0.014±0.006
98	0.011±0.004	0	0.013±0.006	0
100	0.186±0.016	0.427±0.042	0.194±0.020	0.421±0.024
102	0.007±0.003	0.463±0.043	0.043±0.010	0.082±0.013
104	0	0.029±0.015	0.015±0.006	0
106	0.130±0.014	0	0.117±0.016	0.012±0.005
Ho	0.505±0.029	0.529±0.061	0.480±0.036	0.636±0.033
He	0.513±0.020	0.598±0.022	0.687±0.019	0.593±0.012
N	307	68	196	214
Locus *A113*				
200	0	0	0.012±0.005	0.002±0.002
206	0.003±0.002	0	0	0.004±0.003
208	0	0.019±0.011	0	0
210	0.005±0.003	0.019±0.011	0	0.015±0.006
212	0.107±0.013	0.906±0.023	0.171±0.018	0.203±0.019
214	0.003±0.002	0.006±0.006	0.017±0.006	0.011±0.005
218	0.571±0.021	0.013±0.009	0.571±0.024	0.545±0.023
220	0.255±0.018	0.031±0.014	0.124±0.016	0.123±0.015
222	0.009±0.004	0	0.007±0.004	0.021±0.007
224	0.002±0.002	0	0	0.021±0.007
226	0.003±0.002	0	0.033±0.009	0.015±0.006
228	0.035±0.008	0	0.029±0.008	0.021±0.007
230	0	0	0.010±0.005	0.004±0.003
232	0	0.006±0.006	0.007±0.004	0.009±0.004
234	0.007±0.003	0	0.019±0.007	0.006±0.004
Ho	0.521±0.029 *	0.075±0.029 *	0.567±0.034	0.572±0.032
He	0.596±0.017	0.177±0.041	0.626±0.023	0.645±0.020
N	290	80	211	236

Table 5.2 contd. ...

...*Table 5.2 contd.*

Allele size, bp	A. m. mellifera (variants PQQ or PQQQ COI-COII mtDNA, lineage M)	A. m. carpathica (variant Q COI-COII mtDNA, lineage C)	Hybrids based on	
			A. m. mellifera (variants PQQ or PQQQ COI-COII mtDNA)	A. m. carpathica (variant Q COI-COII mtDNA)
Locus *Ap243*				
252	0.014±0.006	0	0.184±0.023	0.395±0.027
255	0.427±0.024	1.000±0.000	0.367±0.029	0.214±0.023
260	0.054±0.011	0	0.126±0.020	0.187±0.021
263	0.330±0.023	0	0.306±0.028	0.157±0.020
266	0.005±0.003	0	0.004±0.004	0.012±0.006
269	0.040±0.010	0	0.011±0.006	0.003±0.003
272	0.094±0.014	0	0.004±0.004	0.021±0.008
275	0.035±0.009	0	0	0.012±0.006
Ho	0.439±0.034 * * *	0.000±0.000	0.252±0.037	0.331±0.037
He	0.694±0.014	-	0.722±0.012	0.738±0.013
N	212	60	139	166
Locus *Ap049*				
120	0.123±0.013	0.047±0.021	0.023±0.007	0.018±0.006
127	0.673±0.019	0.208±0.039	0.645±0.023	0.609±0.023
130	0.175±0.015	0.679±0.002	0.240±0.020	0.144±0.017
139	0.023±0.006	0.006±0.045	0.093±0.014	0.229±0.020
142	0.007±0.003	0.009±0.009	0	0
152	0	0.057±0.022	0	0
Ho	0.447±0.028	0.585±0.068	0.353±0.032	0.418±0.033
He	0.501±0.020	0.490±0.048	0.518±0.021	0.556±0.020
N	309	53	221	225
Locus *SV185*				
260	0	0	0.044±0.011	0.188±0.019
263	0.288±0.021	0.094±0.026	0.310±0.024	0.222±0.020
266	0.117±0.015	0.266±0.039	0.216±0.021	0.290±0.022
269	0.578±0.022	0.359±0.042	0.367±0.025	0.205±0.020
272	0.010±0.005	0.281±0.040	0.052±0.011	0.089±0.014
278	0.006±0.004	0	0.010±0.005	0.005±0.003
Ho	0.527±0.032	0.594±0.061	0.662±0.034	0.676±0.033
He	0.569±0.017	0.712±0.014	0.718±0.010	0.781±0.006
N	243	64	192	207

Table 5.2 contd. ...

...Table 5.2 contd.

Allele size, bp	A. m. mellifera (variants PQQ or PQQQ COI-COII mtDNA, lineage M)	A. m. carpathica (variant Q COI-COII mtDNA, lineage C)	Hybrids based on	
			A. m. mellifera (variants PQQ or PQQQ COI-COII mtDNA)	A. m. carpathica (variant Q COI-COII mtDNA)
Locus *H110*				
152	0	0	0.020±0.007	0
158	0	0	0.015±0.006	0
162	0.789±0.017	0.806±0.038	0.625±0.024	0.806±0.019
166	0.027±0.007	0.176±0.037	0.071±0.013	0.037±0.009
170	0.184±0.016	0.019±0.013	0.270±0.022	0.157±0.017
Ho	0.333±0.028	0.389±0.066	0.559±0.035	0.349±0.032
He	0.343±0.022	0.320±0.049	0.531±0.021	0.325±0.025
N	282	54	204	229
Locus *K0820*				
124	0.044±0.019	0.071±0.031	0.125±0.035	0.104±0.023
126	0	0.357±0.057	0	0.028±0.012
128	0.605±0.046	0.300±0.055	0.761±0.045	0.550±0.037
130	0.053±0.021	0.243±0.051	0.023±0.016	0.170±0.028
132	0.298±0.043	0.029±0.020	0.091±0.031	0.148±0.026
Ho	0.597±0.065	0.400±0.083 * * *	0.432±0.075	0.582±0.052
He	0.540±0.036	0.718±0.022	0.396±0.060	0.635±0.031
N	57	35	44	91
Locus *MRJP3*				
391	0.034±0.014	0.110±0.018	0.077±0.018	0.071±0.021
406	0	0.486±0.029	0.059±0.016	0.103±0.024
437	0.051±0.017	0.021±0.008	0.009±0.006	0.013±0.009
464	0.084±0.021	0.097±0.017	0.023±0.010	0.058±0.019
485	0	0.010±0.006	0.018±0.009	0
495	0	0.003±0.003	0.018±0.009	0.083±0.022
501	0	0.035±0.011	0.009±0.006	0.013±0.009
511	0	0	0.005±0.005	0.096±0.024
518	0	0.197±0.023	0.018±0.009	0.244±0.034
529	0.832±0.028	0.041±0.012	0.766±0.028	0.321±0.037
Ho	0.067±0.027 * * *	0.003±0.002	0.324±0.044	0.615±0.055
He	0.298±0.043	0.006±0.003	0.403±0.041	0.803±0.017
N	89	145	111	78

Note. N—the number of bees studied; He—expected heterozygosity according to Hardy-Weinberg equilibrium; Ho—observed heterozygosity. In the table, the values of allele frequencies and parameters of heterozygosity with a standard error are given. Statistically significant differences in the observed heterozygosity from the expected heterozygosity are marked with (*) (*—$p < 0.05$, * * *—$p < 0.001$).

of morphometric parameters and mtDNA analysis (variants of *PQQ* or *PQQQ* of the *COI-COII* mtDNA locus);

2. *Apis mellifera carpathica* bees (C lineage). This group includes the following bees: (a) honey bees collected from the *A. m. carpathica* bee colonies, obtained from a bee farm of the Carpathian breed (Ukraine); (b) honey bees received from bee colonies imported from the *A. m. carpathica* farm (Ukraine) and living in apiaries in the Tomsk region for several years (zoned breed). All honey bees correspond to the *A. m. carpathica* breed standard according to the analysis of morphometric parameters and mtDNA analysis (a variant of Q of the *COI-COII* mtDNA locus);

3. Hybrid bees obtained from hybrid colonies originating on the maternal line from the *A. m. mellifera* subspecies (variants of *PQQ* or *PQQQ* of the mtDNA *COI-COII* locus were identified);

4. Hybrid bees obtained from hybrid bee colonies originating on the maternal line from subspecies of southern origin (variant Q of the *COI-COII* mtDNA locus was identified), mostly Carpathian breed.

All microsatellite loci studied were polymorphic: the minimum number of alleles was registered for locus *AC117* (4 alleles); the maximum number of alleles was for locus *A113* (15 alleles); the average number of alleles per locus was 8. For *A. m. mellifera* bees, the minimum number of alleles were found for locus *H110* (3 alleles), and the maximum number of alleles was for locus *A113* (11 alleles); for Carpathian bees, the minimum number of alleles was found for the *Ap243* locus (1 allele); the maximum number of alleles (9) was for loci *A008* and *MRJP3*; the average number of alleles per locus for each subspecies was 5.

In addition, the studied loci differed in the variability in honey bees of two subspecies (*A. m. mellifera, A. m. carpathica*). For example, for the *Ap243* locus, only one allele "255' (all the studied individuals were homozygous) was registered in Carpathian bees, whereas there were 8 alleles in the bees of the *A. m. mellifera* subspecies. On the contrary, for the *MRJP3* locus, greater genetic diversity was found in the Carpathian bees (9 alleles were registered), whereas for *A. m. mellifera* bees, 4 alleles with a predominance of the "529" allele (registration frequency more than 0.83) were described (Table 5.2).

Thus, a comparative analysis of the variability of the studied microsatellite loci in bees of two subspecies (*A. m. mellifera, A. m. carpathica*) showed differences in the spectrum and/or frequency of alleles of some loci between studied subspecies. In addition, for some loci (*A008, A043, A113, A024, Ap049,* and *MRJP3*), the different predominant alleles (allele frequency was more than 0.40) were registered in bees of different evolutionary lineages (M and C) (Table 5.2). For example, for the *A008* locus, the predominant allele in *A. m. mellifera* bees (lineage M) was the allele "162" (allele frequency was 0.87), while the predominant allele in *A. m. carpathica* bees (lineage C) was the allele "174" (allele frequency was 0.46). For locus *A113*, the predominant allele in *A. m. carpathica* bees was "212" (allele frequency was 0.91), while the predominant alleles in *A. m. mellifera* bees were alleles "218" and "220" (the sum frequency of these two alleles was 0.83).

An assessment of the heterozygosity of most of the studied loci revealed similar results for the two bee subspecies, namely, lower values of the observed heterozygosity (Ho) compared with the expected heterozygosity (He) (Table 5.2). Only for loci *AC117*, *Ap049*, *H110* (in Carpathian bees), and *K0820* (in *A. m. mellifera* bees), the observed heterozygosity exceeded the expected heterozygosity. At the same time, a statistically significant level of differences between the values of the observed and expected heterozygosity is shown only for some loci: loci *A113* (t = 2.23, p < 0.05), *AC117* (t = 7.67, p < 0.001), *Ap243* (t = 6.94, p < 0.001), and the *MRJP3* locus (t = 4.55, p < 0.001) in *A. m. mellifera* bees; loci *A008* (t = 3.61, p < 0.001), *A043* (t = 4.60, p < 0.001), *K0820* (t = 3.70, p < 0.001), and *A113* (t = 2.03, p < 0.05) in Carpathian bees.

Thus, the analysis of the variability of 11 microsatellite loci in honey bee subspecies (*A. m. mellifera* and *A. m. carpathica*) allowed us to identify diagnostic DNA markers of the nuclear genome to differentiate subspecies of different origin (evolutionary lineages M and C) (Table 5.2). Of particular interest are loci *A008*, *A043*, *A024*, *A113*, and *MRJP3*, which can be considered as subspecies-specific DNA markers differentiating *A. mellifera* subspecies belonging to lineages M and C. For these loci, different predominant alleles were revealed in bees *A. m. mellifera* and *A. m. carpathica*; moreover, the alleles specific for bees of the evolutionary lineage M were registered with a low frequency in bees belonging to the evolutionary lineage C, and vice versa, the genetic variants that dominated the C lineage bees were rare among bees of lineage M.

Assessment of introgression events and bee hybridization in Siberia

The differences in the variability of the studied microsatellite loci detected in honey bees of different origins (subspecies *A. m. mellifera* and *A. m. carpathica*) were used to assess the genetic diversity of hybrid bees obtained from apiaries of the Tomsk region. Based on the analysis of the variability of 11 microsatellite loci, it was found that as in hybrids originating on the maternal line from the *A. m. mellifera* subspecies (variants *PQQ* or *PQQQ* of the *COI-COII* mtDNA locus), and in hybrids derived from breeds of southern origin (variant Q of the *COI-COII* locus), the nuclear genome is more consistent in the *A. m. mellifera* subspecies. For example, for locus *A043*, the allele "128" is specific to the *A. m. mellifera* subspecies (allele frequency is 0.83), whereas in Carpathian bees this allele occurs at a frequency of 0.08; on the contrary, the allele "140" is characteristic (allele frequency is 0.63) of Carpathian bees, whereas in *A. m. mellifera* bees this allele is registered with a frequency of 0.16 (Table 5.2). In hybrids of different origin (on the basis of both the *A. m. mellifera* and the *A. m. carpathica* subspecies), the allele "128" specific to the *A. m. mellifera* is predominant (allele frequencies are 0.84 and 0.64, respectively), whereas the frequency of the allele "140" specific to the *A. m. carpathica* decreased by half (0.29) in hybrids originating on the maternal line from *A. m. carpathica*. A similar picture is observed for most of the studied microsatellite loci (Table 5.2). Consequently, in hybrids on the basis of the *A. m. mellifera* subspecies, introgression of genetic variants of lineage C ("southern" subspecies) occurs in the Tomsk honey bee population. A similar picture is observed in hybrids based on the Carpathian breed: there is introgression of genetic variants of the M lineage, and moreover, this process is more pronounced.

Patterns of genetic variation and admixture, inferred with STRUCTURE software, are shown in Fig. 5.4. When the distribution of the individual admixture proportions in hybrids of different origin was analyzed using the admixture and correlated allele frequency models (based on variability analysis of 11 microsatellite loci), most of the samples corresponded to the *A. m. mellifera* bees.

A principal coordinate analysis (PCA) based on the first two principal coordinates was performed to differentiate subspecies of different evolutionary lineages (M and C) and honey bee hybrids, as well as determine the level of genetic heterogeneity

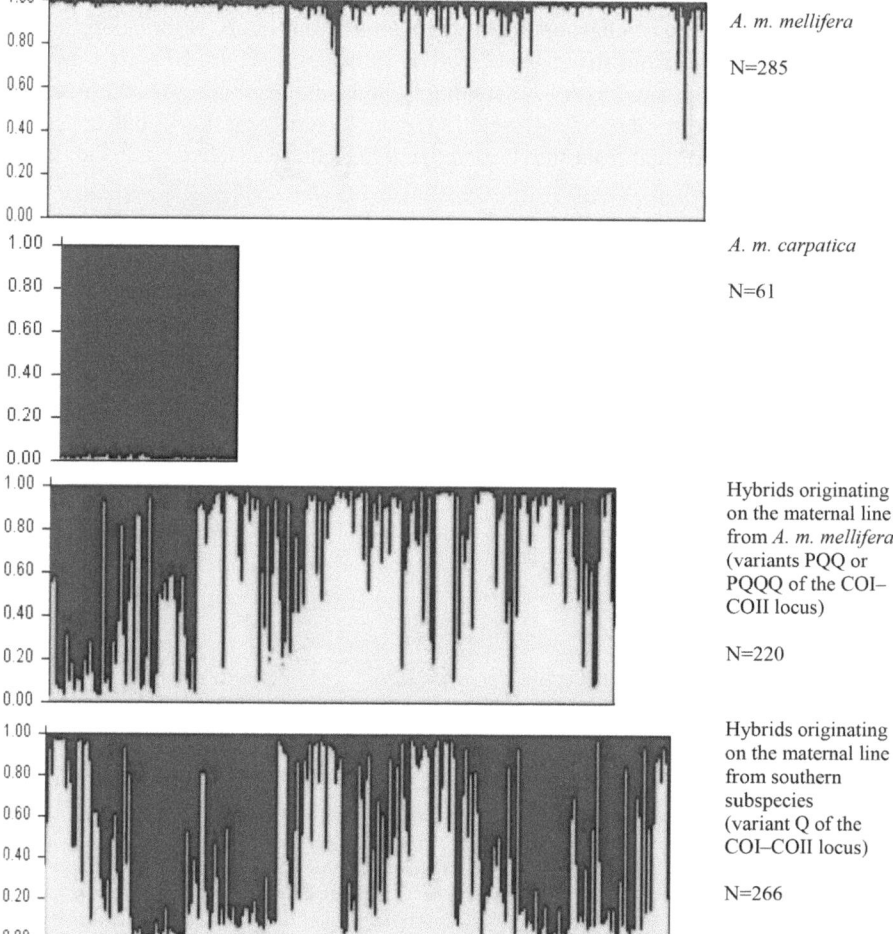

Figure 5.4: Results of STRUCTURE analysis using admixture and correlated allele frequencies models. Individuals are represented by vertical lines, grouped by inferred populations (K = 2). Division of individuals into the colored segment represents the assignment probability of that individual to each of the K groups.

of *A. m. mellifera* samples from Russian populations of different geographic locations (Siberia, Ural) based on the genetic distance among individual samples (Fig. 5.5). To study the maximum number of bee samples of different origin and different geographic localization, the analysis was carried out on the basis of the allele spectrum of only 7 microsatellite loci (*Ap243, 4a110 (= H110), A024, A008, A043, A113,* and *Ap049*). According to the studied DNA markers, the compared bee groups are clearly clustered depending on the origin (subspecies). Coordinate 1 represents 78.0% of the total variability and opposes populations from the M and C lineages (the *A. m. mellifera* samples and the *A. m. carpathica* samples are well differentiated). Coordinate 2 expresses 22.0% of the total variability, and probably reflects the geographic location of the bee samples. Thus, *A. m. mellifera* bee samples clearly differ from each other according to geographical principle (East–West): a subgroup of Siberian populations (the Tomsk region, the Krasnoyarsk Krai, and Altai populations) and a subgroup of Burzyan population (Ural). Hybrids originating on the maternal line from the *A. m. mellifera* subspecies (Hybrid-*PQQ*) were also grouped in an M-lineage cluster. Interestingly, hybrids originating on the maternal line from the southern subspecies (Hybrid-Q) were located near the M-lineage group, but not the group formed from the *A. m. carpathica* samples (lineage C) (Fig. 5.5).

Principal Coordinates (PCoA)

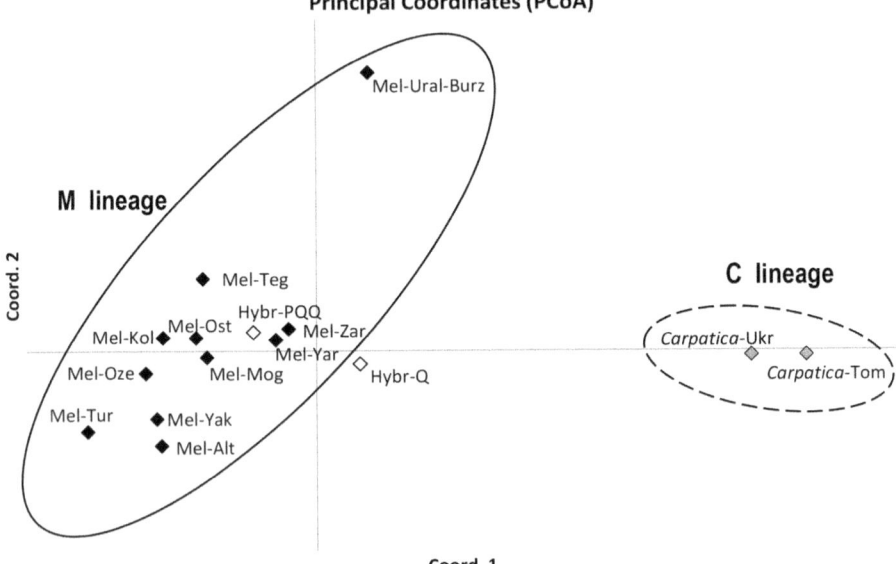

Figure 5.5: Distribution of *A. mellifera* individuals (comparison between bee groups of different origin and honey bee *A. m. mellifera* populations from Siberia and Ural) based on the genetic distance analyzed with principal coordinate analysis (PCA). Note: Coord. 1: Coordinate 1; Coord. 2: Coordinate 2 (see the text for details). Samples of the *A. m. mellifera* bees (Mel) of different populations of Siberia (Teg: Teguldet, Zar: Zarechny, Mog: Mogochino, Ost: Ostyatskoe, Kol: Kolmogorovo, Yar: Yartsevo, Oze: Ozernoe, Tur: Turukhansk, Yak: Yaksha, *ALT*: Altai) and the Urals (Ural-Burz) are marked in black. Samples of the *A. m. carpathica* bees (*Carpatica*) obtained from the bee farm of Ukraine (*Carpatica*-Ukr) and zoned Carpathian bees (*Carpatica*-Tom) are marked in gray. Samples of the hybrid bees are marked in white: Hybrid-*PQQ*-hybrids originating on the maternal line from the *A. m. mellifera* subspecies; Hybrid-Q-hybrids originating on the maternal line from the *A. m. carpathica* subspecies.

Thus, based on the analysis of the variability of 11 microsatellite loci, the genetic diversity of honey bees of two subspecies (*A. m. mellifera*, *A. m. carpathica*) was described, and subspecies-specific microsatellite loci were identified. The differentiation of subspecies of different evolutionary lineages (M and C) was shown. In addition, for the *A. m. mellifera* subspecies, the significance of the geographic component in determining the genetic diversity of bees (a distinct differentiation of Siberian and Ural populations) was revealed. In this regard, for more detailed comparative characteristics of the *A. m. mellifera* bees living in different geographic regions, a comparative analysis of the variability of microsatellite loci in bees of different populations of Russia was performed.

Genetic diversity of the *A. m. mellifera* in Siberia on the microsatellite loci

As a result of the screening of bee colonies of different populations of Siberia (northern and southern areas, forest areas, isolated and closely located apiaries of the Tomsk region, the Kemerovo region, the Krasnoyarsk Krai, the Altai Krai, and the Republic of Altai), a unique *A. m. mellifera* population (the Yenisei population) in the Krasnoyarsk Krai, as well as individual apiaries where honey bees of the *A. m. mellifera* subspecies are bred, were identified in the Tomsk region, the Altai Krai, and the Republic of Altai. We have previously investigated the genetic diversity of honey bees from the Yenisei and Tomsk populations using microsatellite loci; a total of 22 bee colonies were studied (Ostroverkhova et al. 2018a). This paper presents data on the genetic diversity of *A. m. mellifera* honey bees obtained from apiaries of the Tomsk region, the Krasnoyarsk Krai, the Altai Krai, and the Republic of Altai (Figs. 5.6 and 5.7). A total of 35 bee colonies, including 13 previously unexplored bee colonies, were investigated.

Analysis of polymorphism of 11 microsatellite loci in *A. m. mellifera* bees of Siberian populations showed differences in variability of the studied loci (Figs. 5.6 and 5.7). For the majority of loci (*A043*, *A008*, *A024*, *AC117*, *Ap049*, *A113*, and *MRJP3*), one dominant allele was registered with a frequency of more than 40% in all the samples studied (Fig. 5.6). At the same time, for the *K0820* locus, one of the alleles (allele "128") prevailed in the bees of the Tomsk and Krasnoyarsk populations (allele frequencies were 0.61 and 0.51, respectively), but the second allele (allele "132") prevailed in only one of the studied bee samples (in the bees of the Krasnoyarsk population, this allele is registered with the frequency 0.49). For another group of microsatellite loci (*Ap243*, *SV185*, and *H110*), there were some differences in the frequency of the predominant alleles (similar spectrum of alleles was registered) between the bees of different samples (Fig. 5.7).

For the *Ap243* locus, in the bees from the Tomsk sample, two alleles "255' and "263' (the allele frequency was 42.7% and 33.0%, respectively) were more often registered, whereas, in the bees from Altai and Krasnoyarsk populations, only one allele was detected. The allele "257" (56.3%) was predominant in bees from Altai populations; the allele "263" (39.0%) was predominant in bees of the Yenisei population. For the *H110* locus, in the bees of Tomsk and Altai populations, one allele was predominant: the allele "162" (registration frequency was 78.9%) in the Tomsk

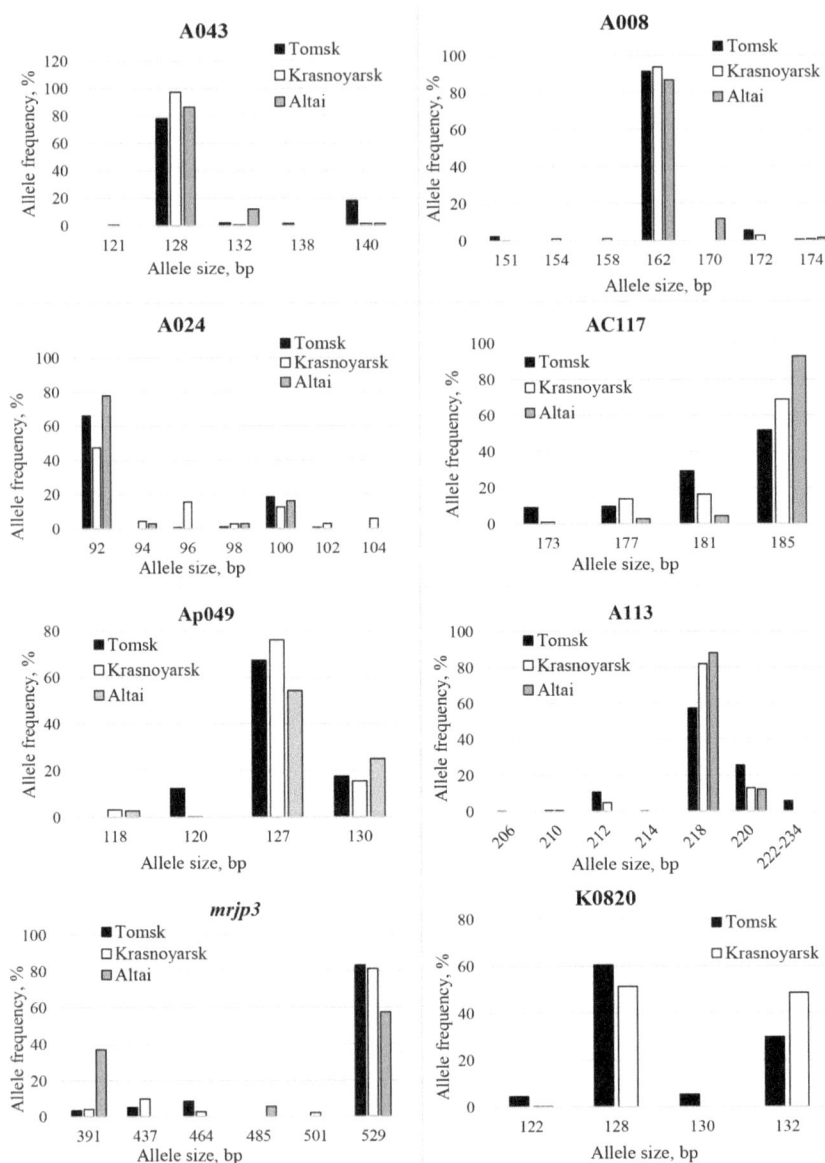

Figure 5.6: Graphs of microsatellite allele frequencies of 8 loci (*A043, A008, A024, AC117, Ap049, A113, K0820*, and *MRJP3*) used to assess the genetic diversity of the *A. m. mellifera* bees of Siberian populations (Tomsk, Krasnoyarsk, and Altai populations). Note: Standard error of the allele frequency is in the range of 0.002–0.067.

bees and the allele "166" (the allele frequency was 72.2%) in the bees from Altai populations. However, there were two predominant alleles "162" (45.2%) and "170" (34.2%) in bees of the Yenisei population. For the *SV185* locus, the allele "269" was most frequently detected in all the studied bee samples (the allele frequency varied from 42.1% to 57.8% in different samples); with a relatively high frequency which

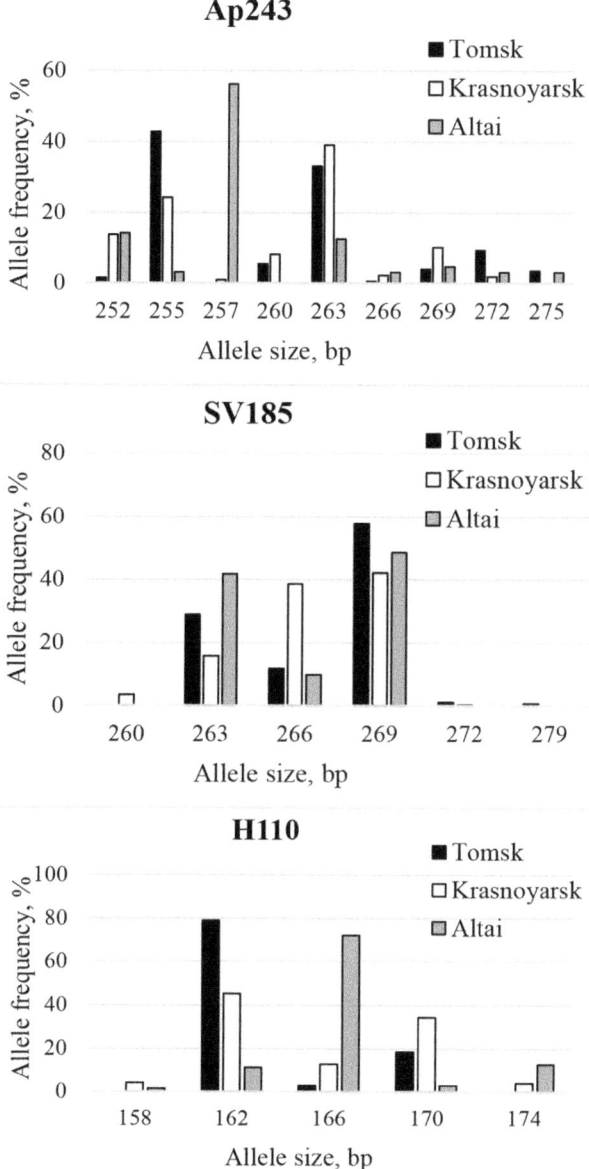

Figure 5.7: Graphs of microsatellite allele frequencies of 3 loci (*Ap243*, *H110*, and *SV185*) used to assess the genetic diversity of the *A. m. mellifera* bees of Siberian populations (Tomsk, Krasnoyarsk, and Altai populations). Note: Standard error of the allele frequency is in the range of 0.003–0.062.

differed between the studied bee groups, alleles "263" and "266" were registered (Fig. 5.7).

The values of the observed heterozygosity (Ho) and expected heterozygosity (He) used to assess diversity in microsatellite loci differed between the bee samples from Siberian populations (Table 5.3). For most loci, lower values of the observed

Table 5.3: Values of the observed heterozygosity and expected heterozygosity (± standard error) on 11 microsatellite loci in the *A. m. mellifera* honey bees from Siberian populations.

Locus		Tomsk region	Krasnoyarsk Krai	Altai populations
A024	Ho	0.505±0.029	0.468±0.022 * * *	0.265±0.076
	He	0.513±0.020	0.722±0.012	0.365±0.065
	N	307	534	34
K0820	Ho	0.597±0.065	0.493±0.030	No data
	He	0.540±0.036	0.501±0.002	
	N	57	280	
A043	Ho	0.279±0.026	0.022±0.007 *	0.030±0.030 * *
	He	0.285±0.021	0.052±0.011	0.239±0.063
	N	305	418	33
A008	Ho	0.226±0.024	0.093±0.014	0.265±0.076
	He	0.243±0.023	0.119±0.015	0.233±0.062
	N	306	451	34
AC117	Ho	0.389±0.028 * * *	0.302±0.021 * * *	0.147±0.061
	He	0.629±0.014	0.482±0.016	0.139±0.055
	N	301	497	34
MRJP3	Ho	0.067±0.027 * * *	0.283±0.029	0.333±0.091 *
	He	0.298±0.043	0.329±0.024	0.530±0.039
	N	89	244	27
Ap243	Ho	0.439±0.034 * * *	0.468±0.028 * * *	0.500±0.088
	He	0.694±0.014	0.753±0.010	0.642±0.058
	N	212	316	32
SV185	Ho	0.527±0.032	0.550±0.030 * *	0.583±0.082
	He	0.569±0.017	0.649±0.009	0.581±0.027
	N	243	278	36
Ap049	Ho	0.447±0.028	0.378±0.023	0.389±0.081 *
	He	0.501±0.020	0.394±0.019	0.611±0.040
	N	309	442	36
A113	Ho	0.521±0.029 *	0.236±0.019 * *	0.000±0.000 * * *
	He	0.597±0.017	0.312±0.017	0.213±0.060
	N	290	509	33
H110	Ho	0.333±0.028	0.421±0.024 *	0.333±0.079
	He	0.343±0.022	0.660±0.009	0.450±0.065
	N	282	439	36

Note. N—the number of bees studied; He—expected heterozygosity according to Hardy-Weinberg equilibrium; Ho—observed heterozygosity. In the table, the values of heterozygosity with a standard error are given. Statistically significant differences in the observed heterozygosity from the expected heterozygosity are marked with (*) (* $p < 0.05$, * * $p < 0.01$, * * * $p < 0.001$).

heterozygosity compared with the expected heterozygosity were found, which may be due to the peculiarity of the reproductive biology of bees. At the same time, only for one microsatellite locus (the locus *A113*), statistically significant differences between the values of the observed and expected heterozygosities in bees of all three groups (Tomsk, Krasnoyarsk, and Altai populations) are shown (t = 2.24, p < 0.05, t = 2.97, p < 0.01, and t = 3.54, p < 0.001, respectively). Statistically significant differences between the values of the observed and expected heterozygosities were also found for loci *AC117*, *Ap243*, and *MRJP3* in a sample of Tomsk bees and for loci *A043*, *Ap049*, and *MRJP3* in a bee sample from Altai populations. In the bees of the Yenisei population (Krasnoyarsk Krai), statistically significant differences were shown between the values of the observed and expected heterozygosities for a larger number of microsatellite loci (except loci *A008*, *Ap049*, *K0820*, and *MRJP3*), which may be due to the isolation of apiaries.

For some loci, differences in the ratio of the observed heterozygosity and expected heterozygosity were found between the bees of different Siberian populations. For example, for the *K0820* locus, the value of the observed heterozygosity was higher than the value of the expected heterozygosity in bees of the Tomsk population, in contrast to the bees of the Yenisei population. Finally, for bees from Altai populations, for some loci, either close values of observed and expected heterozygosities (the *SV185* locus) were detected, or the value of the observed heterozygosity exceeded the values of the expected heterozygosity (loci *A008* and *AC117*). Thus, according to the observed and expected heterozygosities, the studied bees from Altai populations differed from all other studied populations.

Thus, the study of the genetic diversity of the *A. m. mellifera* bees living in different regions of Siberia using a complex of microsatellite loci showed the significant similarity of the spectrum and frequencies of alleles for most of the studied DNA markers. This may indicate the existence of the Siberian ecotype of the *A. m. mellifera* subspecies on the territory of Siberia. To test this assumption, a comparative analysis of the genetic characteristics of honey bees of the *A. m. mellifera* subspecies of Siberian populations and honey bees of Ural *A. m. mellifera* populations (the Burzyan dark-colored forest bee population located in the reserve "Shulgan-Tash", the Republic of Bashkortostan, Ural) was carried out (Ilyasov 2016, Ilyasov et al. 2016).

Comparative characteristics of the genetic diversity of the honey bees of the *A. m. mellifera* populations in Russia based on the microsatellite loci

The original range of *A. m. mellifera* extends from the Alps and Carpathians to the latitude of 60°N, from the Atlantic seashore of western Europe eastward to the Urals and beyond (Meixner et al. 2010). It is assumed that the dark-colored forest bee *A. m. mellifera*, which lives in the vast territory of Eurasia, cannot exist with a similar structure of the gene pool in all local populations. There are probably ecological groups (ecotypes) that differ from each other both in genetic features and in behavioral, physiological, and morphological characteristics at a level below the subspecies. The health of honey bee colonies cannot be understood without taking account of the genetic variability of bee populations, and their adaptation to regional environmental factors, such as climate and vegetation, prevailing diseases, and other aspects of regional importance (Ilyasov et al. 2007, Meixner et al. 2013).

To identify the genetic characteristics (specificity) of bees *A. m. mellifera*, living in different geographical regions of Russia, and determine whether there is a Siberian ecotype of the dark-colored forest bee *A. m. mellifera*, a comparative analysis of the variability of microsatellite loci of different samples of Siberian (the Tomsk region, the Krasnoyarsk Krai, and Altai populations) and Ural (the Burzyan population) bee populations was carried out. For this analysis, the data of a scientific publication on the variability of microsatellite loci in a dark-colored forest bee *A. m. mellifera* of the Burzyan population (Ural, Russia) was used (Ilyasov 2016, Ilyasov et al. 2016).

A comparative analysis of the genetic diversity of *A. m. mellifera* bees of different populations of Russia (Siberian and Ural populations) on 7 microsatellite loci (common loci from the studied loci in bees of different populations) showed the following:

1) Honey bees of the Siberian population (the Tomsk region, the Krasnoyarsk Krai, and Altai populations) are close to each other in the spectrum and frequency of alleles of most studied microsatellite loci (Fig. 5.6);

2) Honey bees of the Burzyan population (Ural) located to the west of the Siberian region differ from the bees of the Siberian populations in some loci (Table 5.4). So, differences in the spectrum of alleles of loci *A008* and *H110*, in the frequency of alleles of loci *A113*, *Ap049*, and *Ap243*, and both in the spectrum and in the frequency of alleles of locus *A024* were registered.

Only for one locus *A043*, the similarity in the spectrum and frequency of alleles in *A. m. mellifera* bees from different populations of Russia were found, which allows us to consider this microsatellite locus a subspecies-specific DNA marker. This assumption is also supported by the data that locus *A043* differentiates the *A. m. mellifera* and *A. m. carpathica* subspecies (Table 5.2).

These sets of data were corroborated by using the principal coordinate analysis (PCA) to visualize the genetic distances of the *A. m. mellifera* samples of different populations of Russia (Siberia, Ural) between themselves (Fig. 5.5). According to the studied DNA markers (7 microsatellite loci), the compared bee groups (Siberian and Ural populations) are distanced from each other and clearly clustered depending on the geographic location (East–West). At the same time, some heterogeneity in the genetic diversity of the bee samples of Siberian populations was shown. The bee samples of the Tomsk region (s. Mogochino, p. Zarechny) and the Krasnoyarsk Krai (s. Ostyakoe, s. Kolmogorovo, and s. Yartsevo) were closest to each other, while the Altai populations and some Krasnoyarsk samples (s. Turukhansk and s. Yaksha) were the most distant, which may be due to the isolation of apiaries.

It should be noted that a comparative analysis of the spectrum of alleles in bee samples of different geographic locations (Siberian and Ural) for some loci revealed alleles that were not identical, but close in the number of nucleotides. This situation, in particular, was observed for loci *H110* and *A113*, whose alleles differed by 2 nucleotides in bees of Siberian and Ural populations. The most likely reason for this situation may be the peculiarities of the methodological approaches used in

Table 5.4: Parameters of the genetic diversity of 7 microsatellite loci in the *A. m. mellifera* samples of different populations in Russia.

Locus	Parameter		Allele frequency			
			Siberia			Ural
			Tomsk region	Krasnoyarsk Krai	Altai populations	Burzyan population
Ap049	N		149	371	36	326
	NA		8	6	4	3
	Min/max		117/142	120/152	117/152	129/142
	Allele (bp) *	127	0.71	0.76	0.54	0
		130	0	0	0.25	0.78
A113	NB		149	367	33	326
	NA		5	7	2	4
	Min/max		212/228	212/232	218/220	216/228
	Allele (bp)	218	0.63	0.80	0.88	0.09
		220	0.30	0.15	0.12	0.85
H110	NB		144	376	36	326
	NA		3	7	5	3
	Min/max		162/170	158/170	156/174	160/168
	Allele (bp)	160	0	0.04	0	0.68
		162	0.73	0.48	0.11	0
		166	0.19	0.13	0.72	0
		170	0.09	0.33	0.03	0
Ap243	NB		109	203	32	326
	NA		6	9	8	3
	Min/max		257/275	254/284	254/275	254/260
	Allele (bp)	254	0	0	0.14	0.62
		257	0.47	0.30	0.56	0.32
		263	0.27	0.55	0.13	0
A008	NB		145	295	34	326
	NA		4	7	3	3
	Min/max		151/173	151/173	163/173	154/158
	Allele (bp)	154	0	0	0	0.87
		162	0.91	0.91	0.87	0
A043	NB		76	236	33	326
	NA		4	3	3	3
	Min/max		128/140	121/140	128/140	128/140
	Allele (bp)	128	0.78	0.98	0.86	0.76
A024	NB		148	376	34	326
	NA		6	5	4	3
	Min/max		92/106	92/102	92/100	98/108
	Allele (bp)	92	0.29	0.67	0.78	0
		94	0.35	0	0.03	0
		98	0	0	0.03	0.63

Note. NB—the number of bees studied; NA—the number of registered alleles studied; Min/max—the minimum/maximum size of alleles (bp). In the table, alleles with a frequency of registration of more than 30% are shown. Data on the Burzyan population (Ural) is given according to Ilyasov (2016).

genotyping. Therefore, the most important task of studying the genetic diversity of bees is the development of a standard allelic ladder for microsatellite loci.

In this study, we presented a description of the subspecies composition of honey bees living in Siberia (the Tomsk region, the Kemerovo region, the Krasnoyarsk Krai, the Altai Krai, and the Republic of Altai) and a characteristic of their genetic diversity using a complex of microsatellite loci. This is the first stage of research on the honey bee, necessary for understanding the consistent pattern of formation of the population structure and the characteristics of the biodiversity of the honey bee populations.

We found that despite the fact that the majority of bee colonies in Siberian apiaries originate from the *A. m. mellifera* subspecies on the maternal line, most studied bee colonies are hybrids, mainly based on the *A. m. mellifera* and *A. m. carpathica* subspecies. We have shown that the process of bee hybridization in Siberia proceeds in a peculiar manner, negatively affects the state of the honey bee populations, and is probably irreversible: according to the analysis of the variability of microsatellite loci, the nuclear genome is more consistent with the *A. m. mellifera* subspecies both in hybrids based on the *A. m. mellifera* subspecies and hybrids based on the southern subspecies. On the one hand, the bee hybridization for the *A. m. mellifera* colonies are characterized by the loss of pure breed; on the other hand, the *A. m. carpathica* colonies, introduced into Siberia, also undergo hybridization and lose the "purity" of the breed (the mitochondrial genome corresponds to the southern subspecies, while the nuclear genome more closely matches the *A. m. mellifera* subspecies). Finally, the process of introgression of the M-lineage genes into the C-lineage in *A. m. carpathica* bees proceeds more extensively compared with the introgression of the lineage C into the lineage M (the *A. m. mellifera* subspecies), since there is a greater percentage of nuclear M-lineage genes than the C-lineage genes in bees originating from the southern subspecies.

Perhaps this situation is due to the fact that for the majority of studied microsatellite loci, the similarity in the allele spectrum is observed between these two subspecies (*A. m. mellifera* and *A. m. carpathica*). When hybridizing bees of southern origin, the presence of often registered alleles can give them certain adaptive advantages. Such an assumption indirectly confirms the data on the presence of the Siberian *A. m. mellifera* ecotype, which differs from the individuals of this subspecies inhabiting other territories, according to the genetic characteristics of some microsatellite loci. However, this assumption needs confirmation.

However, in Siberia (the Tomsk region, the Krasnoyarsk Krai, the Altai Krai, and the Republic of Altai), the *A. m. mellifera* populations were identified. A comparative analysis of the genetic characteristics of the *A. m. mellifera* bees of different populations of Russia (Siberia, Ural) indicates the existence of the Siberian ecotype of the *A. m. mellifera* subspecies.

In conditions of widespread bee hybridization, it is necessary to improve the genetic methods for controlling the purebreed of bee colonies. These studies require the use of highly informative and reliable molecular approaches. At present, both morphometric and molecular genetic methods are used to study the honey bee origin (subspecies), however, the results achieved using different techniques are oftentimes not congruent. There is now a critical need to integrate databases, make reference data sets generally accessible, and harmonize procedures to achieve sound and generally recognized "consensus reference data sets" (Meixner et al. 2013). The development of an accessible database of European subspecies and ecotypes (both in Europe and in other regions where they live) as reference material is a necessary task for further research on the honey bee. The accumulation of data on the variability of different marker systems in individuals of different bee subspecies (races, breeds, lines) living in different climatic and ecological conditions will reveal the subspecies-determining traits and genes causing adaptation to different geographical and environmental conditions, identify the markers underlying the formation of differences in bee colonies by a set of economically significant traits, etc. In addition, the study of the biological and genetic diversity of honey bees is a scientific basis for carrying out selection work to preserve and restore the gene pools of unique honey bee subspecies, for example, the *A. m. mellifera* subspecies.

Conclusion

Despite the active study of the honey bee all over the world, key issues, such as the mass death of bee colonies and interbreed bee hybridization, as well as some more specific problems, remain unresolved. The causes and/or consequences of these negative processes are not fully understood, and the results of studies by various authors are contradictory and ambiguous. For example, the main causes of the bee collapse are certain diseases—varroosis caused by the mite *Varroa destructor* and type C nosemosis caused by the pathogen *Nosema ceranae* (Guzmán-Novoa et al. 2010, Dainat et al. 2012a,b, Dietemann et al. 2012). However, if this is true for some geographic regions, then for other territories it is not proven. For example, the *N. ceranae* pathogen is often considered to be the cause of the mass colony deaths in Spain (Higes et al. 2008, Botías et al. 2013, Goblirsch et al. 2013), but in Germany and Siberia (Russia), in the honey bee populations, this parasite does not predominate and does not displace the traditional *N. apis* species (Gisder et al. 2017, Ostroverkhova et al. 2019a). Consequently, the bee collapse as a result of invasion by the parasite *N. ceranae* is more likely to be a regional, but not a global problem; therefore, the automatic transfer of the results obtained in some territories to other geographical areas and climatic habitats of honey bees is unreasonable.

Similarly, it can be assumed that the results obtained for one subspecies of the honey bee (for example, assessment of biological and genetic characteristics, infestation with parasites and pathogens, economically significant indicators, etc.) inhabiting different climatic conditions, and even more so for different subspecies, will be different. Thus, when studying the variability of the microsatellite *MRJP3* locus in Siberian honey bees, we showed that this locus is more likely to be a subspecies-specific DNA marker, but does not define economically significant indicators (Ostroverkhova et al. 2018b). Our data does not agree with the results obtained on Africanized bees, according to which the significance of some alleles and genotypes of the microsatellite *MRJP3* locus is shown for the selection of bee colonies for the productivity of royal jelly (Baitala et al. 2010). In addition, subspecies *A. m. mellifera*, which lives in different regions, has a high genetic heterogeneity; therefore, extrapolation of the results obtained for one population to the subspecies as a whole is possible only after they are confirmed by other studies conducted on representatives of other populations (Lebedinskiy et al. 2016). Finally, it is likely that with the introduction of the same honey bee subspecies into different regions where different aboriginal populations live, in each case the specific features of the hybridization process (course, speed, level, direction) will be revealed.

Thus, only the accumulation of data on controversial and topical issues of studying the honey bee, taking into account the climatic features of the location of apiaries, the geographical localization of populations of honey bees, the origin of bee colonies (evolutionary lineage, subspecies and/or ecotype), etc., will highlight general (universal) processes and mechanisms and specific details on the above problems.

CHAPTER 6

Current Drivers of Taxonomic Biodiversity Loss in Asian and European Bees

Hatjina, F.,[1,]* *Gajda, A.*[2] and *Dar, S.A.*[3]

Introduction

Bees and their ancestors constitute a very old group- almost 120 million years old (Michener 2007, Michez et al. 2012, Danforth et al. 2013). Today over 16,000 species have been described worldwide, and undoubtedly they constitute a major part of the food chain, as well as the ecosystem's sensitive equilibrium through their service towards the pollination needs of the flowers, trees, and vegetables (Klein et al. 2007, Potts et al. 2016). However, this is a mutual relationship, as the bees rely on plants for their protein and carbohydrate nutrients, as well as on water and resins. The increasing human activity (habitat destruction, fragmentation, land use intensification, insecticides application), pathogens, and alien species threaten the diversity of life as a whole, and subsequently of bees around the world (Osborn et al. 1991, Banaszak 1995, Williams 1986, Steffan-Dewenter and Westphall 2008, Alaux et al. 2010a, Vidau et al. 2011) (Figs. 6.1 and 6.2).

Climate change also poses many risks for agricultural crops (Lobell et al. 2011, Rosenzweig et al. 2014), and eventually for pollinators (Freitas et al. 2009, Burkle et al. 2013, Vanbergen et al. 2013, Giannini et al. 2017). Hooper et al. (2012) suggested that species loss affects primary production and decomposition of plant

---

[1] Division of Apiculture in Institute of Animal Sciences, Nea Moudania, 63200, Greece.
[2] Laboratory of Bee Diseases, Department of Pathology and Veterinary Diagnostics, Institute of Veterinary Medicine, Warsaw University of Life Sciences, 8 Ciszewskiego Street, 02-786 Warsaw, Poland, E-mail: anna_gajda@sggw.pl.
[3] Entomology in Sher-e-Kashmir University of Agricultural Sciences and Technology of Kashmir, Jammu and Kashmir, 192401, India, E-mail: sadar@skuastkashmir.ac.in.
* Corresponding author: fhatjina@instmelissocomias.gr

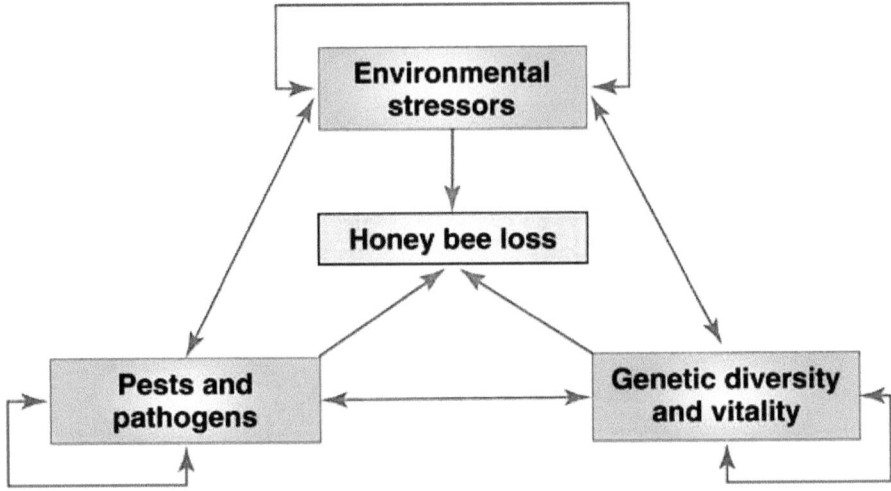

Figure 6.1: Interactions among multiple drivers of honey bee loss. Blue boxes represent the three main groups of drivers associated with honey bee loss; red arrows represent direct pressures on honey bees from drivers; green arrows represent interactions between drivers; black arrows represent interactions within drivers. (Retrieved from Potts et al. 2010b–TRENDS in Ecology and Evolution).

Color version at the end of the book

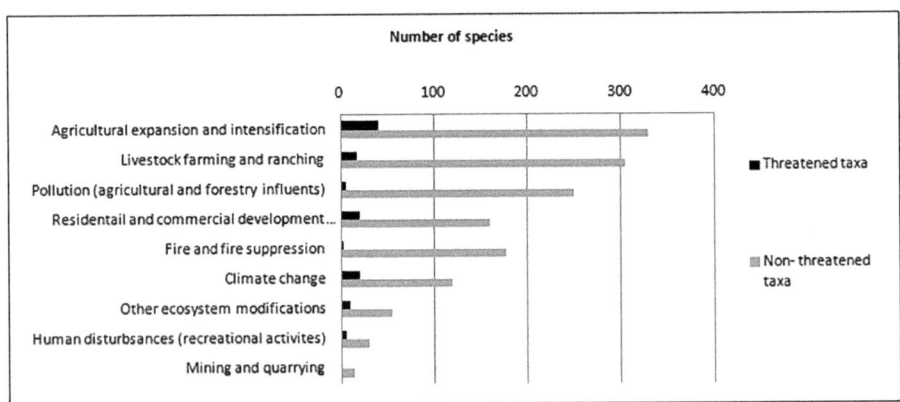

Figure 6.2: Drivers reducing biodiversity (Retrieved form Nieto et al. 2014).

species, and that biodiversity loss in the 21st century could be one of the major drivers of ecosystem change. The International Union for Conservation of Nature (IUCN) predicts a global loss of 20,000 flowering plant species within the next few decades; which will lead to the decline of dependent pollinators, such as honey bees (Heywood 1995). Also, the pollinator loss ultimately results in the decline of plant biodiversity (Thomas et al. 2004, Biesmeijer et al. 2006, Carvell et al. 2006, Pauw and Hawkins 2011).

Especially for the honey bees, there are reports of declines of their managed populations throughout the world (Potts et al. 2010b, vanEngelsdorp and Meixner 2010). Similarly, elevated colony losses have been reported from

Europe (Crailsheim et al. 2009, Potts et al. 2009), including countries, such as Austria (Brodschneider and Crailsheim 2010), Denmark (Vejsnaes et al. 2010), England (Aston 2010), France (Chauzat et al. 2010), Greece (Bacandritsos et al. 2010, Hatjina et al. 2010), Italy (Bortolotti et al. 2010), the Netherlands (van der Zee 2010), Norway (Dahle 2010), Poland (Topolska et al. 2010), Scotland (Gray et al. 2010), Bulgaria (Ivanova and Petrov 2010), Croatia (Gajger et al. 2010), Bosnia and Herzegovina (Santrac et al. 2010), Canada (Currie et al. 2010) and the USA (Ellis et al. 2010a, vanEnglesdorp and Meixner 2010), and the Middle East (Haddad et al. 2009a).

Environmental factors, such as weather conditions, availability of nesting sites, food sources, and chronic exposure to insecticides might cause CCD-like symptoms (Oldroyd 2007). Similarly, the pathogens and parasites which have been demonstrated to be involved in colony losses in different regions of the world, are considered current threats to honey bees and beekeeping (Genersch et al. 2010). In Fig. 6.2, we see the main drivers for bees decline, as reported by Nieto et al. (2014). Furthermore, in the most recently published review, we read "The main drivers of taxonomic insect biodiversity appear to be in the order of importance: (i) habitat loss and conversion to intensive agriculture and urbanization; (ii) pollution, mainly that by synthetic pesticides and fertilizers; (iii) biological factors, including pathogens and introduced species; and iv) climate change" (Sánchez-Bayo and Wyckhuys 2019). Concerning the bees (Apidae), the same authors report that almost 15% of bee species are extinct, 8% are threatened, and 42% are in a vulnerable condition.

Honey bee colony losses in general

Loss of taxonomic biodiversity of honey bees can partly be attributed to colony losses. Elevated losses of *Apis mellifera* colonies have been observed since over a decade now in numerous countries (Neumann and Carreck 2010). They were first observed in 2006 in North America (vanEngelsdorp et al. 2007, Ellis et al. 2010a), and were estimated to be 31.8 percent. Shortly after that, high loss was observed in Canada (36% and 35% in the respective winters of 2006/2007 and 2007/2008). Since then, many other countries started observing this trend, starting from the winter 2007/2008 (Brodschneider et al. 2010, Charriere and Neumann 2010, Gray et al. 2010, Hatjina et al. 2010, Ivanova and Petrov 2010, Vejsnæs et al. 2010, van der Zee 2010). In some regions of Italy, they reached 40% (Mutinelli et al. 2010), in Poland it was 15.3%, but it was still a big increase from 9.9% loss rate from the previous winter (Topolska et al. 2008, Topolska et al. 2010), which shows the scale of this phenomenon.

At some point, it became apparent, that individual countries are not going to solve this problem, so in 2008, an international network of honey bee experts-COLOSS—Prevention of Honey bee Colony Losses—was formed under COST funding (FA0803). Since then the network is working to identify the drivers of those elevated losses, as well as to estimate how high the losses were in different countries, and if there is any pattern to it. It has been observed, that the losses vary greatly across space and time. The "acceptable" loss level has been set at 10%, and anything that is higher than that indicates elevated colony losses.

A first comprehensive report from Europe was looking at the winter of 2008/2009, when the overall (for all participating countries) winter loss was 12.3%, which already exceeded the acceptable rate. Although looking at respective countries, the losses varied greatly, yet there were many, where the loss rate was almost double the acceptable limit (the Netherlands and Ireland 21.7%). The next winter (2009/2010) brought even bigger losses. For instance, in Italy, they were almost 30%, and in the Netherlands 29.3% (van der Zee et al. 2012). Since then, most countries still experience elevated colony losses (van der Zee et al. 2014, Brodschneider et al. 2016).

The most recent data shows an overall loss rate of 20.9% during the winter of 2016/2017 for Europe, yet, when looking at separate countries, the situation seems even worse. Overall losses noted in Spain were 27.6%, in Serbia 24.1%, Poland 21.8%, just to name the highest (Brodschneider et al. 2018).

In the US, the elevated losses still continued (and still do), and since 2008 they were reported to increase to 35.9% (vanEngelsdorp et al. 2008), and haven't gotten under the desired 10% since (Ellis et al. 2010b, Spleen et al. 2013, Steinhauer et al. 2014, Bruckner et al. 2018).

A report by FAO (2013) indicates that in the period between 1961 and 2007, the number of managed colonies decreased in Europe (26.5%), North America (49.5%), Asia (42.6%), Africa (13.0%), South America (86%), and Oceania (39%) (Fig. 6.3).

In Asia, colony losses of managed honey bees have raised a major concern, and surveys of colony losses were conducted around the globe to understand the apicultural situation. Till date, most studies have focused on the mortality of the western honey bee—*Apis mellifera*; however, little is known about the mortality of its eastern counterpart—*Apis cerana*. Here, we report the survey results of *A. cerana* colony losses in three consecutive years (2011–2012, 2012–2013, and 2013–2014) in China, which were 12.8%, 95%, and 11.9–13.7%, respectively, but varied among years, provinces, and types of apiaries (Chen et al. 2017).

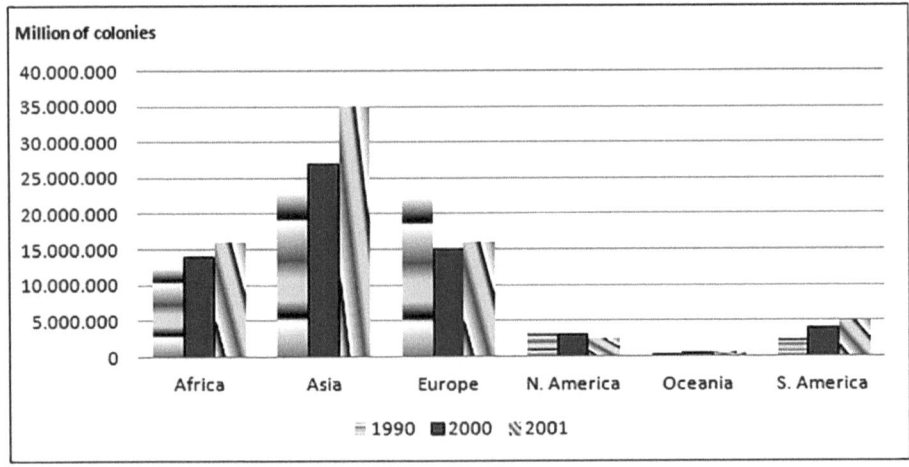

Figure 6.3: Honey bee colony declines over the years (Retrieved from FAO 2013).

Much thought and research has been given to determine the causes of those nearly catastrophic losses all over the world and it has been concluded that pathogens play a key role in this global decline. The intensity may vary between continents or countries, but *Varroa destructor* with viral infections that it vectors and activates has been unanimously named as the main cause of global colony losses (van der Zee et al. 2015). More recently, *Nosema ceranae* has been described as the second most important cause of mentioned losses (Higes et al. 2009).

Pathogens of honey bees

The drivers of colony losses (that add to the loss of taxonomic biodiversity of bees) are most often pests and pathogens, especially the (relatively) newly introduced ones. It is widely known, that an introduction (or invasion as a matter of fact) of an alien species into an ecosystem is one of the most important sources of biodiversity loss and may result in host eradication (Deredec and Courchamp 2003). Invasive alien species is an alien species whose introduction and/or spread threatens biological diversity. Introduced pathogens and parasites are often able to switch hosts, therefore posing new threats to native species which lack innate abilities to fight the infestation/infection/invasion (Cuthbertson et al. 2013).

There is also the matter of biodiversity-disease relationship. There are many hypotheses on how these relations' dynamics work, but the most plausible is the "dilution effect" hypothesis, which puts weight on the fact that increased diversity will actually decrease the disease transmission, which means, that saving biodiversity of bees will help prevent diseases in the long perspective, as a sort of upward spiral.

Invasive alien honey bee pest and pathogen species are impacting global bee biodiversity, which is a major concern because biodiversity is essential in the functioning of healthy ecosystems. Those alien species are steadily increasing in numbers as a result of world trade expansion and improvement of transport. Also, global warming makes it possible for pests, which are native for tropics, to move, and settle in a more temperate climate. It is believed that wild populations are not normally threatened by the parasites and pathogens with which they co-evolved. However, adverse effects of pests and diseases may arise when wild populations are stressed by environmental degradation. Allen et al. (1990) found a Nepalese population of *A. laboriosa* that was severely infected with European foulbrood (*Mellisococcus plutonius*), which they attributed to environmental stress brought on by deforestation.

The *Varroa* mite

Regarding the honey bees, the most dangerous introduced pathogen is (still, after over 3 decades) *Varroa destructor* (Fig. 6.4 A).

Originally it was a parasite of *Apis cerana*, but due to the export/import of bees, which provided contact between Western and Eastern honey bee colonies, it had the opportunity to broaden its host range to *Apis mellifera*, which was far less resistant to the infestation and its consequences. As a result, the mite has spread almost worldwide. Till date, only Australia is free of this parasite (AQIS, Australian Government: http://www.daff.gov.au/aqis/quarantine/pests-diseases/honey bees).

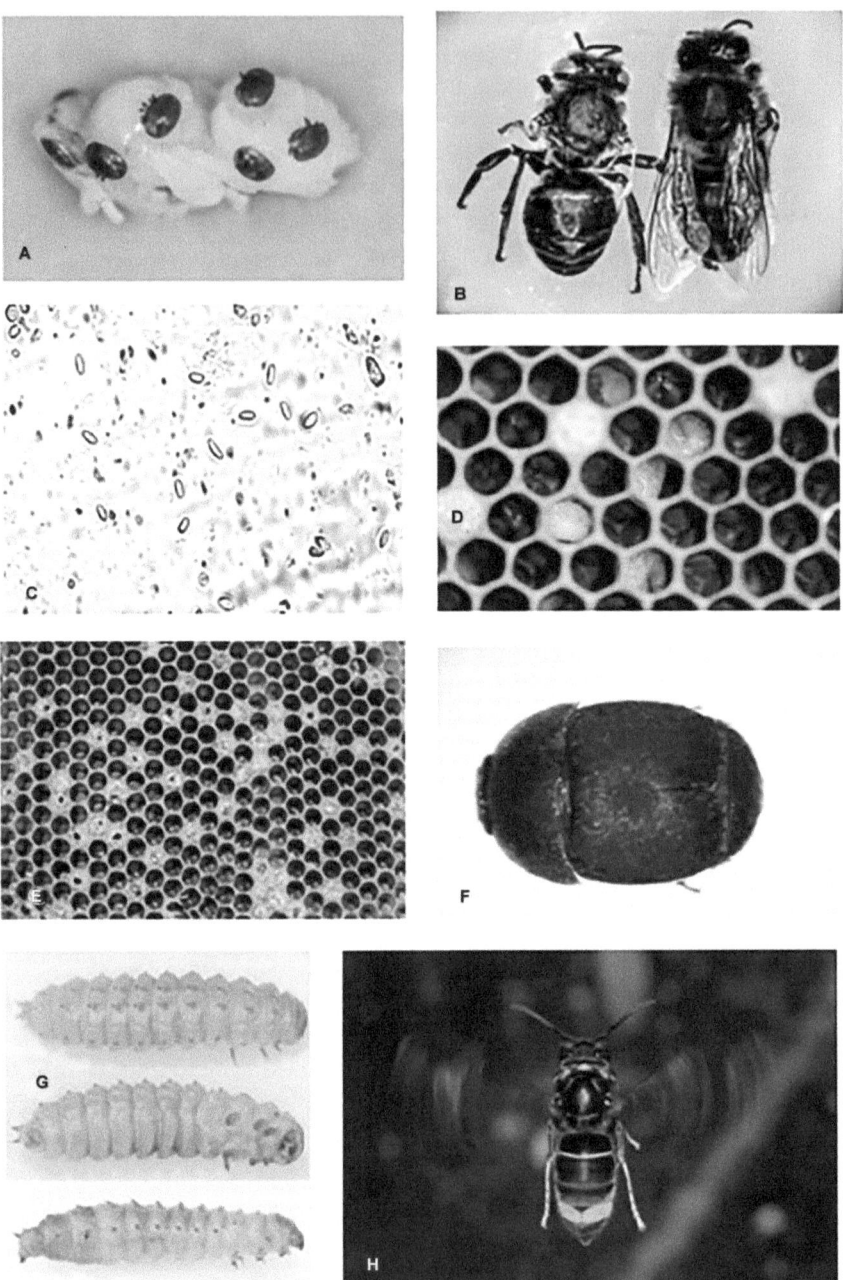

Figure 6.4: (A) The *Varroa* mite females feeding on a honey bee pupa. (B) A bee with deformed wings and shortened abdomen due to the DWV infection. (C) *Nosema ceranae* spores seen in 400x magnification. (D) A brood comb with twisted, diseased larvae attacked with *Melissococcus plutonius*. (E) a brood comb with brood cappings clearly punctured and scattered brood pattern. (F) The small hive beetle imago. (G) larva. (H) The yellow-legged hornet (Photos A, B, C, D, E, F, G: A. Gajda, Photo H: Per Kryger).

Color version at the end of the book

V. destructor female feeds on the fat body and hemolymph by piercing the cuticle of brood and bees. With low and medium levels of colony infestation, the symptoms in individual bees are not visible, however, adult bees emerging from infected cells have lowered immunity and their life span is significantly shortened. Feeding females cause physical injuries to the bees and brood, which lowers their protein, fat, and carbohydrate levels, and interferes with organ development (Bowen-Walker and Gunn 2001).

The mite is also known to be a vector and activator for some bee viruses, such as Kashmir Bee Virus (KBV), Acute Bee Paralysis Virus (ABPV), Israeli Acute Paralysis Virus (IAPV), and Deformed Wing Virus (DWV) (Tentcheva et al. 2004, Boecking and Genersch 2008). Those viruses were long present before the occurrence of *V. destructor* in *A. mellifera*, but haven't been observed to cause major problems (Bailey and Ball 1991, Bowen-Walker et al. 1999). It is thought, that the direct injection of the virus to the hemolymph provokes typical disease symptoms. The most widely researched is DWV infection. It results in deformed wings (Fig. 6.4 B) and shortened abdomens in heavily infested honey bee colonies (Boecking and Genersch 2008, De Miranda and Genersch 2010).

Initially, when the *Varroa* mite arrived in its non-native countries, it had a tremendous effect on wild bee populations (Kraus and Page 1995), and massive honey bee losses were documented, for example in Greece, where 20–25% losses were reported (Santas 1983). The parasite was "unknown" for the beekeepers, as there was negligible knowledge about its biology, its behavior, and there were no treatment methods. It took many years until beekeepers started to deal with this mite in a way that they could reduce the loss of their colonies and the reduction of their income (Emmanouel et al. 1984, Boecking and Genersch 2008). At the same time, beekeepers started to replace their lost colonies with populations from other places, from the same or different countries, resulting in introgressive hybridization of the native subspecies and ecotypes (Jensen et al. 2005, Ivanova et al. 2007, Soland-Reckeweg et al. 2009, Martimianakis et al. 2011), or they even stopped beekeeping completely.

An optimistic message has recently come from different parts of the world: there are many reports of colonies resistant to the mite infestation (Büchler et al. 2010, Danka et al. 2012, Locke et al. 2012, Kirrane et al. 2015), but with the current state of worldwide beekeeping, where many colonies have extremely high infestation levels due to poor treatment, the resistant colonies are simply attacked too strongly when introduced to an environment, and they automatically are in a losing position simply because they are "introduced" (Büchler et al. 2014). *Varroa* resistant colonies are, however, thought to be the best solution in eliminating the problem of colony losses due to this parasite, but it seems that global beekeeping and *Varroa* management need to be controlled and advised in a way that will allow for the resistant bees to thrive. This will, of course, mean further loss of diversity in favor of resistant genotypes, yet this seems to be the only plausible solution, allowing the end of the great colony losses observed due to this widespread and dangerous parasite. We also need to note that 30 years is simply not enough time for the host-parasite relationship to co-evolve in a way that would create some sort of homeostasis between them and allow them to co-exist, as it is observed in *A. cerana* colonies, for which the parasite doesn't pose such a big threat. Therefore, steps must be taken towards the most sustainable solution in saving honey bee diversity.

The Microsporidium *Nosema ceranae*

The second most dangerous pathogen causing colony losses is the Microsporidium *Nosema ceranae* (Fig. 6.4 C). Much like *V. destructor*, *N. ceranae* was previously parasitizing only *Apis cerana*. In 2005, it was found in *Apis mellifera* for the first time, but soon it turned out that it had made a host switch to *A. mellifera* at least 10 years before (Paxton et al. 2007, Chen et al. 2008, Topolska et al. 2008). By now, *N. ceranae* is spread worldwide, and causing major colony losses in many countries (Higes et al. 2006, Huang et al. 2007, Klee et al. 2007, Chen et al. 2008, Giersch et al. 2009, Higes et al. 2009, Invernizzi et al. 2009), including Canada and USA (Williams et al. 2008).

The microsporidia infect the epithelial cells of the midgut of adult bees, causing digestive disorders, which leads to a shorter life span. Just one infected cell is enough for the disease to develop, moreover, constant intake of spores is also not necessary (Meana 2009). The parasite damages not only the epithelial cells, but also the regenerative crypts of the midgut, which leads to irreversible damage (Kasprzak and Topolska 2007). The death of the bee is a result of a prolonged hunger state, which leads to energetic stress (Mayack and Naug 2009). This state of hunger is the reason bees are actively seeking others, who will potentially feed them (via trophallaxis), which leads to a rapid spread of the disease in the colony.

The threat of this disease is mostly attributed to the fact that it can kill quickly and there are no symptoms right until the moment when the level of infection is much too high to fight anymore. Beekeepers often find their colonies dead at the bottom board after the winter period, or "disappeared" if the infection level is already too high before the winter. In this case, it is thought that sick bees are leaving the hive to find food, and never come back due to energetic stress. As a consequence beekeepers will need to replace their lost stock with others, sometimes from foreign genotypes, resulting in an even higher dwindling of local biodiversity.

It seems that warmer climate favors the disease, while in colder regions of the world *N. ceranae* doesn't seem to be a problem (Forsgren and Fries 2013). In Kashmir, Dar and Ahmad (2013) also observed the highest infection rate of 41.1% in spring and 13.3% in summer, however, the infection rate was low to none in autumn. The mountainous climate of Kashmir with humid springs and summers favors *Nosema* spread. This usually leads to higher colony losses due to nosemosis in the warmer climate, meaning that the global diversity of honey bees is already dwindling, since the pathogen is wiping out bees in some parts of the world and beekeepers might seek "more resistant" bees from other countries. It is important to note here that even the scientific approach of seeking bee genotypes resistant to *Nosema* spp. (Chaimanee et al. 2012), might lead to a reduction of honey bee biodiversity.

Bacterial pathogens of honey bees

There are two main bacterial pathogens of honey bees, and both are pathogenic to larvae but not to adult bees: *Melissococcus plutonius*, causing European foulbrood (Bailey 1956, 1957) (Fig. 6.4 D) and *Paenibacillus larvae*, causing American foulbrood (Fig. 6.4 E) (Genersch et al. 2006, Genersch 2010). *P. larvae* is a spore-forming bacteria, which makes its control more difficult than *M. plutonius* (which does not form spores). The *M. plutonius* enters the larvae through ingestion and proliferates in the larval midgut, assimilating much of the larval food and the infected larvae die from starvation (Bailey 1983). There are few reports of colony losses due to bacterial pathogens around the world. Allen et al. (1990) found a Nepalese population of *A. laboriosa* that was severely infected with *M. plutonius*, which they attributed to environmental stress brought on by deforestation.

American foulbrood (AFB) is a notifiable disease in many countries. It is highly contagious, easily and rapidly spread within a colony and among colonies. Such colonies with AFB should be destroyed to prevent the disease from spreading further. AFB has had big impact on the beekeeping industry. In 2000, the annual economic loss attributed to AFB infection in the U.S. was $5 million (Eischen et al. 2005). It is considered to be more virulent than European foulbrood.

European foulbrood (EFB) has been reported from across every continent that honey bees inhabit (Matheson 1993), and currently appears particularly prevalent and dramatically increasing in the UK (Wilkins et al. 2007, Tomkies et al. 2009) and Switzerland (Belloy et al. 2007, Roetschi et al. 2008, Forsgren 2010), but EFB it is not believed to be a major factor to explain widespread colony losses.

The small hive beetle and the yellow-legged hornet

In recent years, two more invasive alien species were introduced to Europe, namely: small hive beetle *Aethina tumida* and the yellow-legged hornet *Vespa velutina*. The small hive beetle *A. tumida* is a parasitic pest and scavenger of social bee colonies (honey bees, bumble bees, stingless bees) in sub-Saharan Africa. In its original ecosystem, it rarely inflicts damage on colonies, which co-evolved with the pest and developed mechanisms to keep the invasion on non-threatening levels. However, *A. tumida* has since "escaped" from its native environment and has invaded North America and Australia, where it impacted the apiculture industry greatly. A decline in native bees due to infestation by small hive beetles will potentially have a negative impact on bee biodiversity (Cuthbertson and Brown 2009).

Adult small hive beetles (Fig. 6.4 F) measure on average 5.7 mm × 3.2 mm (Ellis et al. 2002). The size of beetles can vary due to food availability and variations in climate (Ellis 2004). They can fly several kilometers (Somerville 2003), which makes their spread fast and quite easy. But the real threat is the larvae (Fig. 6.4 G), which can destroy the honey bee nests nearly completely by feeding on brood, honey, and pollen, and destroying combs in the process. As it is a very new, and potentially big threat to European beekeeping, the infestation is compulsorily notifiable in the European Union. After a case is confirmed, Veterinary Services take necessary measures to stop its further spread. A decline in native bees due to infestation by small hive beetles will potentially have a negative impact on bee biodiversity (Cuthbertson and Brown 2009).

The invasive yellow-legged hornet (*V. velutina*) is a predator of honey bees introduced into Europe from Asia (Fig. 6.4 H). It poses an additional threat to honey bees and other pollinators in light of current, already high, colony (and generally pollinator) losses. It was first observed in France in 2004 (Rortais et al. 2010), and has rapidly spread to Spain, Italy, Belgium, Portugal, Germany, and Great Britain. Other areas of introduction are South Korea and Japan. The diet of *V. velutina* may be comprised of honey bees and other Apoidea in 70 percent. Besides honey bees, *V. velutina* preys on other Hymenoptera, including various species of wild bees, but also preys on Vespidae, Diptera, and other insects. Therefore, the hornet might have a negative effect on insect communities, reducing biodiversity.

All the above mentioned invasive species have a great (but not only) impact on the honey bee biodiversity, as they are adding to the colony losses that are already high in recent years. However, research shows that certain genotypes of bees may be resistant, while others remain susceptible to different pathogens or pest invasions. Studies show that no matter the disease/invasion, locally adapted bees will always do better when exposed to certain pests and pathogens in their own environment, than the introduced ones (Meixner et al. 2015), which gives a strong hint as to how to manage bee populations.

Interestingly, many microparasites, including *Nosema*, viruses, and fungi which affect honey bees were detected also in solitary bees in the vicinity of the apiaries, suggesting "that beehives represent a putative source of pathogens for other pollinators. Similarly, solitary bees may act as a reservoir of honey bee pathogens" (Ravoet et al. 2014).

Agricultural intensification and pesticides

Land Use. As already mentioned in the Introduction, intensification of agriculture has a number of impacts on the availability of resources for wild and managed bees at a landscape scale (Feon et al. 2010), and is widely regarded as the primary driver of bee declines across Europe (Kuldna et al. 2009). Discussing about bee losses, we do not refer only to actual number of insects or honey bee colonies (Aizen and Harder 2009, Le Conte et al. 2010, Neumann and Carreck 2010, van der Zee et al. 2012, van der Zee et al. 2014, Brodschneider et al. 2016), but also to the loss of bee biodiversity, loss of taxa, loss of honey bee subspecies, and other pollinators (Biesmeijer et al. 2006, Moritz et al. 2007, Goulson et al. 2008, Williams and Osborne 2009, Settele et al. 2008, Meixner et al. 2010). The European Red List of Bees provided information on the status of nearly 2,000 species of bees in Europe, including *A. mellifera*. According to the Red List of Bees 7.7% (150 species) of European bees have declining populations, while 9% of bees in Europe are threatened with extinction, mainly due to habitat loss as a result of agriculture intensification (Nieto et al. 2014).

Deforestation during the last 25 years (Sodhi et al. 2004) and an extremely high human population density in some countries, such as Pakistan, Nepal, and Bangladesh inevitably causes increased pressures on natural ecosystems. Broad-scale conversion of primary forest to short-cycle forestry, rubber and oil palm plantation, agriculture, and urban areas (Kevan and Viana 2003, Sodhi et al. 2004) are a major concern for honey bee conservation. Sodhi et al. (2004) outline the depressing reality of deforestation in Southeast Asia. This region has the highest rate of tropical deforestation in the world, and is predicted to lose three-quarters of its original forest and 42% of its biodiversity in the next hundred years.

In most industrialized countries, intense land use has led to a progressive reduction of habitats suitable for honey bees with a negative impact on feral and wild populations (Biesmeijer et al. 2006, Kremen et al. 2007, Flynn et al. 2009). Agricultural intensification and forestry have been shown to reduce the diversity and abundance of native bees in the US, diminishing their pollination services by 3 to 6-fold (Kremen et al. 2007). In Europe, commonly practiced intense land use is likely to reduce not only the availability of floral resources, but also nesting sites suitable for honey bees (Biesmeijer et al. 2006, Murray et al. 2009). According to Nieto et al. (2014), "Europe seems to have the most highly fragmented landscapes of all continents, and only a tiny fraction of its land surface can be considered as wilderness".

In areas of intense land use, facilitating the establishment of native honey bees in nature reserves may be a strategy to reestablish wild populations. Many European governments (e.g., Belgium, Netherlands, and the UK) however, have implemented national policies to warrant the legal exclusion of managed pollinators from Protected Areas (Sections 14 and 16 of the UK's Wildlife and Countryside Act 1981), based on questionable evidence on competition between honey bees and other native pollinators (Huryn 1997, Paini 2004). Based on the same evidence, other countries, such as Germany and Austria, permit apiculture within designated Protected Areas.

Fragmentation of habitats is responsible for changing pollinator populations (Thomas et al. 2004); it causes genetic erosion by reducing gene flow and increases the chance of extinction of populations and species (Barrett and Kohn 1991), as well as decreasing food and nesting resources (Hines and Hendrix 2005, Potts et al. 2005). Habitat loss might be one of the biggest factors impacting honey bee decline and the agricultural landscape changes after the Second World War (Winfree et al. 2009).

Use of Pesticides. During the last 5 years, an exponential increase in the interest of scientists to determine the effects of all types of pesticides on bees' behavior, immunity, biodiversity, and general well-being is observed. It is due to a massive expansion in the use of chemicals active at very low doses, which do not really kill the individuals, but they jeopardize their well-being. According to the European Environment Agency, the agricultural sector is one of the mainland users in Europe. Farming systems in Europe, as well as around the world have been strongly intensified over the past 50 years, and have led to the homogenization of agrosystems. Eventually, higher amounts and new chemical compounds are used every year to protect the crops (Fig. 6.5).

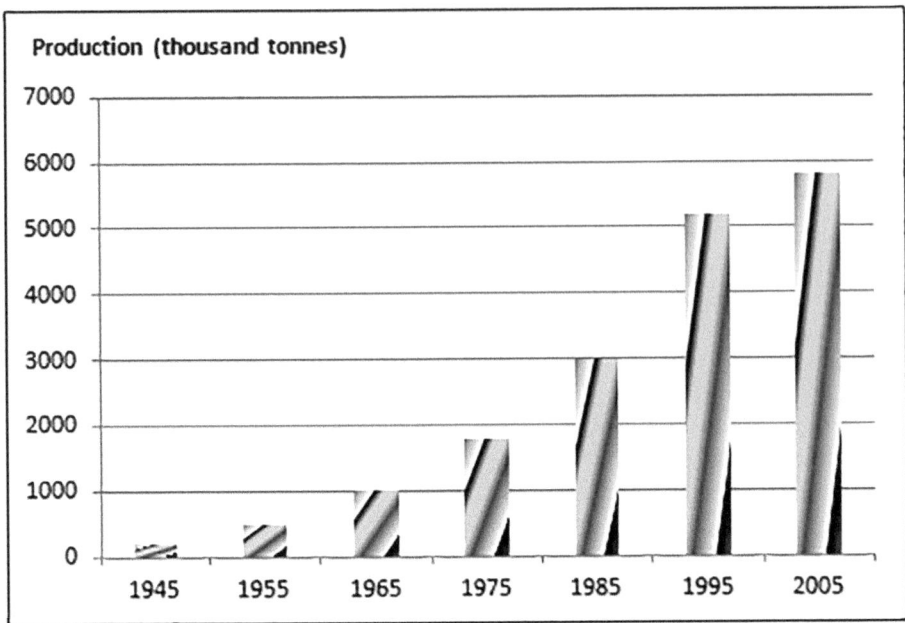

Figure 6.5: World production of pesticides. (Retrieved from Agrochemical Service 2000 in Carvalho 2006).

A recent update from the Worldwide Integrated Assessment on the Impact of Systemic Pesticides on Biodiversity and Ecosystems has evaluated 500 scientific evidence published since 2014 and confirms the high risk posed by these substances not only to insects but also to vertebrates and wildlife in general. Almost 90% of these publications demonstrate the devastating effects of pesticides on biodiversity of bees around the world. The reduction is referring to species richness as well as to population numbers of all: solitary bees, bumble bees, and honey bees (Simon-Delso et al. 2014, Goulson et al. 2015, Raine and Gill 2015, Van der Sluijs et al. 2015, Kerr 2017, Lundgren 2017, Rortais et al. 2017).

Honey bees and non-*Apis* bees may differ in their susceptibility and exposure to pesticides because they differ significantly in terms of morphology, physiology, life cycle, survival needs, and behavior (Arena and Sgolastra 2014, Heard et al. 2017). Moreover, non-*Apis* bees may also experience higher pesticide exposure than honey bees due to their life cycle, which makes them more susceptible (Arena and Sgolastra 2014, Heard et al. 2017, Gradish et al. 2019).

When the effect of pesticides is high, for example when there is direct oral or contact exposure to high concentrations, the population numbers of all species are affected. However, when the exposure is through a continuous low concentration poisoning, then the behavior and the immune system is altered, and the effect could even be on a species level, as most of the species are oligolectic and have a short lifespan (Arena and Sgolastra 2014). Pesticides, therefore, could be regarded as the number one stressor on species decline of European and World bees, although this might not apply to honey bees. Nonetheless, the synergistic or combination effect of pesticides together with honey bee diseases (e.g., nosemosis) should not be overlooked, as it can wipe out vast numbers of managed bees (Vidau et al. 2011, Pettis et al. 2012).

Hunting pressure of honey bees (Asia). Asian people have been hunting honey bees for more than 40,000 years (Crane 1999), and bee hunting is still widely practiced throughout the continent. To take an A. florea or *A. andreniformis* colony, the hunter merely shakes the bees off, snips the branch holding the colony, and carries the comb home. It is assumed that provided there is plenty of food available, the colony recovers from the theft of its comb more often than not. Hunting *A. dorsata* and *A. laboriosa* is much more brutal and often involves burning the bees with a smoldering torch of the tightly-bound brush (Valli and Summers 1988, Lahjie and Seibert 1990, Nath et al. 1994, Crane 1999, Tsing 2003). Some harvested colonies may be able to regroup, especially if the hunt occurs in daylight. Often, however, the hunt is conducted in the darkness. The hunter bangs his torch on the branch supporting the colony to create a shower of sparks. The bemused bees follow the sparks to the forest floor (Tsing 2003) where they crawl, often with burned wings. Many queens are lost during these harvests, and their colonies perish along with them. Night hunting is preferred by many hunters because it reduces the number of stings received. This method of hunting kills many, if not most, colonies.

The level of hunting pressure is most likely increasing in many areas. Even the poorest communities (who are more likely to engage in hunting than landowners) have increasing access to motorized transport so that they can access nests over a broad area. Conversion from a barter/subsistence economy to a cash-based economy increases the incentive to produce a high value, an easily-transported product like honey (Nath et al. 1994, Tsing 2003, Nath and Sharma 2007). Increasing population in the cities and rural towns may increase the demand for wild honey, which is perceived as being more natural, pesticide-free, more healing and delicious than honey produced from domestic colonies. Finally, decreasing areas of forested land increase the hunting pressure on the remaining forested pockets (Nath et al. 1994). Consequently, hunting pressure might lead to the extinction of local species across extensive areas of Asia, although a complete extinction of any honey bee species is unlikely to occur.

Beekeeping practice, breeding, and hybridization

Europe. Fifteen (15) subspecies of *A. mellifera* are recognized today in Europe (Ruttner 1988, Garnery et al. 1992, De la Rúa et al. 2004, De la Rúa et al. 2009, Meixner et al. 2013), and it seems that species richness for both *Apis* and non-*Apis* bees is increasing from the North to the South of Europe. According to Ruttner (1988), the present distribution of European honey bee subspecies has been influenced by the Last Glacial Maximum (LGM), when the mountain chains of the Pyrenees, the Alps, and the Balkans acted as geographic barriers in maintaining the European populations in isolation (Fig. 6.6).

As can be seen in Fig. 6.6, the subspecies with the largest natural area of distribution is *A. m. mellifera* and the ones with the smallest are *A. m. siciliana*,

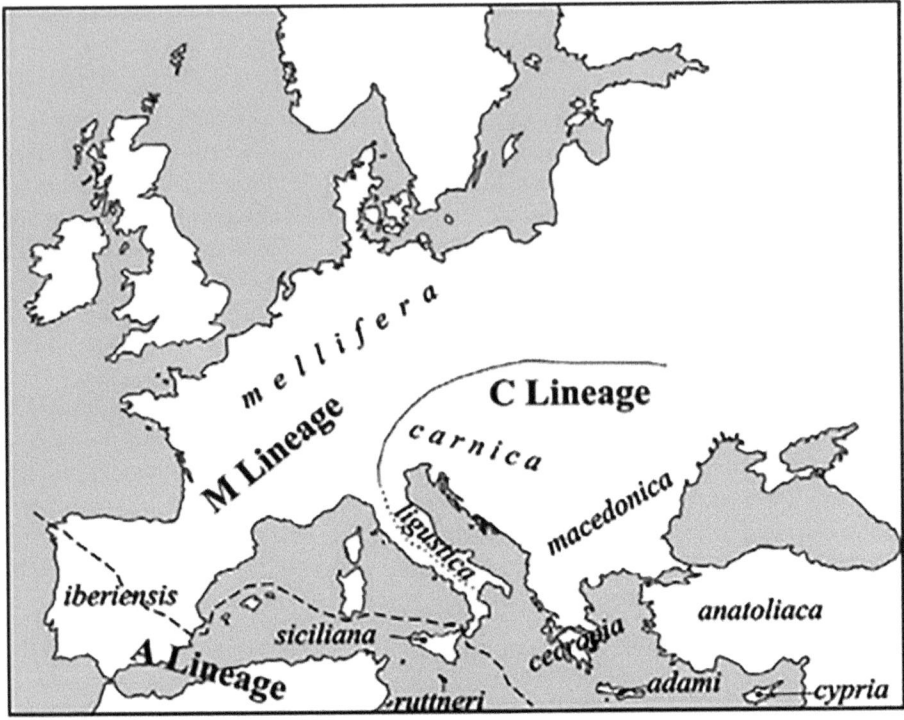

Figure 6.6: Approximate distribution of the *Apis mellifera* evolutionary lineages and subspecies in Europe (Source: De la Rúa et al. 2009).

A. m. ruttneri, and *A. m. adami*. Therefore, it is not surprising that the species with the smallest distribution suffers the most effects of all threats discussed in this chapter (Fontana et al. 2018).

The Italian bee (*A. m. ligustica*), the Carniolan bee (*A. m. carnica*), and the Caucasian bee (*A. m. caucasica*) are regarded as the "favorite" subspecies kept by many beekeepers around the world. The breeding activities and intense dissemination of these particularly docile and productive "superior" honey bees throughout the European continent have resulted in introgressive hybridization of the native subspecies and ecotypes (Jensen et al. 2005, Ivanova et al. 2007, Soland-Reckeweg et al. 2009, Martimianakis et al. 2011). In other areas, a complete replacement of the local subspecies has occurred. For example, the *A. m. mellifera* in central European countries, such as Germany. As a result of beekeeping activities, the distribution map described originally by Ruttner (1988), as shown in Fig. 6.6, has been changed. In other words, the "introduction" of the "superior" subspecies outside their natural range, and in regions where other subspecies of *A. mellifera* are endemic (Europe, Africa, and western Asia) or where *Apis* species are not endemic (America, Australia), could be regarded as an invasion, and all "invasion" types have resulted in loss of biodiversity (Moritz et al. 2005).

It has also been claimed that the consequence of the introduction of particular *A. mellifera* subspecies in a certain region is the extinction of its native honey bees

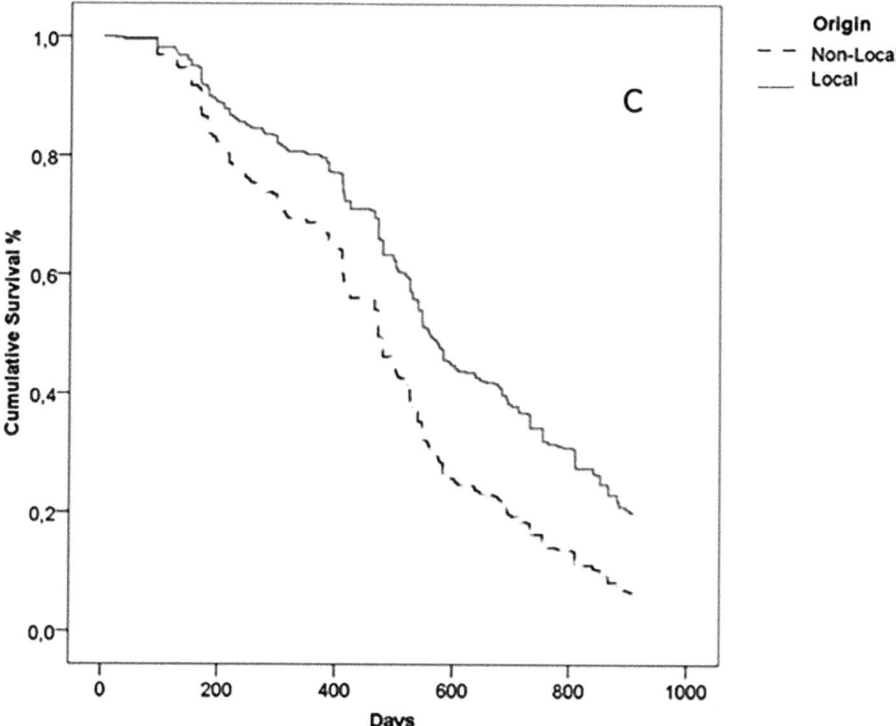

Figure 6.7: Colony survival in relation to the origin of the queens (Source: Büchler et al. 2014).

(Ruttner 1969, Ruttner 1988). However, recent studies fail to document a complete extinction, and autochthonous subspecies or genotypes can still be identified (De la Rúa et al. 2004, Martimianakis et al. 2011). Certainly, a racial admixture has occurred and valuable characteristics might have been lost.

Commercial breeding activities often overlook the importance of local adaptation. The long coexistence and adaptation of honey bees with plants and the environment, in general, is recognized by distinct genotype-environment interactions (GEI), and is expressed through the degree to which different genotypes are affected by the environmental conditions (Falconer and Mackay 1996). It has well been documented through a recent Genotype-Environment Interaction experimentation, that local populations/subspecies/genotypes are better adapted, longer survived, and more productive (Fig. 6.7) (Büchler et al. 2014, Hatjina et al. 2014, Meixner et al. 2014, Uzunov et al. 2014b, Meixner et al. 2015).

Asia. East Asia is home to at least 9 indigenous species of honey bees. Furthermore, the Asian honey bee *A. cerana* has been raised in indigenous cultures across the Asian continent for around two thousand years (Crane 1995). At the beginning of the 20th century, however, beekeepers across Asia began to import the European honey bee, *A. mellifera*, a species that lives in significantly larger colonies (30,000–50,000 vs. 2,000–20,000 in *A. cerana*). For example, some of the first European honey bees

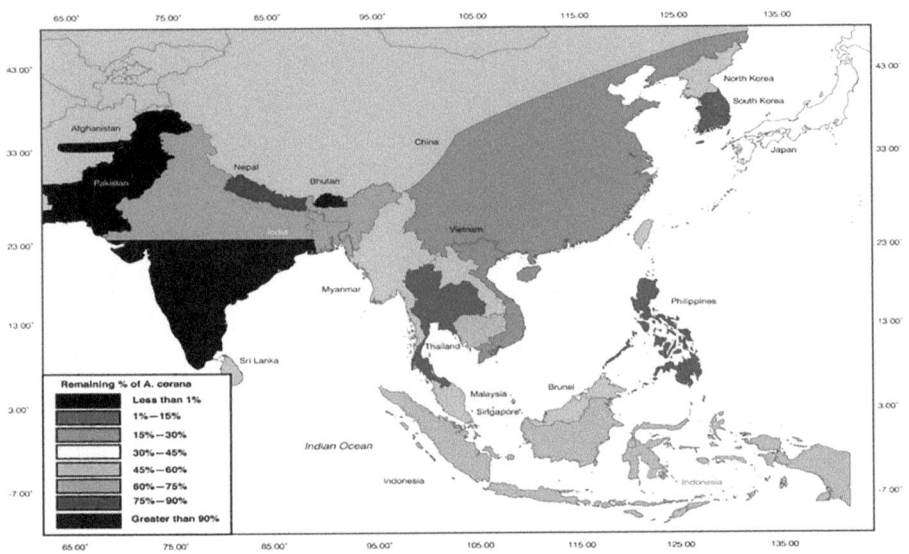

Figure 6.8: Map of the estimated remaining percentage of *A. cerana* compared with *A. mellifera* based on 2014–15 of 31 apiculturists in 16 Asian countries. Due to lack of data about wild honey bee species, only the commercially kept species (*A. cerana* + *A. mellifera*) could be taken into account. The sum of both species represents 100%, while the map details the proportion of *A. cerana* still extant (Source: Jones and Bienefeld 2016).

Color version at the end of the book

arrived in India in 1880 and were firmly established by 1960 (Abrol 2013, Crane 1995).

A. cerana was established in Vietnam in the 1960s (Tan and Binh 1993), and in Pakistan by the late 1970s (Sivaram 2012), but only reached Nepal and Bangladesh in 1990 and 1995, respectively (Sivaram 2012). *A. mellifera* is considered to be more productive than its Asian counterpart, and therefore more suitable for commercial beekeeping (Abrol 2013, Atwal 2000). Whether or not *A. mellifera* is profitable, its introduction into the native range of *A. cerana* is often problematic, for a myriad of reasons, i.e., competition for floral resources, interference with *A. cerana* mating by *A. mellifera* drones, who will pursue *A. cerana* queens, the introduction and exchange of pests and diseases. All this has contributed to a drastic decline in *A. cerana* health and numbers (Fig. 6.8).

The decline of *A. cerana* was observed in Afghanistan, Bhutan, China, India, Japan, South Korea, Myanmar, Pakistan (Verma 1998, Abrol 2013), the Philippines (Mojica 2011, Wendorf 2002), Taiwan, and Vietnam (Koetz 2013a) on an average of about 55 percent (Jones and Bienefeld 2016). It is now a reality across all of Asia. The essential drivers of this decline are lower productivity per hive, the impact of habitat destruction, alien species invasion, monoculture, chemical use, and the management challenges of absconding behavior (Akratanakul et al. 1990, Abrol 2013, Koetz 2013b).

China has six million bee colonies, About 200,000 beekeepers in this region raise Western honey bees (*A. mellifera*) and Eastern honey bees (*A. cerana*). In recent years, Chinese beekeepers have faced several complex symptoms of colony losses in both *Apis* species. Certain losses are known to be caused by *Varroa* mites (*A. mellifera*), Sacbrood virus (*A. cerana*), and *Tropilaelaps* mites (both species). However, other factors and mechanisms are being investigated, although no data has been published till date (UNEP 2010).

The red dwarf honey bee, A. florea, in the Asian subcontinent is actually expanding its range into the Middle East (Mossagegh 1993), and the Eastern honey bee, *A. cerana*, into New Guinea (Anderson 1994). In Hong Kong, one of the most urbanized and altered landscapes on the planet, *A. cerana*, remains common and is an important pollinator of remnant vegetation (Corlett 2001).

Nonetheless, there are obvious signs of threatening processes at work in some species in some areas, and it is suspected that these processes either have or soon will drive local extinctions. Perhaps this has already occurred in the dwarf bees on the island of Hong Kong, where they are apparently absent (Corlett 2001). The red honey bee, *A. koschevnikovi*, is now extremely rare on peninsular Malaysia and the south of Thailand (Otis 1996).

Take home message. The loss of taxonomic biodiversity in honey bees could somehow be controlled by breeding of local bee strains rather than imported ones. At the same time, conservation efforts also need to be reinforced and increased, ensuring that endangered populations will be protected from uncontrolled introgression of imported strains, and genetic richness will be available for the future generations.

CHAPTER 7

The Loss of Taxonomic Biodiversity of Honey Bees *Apis mellifera* and Main Breeds in Russia

Berezin, A.S.,* Borodachev, A.V.,* Borodachev, V.A.,
Mitrofanov, D.V.* and Savushkina, L.N.

The breeds of honey bees and their selection in Russia

Honey bee (*Apis mellifera* L.) belongs to agricultural animals, as it is of considerable interest and is directly related to the products of agroindustrial complex. Breeding of honey bees provides an increase of productivity of entomophilous crops, receiving of dietary food and medical preparations for apitherapy, getting raw materials for a national economy, preservation of biodiversity, and the solution of problems of ecological monitoring of environment and food security of the country. Groups of bees were formed in the process of long evolution under the influence of different natural climatic conditions. Despite a certain human effect in the breeding of bees, they are still at the stage of semi-domestication and are able to exist independently in their natural habitat. The Russian Federation has rich honey resources (4.9 million tons) and a valuable gene pool of honey bee species and populations (*Apis mellifera* L.) (Borodachev et al. 2016). The extensive territory of Russia, due to the diversity of natural climatic conditions, is recommended for breeding Middle Russian, Bashkirian, Carpathian, Gray Mountain Caucasian, Far-Eastern bees, characterized

Federal state budgetary scientific institution "Federal beekeeping research centre", Pochtovaya street, 22, Rybnoe, 391110, Russia.
* Corresponding authors: mellifera@yandex.ru; rybnoe-bee@mail.ru; dima-mitrofanoff2012@yandex.ru

by a specific set of biological and economic characteristics. Breed types of Middle Russian: "Priokskiy", "Orlovskiy", "Tatarskiy", "Burzyanskaya Bortevaya", Carpathian: "Maikopskiy", "Moskovskiy", and "Krasnopolyanskiy" of Gray Mountain Caucasian were included in the state register of breeding achievements of the Russian Federation. Successful breeding of bees largely depends on the effectiveness of the protection and rational use of the gene pool of existing species and types of the honey bee (Borodachev and Savushkina 2012).

Middle Russian breed (*Apis mellifera mellifera* L.). The appearance of Middle Russian bees or dark forest bees due to evolution was at the Northern border of the range of *Apis mellifera* L. It is a native breed for Russia. Among the populations of this breed are widely known breed types, such as Bashkirskaya, Burzyanskaya, Vologodskaya, Gorno-Altaiskaya, Krasnoyarskaya, Permskaya, Tatarskaya, Uralskaya, etc. These bees are dark in color, have a body weight of 110 mg, the width of the 3rd tergitum is 4.8–5.2 mm, but they have the shortest proboscis (5.9–6.35 mm), cubital index 60–65%, tarsal index 50–55%, negative discoid displacement, the straight shape of the back border of the wax plate of the fifth sternite. The weight of the virgin queen is not less than 195 mg, fertile queen is 220 mg, drone is 260 mg.

One of the characteristic features of these bees is aggression. Therefore, when beekeepers open the nest, especially in the absence of honey flow and lack of responsible care with bees, they will fly and attack surrounding persons massively. During the inspection of the honey frames, bees long to go away opposite to the light side, run down and hang in clusters on the bottom bracket. They do not build wax walls between the combs, which facilitates sorting of the nest. After that, the flight activity of a bee colony can be reduced to 90 percent. Honeycombs are sealed with a light capping. Frames with such capping are the most suitable in the production of the honey section. The Middle Russian bees are worse than other breeds guided in space, so they have more frequent exchanges of specimens from one bee colony to another. They do not protect their nests from other bees and are prone to theft to a lesser degree.

During the honey yield, the bees use nectar and pollen to increase the growth of brood and compose it, first, in the honey chamber, and then in the brood chamber. Composing honey in the chamber does not limit the queen bee in the laying of eggs in the lower chamber and ensures the honey harvesting by whole bodies and supers. Mobilization activity of bees starts only at 20% concentration of sugar in nectar. In these bees there are no "silent" change and cohabitation of queen bees. Bees accept introduced queens significantly better than others in the absence of the queen. Middle Russian bees propolize nest moderately. They are able to harvest a large amount of pollen in the brood part of the nest, which can reduce the area for laying eggs by the queen bee. They keep their brood nests clean too. Considering biological features, we need bee colonies of Middle Russian bee in the hives with the vertical placement of the bodies and chambers.

The bees are characterized by high winter hardiness. They are well adapted and remain viable for a long winter for up to 6 months without any flights. Usually, queen bees of this breed after the honey flow are preparing for winter time. Bee colonies go into the winter with up to 2.5 kg of offsprings, resulting in less depletion the individual specimens. During the wintering period, bee colonies maintain a high level of carbon dioxide (up to 4 percent). They react less to adverse conditions, while in a deeper rest, at the same time. In the spring, queen bees start to lay eggs almost a month later. These bees are characterized by increased resistance to *Nosema* disease, European foulbrood, and honeydew toxicosis, surpassing southern breeds (Krivtsov and Grankin 2004). Spring development begins relatively late, but passes more intensively and ends in mid-June. During the period of intensive growth of the bee, weak colony and medium honey harvesting enhances the cultivation of bee brood.

The queen bees are the most prolific; daily egg laying consists of 1,500–2,000 pcs. Often the weight of eggs laid during the day is higher than the weight of the body of the queen. This egg-laying capacity of queen bees in spring and summer allows the growth of strong bee colonies with a body weight of 5–7 kg (Vereshchaka and Grankin 2011). These bees are characterized by increased swarming. In some years, up to 90% of bee colonies have swarming impulse in the apiary. Being in the swarm impulse, bees reduce the cells building, brood growing, honey flow, which negatively affects their productivity. The average number of laid eggs by swarm queen is 15–30 pcs. Destruction of cells or expansion of the nest does not stop, but only delays the period of swarm impulse. Usually, the colony shifts from the swarm impulse to a working state with the start of the honey flow harvesting at least 3.0 kg honey per day.

Middle Russian bees most effectively use strong honey yield in late summer. In conditions of weakness in unstable honey harvesting, they are inferior to the southern bees. Honey productivity in different regions of Russia ranges from 20 to 50 kg per bee colony. The wax productivity is higher than the other breeds. They have better-developed wax glands and therefore they rebuild a large number of cells with large bee cells (5.56 mm). The bees are irreplaceable in regions with severe climatic conditions and are recommended for cultivation in 52 regions of the Volga, Ural, Siberian, Northwest, Central Federal districts, and Middle Russian bees comprise 60% of a total number of bee colonies in the country. Due to stable selection work in recent years based on the selected source of material, Bashkirian breed and breed types of Middle Russian: "Priokskiy", "Orlovskiy", "Tatarskiy", and "Burzyanskaya Bortevaya" were obtained, characterized by increased productivity and other valuable traits.

Bashkirian breed (*Apis mellifera*). The Bashkirian breed was bred in the Bashkir research center for beekeeping and apitherapy based on the Bashkir population of *Apis mellifera mellifera* L. (patent No. 3206 from 12.10.2006). Bees are dark in color, have a body weight of 113 mg, proboscis length of 6.15 mm, width of the third tergite of 5.02 mm, cubital index of 63%, tarsal index of 55.1%, discoidal shifts is the negative, and straight shape of the back border of the wax plate of the fifth sternite (Ishemgulov 2006). The weight of virgin queen is not less than 195 mg, fertile queen is 220 mg, drone is 240 mg. The bees are less aggressive and unlike the starting population, run away from the cells during the inspection and have light capping. They do not have "quiet" shifts and cohabitation of queen bees. These bees have the best winter hardiness and they are more resistant to *Nosema* disease, European foulbrood, and honeydew toxicosis bee. Their spring development starts late, but intensively. Daily egg production in the period of buildup bees to the honey harvest is more than 2,000 pcs. The swarming is 20–25% lower than in the starting population. The bees are able to harvest a large amount of pollen and beebread, which can be selected for sale. The best plans for honey gathering are lime tree and buckwheat. Honey productivity is 40–70 kg per bee colony. Wax productivity is high (Abdulov et al. 2008). They are recommended for breeding in the regions of the Volga, Ural, and Siberian Federal districts. The number of bee colonies is more than 200,000. The breeding base is two farms: GBU Bashkir research center for beekeeping and apitherapy, and LLC "Nectar" located in the Republic of Bashkortostan.

Breed type "Priokskiy" of Middle Russian breed. Breed type "Priokskiy" of Middle Russian breed (License 5818 dated 21.10.1992) was bred at the Federal beekeeping research center on the base of cross-breeding of Middle Russian and Gray Mountain Caucasian breeds for keeping in the Central Federal district of Russia (Borodachev et al. 2017). Bees are gray, have a body weight of 105 mg, 6.6–6.9 mm proboscis length, 4.6–5.0 mm is the width of the third tergite, 55–60% cubital index, the discoidal shift is the negative and straight shape of the back border of the wax plate of the fifth sternite. The body weight of the virgin queen is at least 190 mg, fertile queen is 215 mg, drone is 235 mg. These bees are more peaceful than other breeds and continue to work on the honey frame even after being removed from the nest. Honey capping is predominantly light. The bees are guided in space well. "Quiet" change and cohabitation of queens are rare. They protect their nests from other bees well, but they are prone to theft (Savushkina and Borodachev 2014). Their winter hardiness is equal to Middle Russian bees. Resistance to diseases is higher than *Apis mellifera caucasica* Gorb. Daily egg production of queen bees before the main honey yield is from 1,600 to 2,000 pcs. The swarm intensity is twice lower than the average. In the spring, they grow more on 15%, swarming 2 times less *Apis mellifera mellifera* L. They productively use herbs for both weak and strong honey gathering and lime tree or buckwheat for strong honey gathering. These bees pollinate crops, including legumes effectively. Honey productivity per bee-family is 30–50 kg. Wax productivity is high (Borodachev et al. 2017). The bees propolize their nests moderately. There are 20,000 bee colonies of "Priokskiy" breed type. There are two breeding farms: enterprise "Bortniki" located in Rybnovskiy district of the Ryazan region and LLC "RegionAgro" of Aleksinskiy district located in the Tula region.

Breed type "Orlovskiy" of Middle Russian breed. Breed type "Orlovskiy" of Middle Russian breed (patent No. 4110 dated 23.06.2008) was developed at the Orel experimental station of beekeeping on the basis of crossing 11 populations of Muddle Russian breed (Grankin 2008). Bees are dark-gray, have a body weight of 104 mg, 6.1–6.3 mm proboscis length, 4.9–5.1 mm is the width of the third tergite, 60% cubital index, 55.8% tarsal index, negative discoidal shift, and straight shape of the back border of the wax plate of the fifth sternite. The body weight of the virgin queen is not less than 200 mg, fertile queen is 220 mg, drone is 250 mg. Unlike the Middle Russian, the bees are less aggressive. During the inspection of the honey frames, they run down and hang in clusters on the bottom bracket. The capping is light. They protect their nests badly from other bees and are not prone to theft. They do not have a "quiet" shift and cohabitation of queens. These bees are characterized by high winter hardiness and raised stability to diseases. Their spring development starts late, but intensively. Daily egg production in the period of buildup bees to the honey harvest is 1,800–2,000 pcs, and the maximum is 3,000 pcs. During intense honey yield, bees have low swarm intensity. The best plants for honey gathering are lime tree, rosebay, willowherb, and buckwheat. Honey productivity of the bee colony is equal to 40–70 kg. Wax productivity of bees is high. The bees are able to harvest a large amount of pollen and bee bread. There are 20,000 bee colonies of "Orlovskiy" breed type. They are recommended for breeding in the regions of the Central, Volga, Ural, and Siberian Federal districts. The base breeding farm is National park "Orel Polesie", which is located in Orel region.

Breed type "Tatarskiy" of Middle Russian breed. Breed type "Tatarskiy" (patent No. 5476 dated 28.07.2010) was bred by researchers of the Scientific and production center of selection "Tatarskiy" and the Federal beekeeping research centre on the basis of the Tatar population of Middle Russian breed using the methods of pure breeding (Safiullin et al. 2011). Bees are dark gray, have a body weight of 110 mg, 6.1–6.4 mm proboscis length, 4.8–5.2 mm is width of the third tergite, 60.6% cubital index, 55.2%—tarsal index, negative discoidal shift and straight shape of the back border of the wax plate of the fifth sternite. The body weight of the virgin queen is not less than 200 mg, fertile queen is 220 mg, drone is 250 mg. The bees are less aggressive unlike the starting population, and run away from the cells during the inspection and have light capping. They are not prone to theft, but badly protect their nests from other colonies bees. They do not have a "quiet" change and cohabitation of queen bees. They quickly learn the upper honey chamber during the nest expansion (Safiullin and Savushkina 2011). The bees are characterized by high winter hardiness and increased resistance to diseases. Spring development is relatively late, but intense. Daily egg production of queen bees in the period of buildup bees to the honey harvest is more than 1,995 pcs. The swarm intensity is lower than the starting population. Bee colonies most successfully use late-summer honey gathering from lime tree, rosebay, willowherb, and buckwheat. Honey productivity is 40–60 kg per bee colony. Their pollination of entomophilous agricultural crops takes place as well. Wax productivity is high. The bees moderately propolize their nests. The bees are able to harvest a large amount of pollen and bee bread. There are 130,000 bee colonies of "Tatarskiy" breed type. These bees are recommended for breeding in the regions of the Volga, Ural, and Siberian Federal districts. There are four base breeding farms located in the Republic of Tatarstan: LLC "Rasset" Bugulminskiy district, LLC "Rapeseed", and LLC named after Timiryazev Baltasinskiy district, LLC "Nyrty" Sabinskiy district (Safiullin et al. 2013).

Breed type "Burzyanskaya Bortevaya" of Middle Russian breed. Breed type "Burzyanskaya Bortevaya" of Middle Russian breed (patent No. 5956 dated 14.06.2011) was bred by researchers of FGBI "State nature reserve "Shulgan-Tash" of the Republic of Bashkortostan (Kosarev et al. 2011). Bees are dark gray, and have a body weight of 108 mg, 6.2 mm proboscis length, 4.9 mm is the width of the third tergite, 59.2% cubital index, 57% tarsal index, a negative discoidal shift and the straight shape of the posterior border of the wax plate of the fifth sternite. The body weight of the virgin queen is not less than 200 mg, fertile queen is 220 mg, drone is 240 mg. The bees are more aggressive unlike the starting population, and run away from the cells during the inspection and have light capping (Kosarev et al. 2012). These bees have exceptional winter hardiness and they are more resistant to *Nosema* disease, European foulbrood, and toxicosis honeydew flow. These bees are characterized by late intensive spring development. Egg production in the period of buildup bees to the honey harvest is 1,935 pcs, and the maximum is 2,235 pcs per day. Bee colonies most successfully use late-summer honey gathering from lime tree, buckwheat. Honey productivity of the bee colony in wild hives is 20–30 kg, and 50–60 kg in hives. Wax productivity is high. There are 10,000 bee colonies of "Burzyanskaya Bortevaya" breed type. They are characterized by an increased collection of pollen and propolizing.

The bees are recommended for breeding in the regions of the Volga, Ural, and Siberian Federal districts. The breeding base farm is FGBI "State nature reserve "Shulgan-Tash" of the Republic of Bashkortostan (Nikolenko et al. 2010). Important work of preservation and reproduction Middle Russian bees of the permskiy population is carried out in Perm region. Results in 2016 researches indicate that permskie bees have proboscis length of 6.0 mm, C_v= 1.7% (lim 5.8–6.3 mm), width of the third tergite is 4.9 mm, C_v= 2.7% (lim 4.5–5.3 mm), cubital index is 60.8%, C_v= 14% (lim 40–93%), and was correspondent to standard parameters target of Middle Russian breed (Simankov et al. 2017). In the region, there are two functioning breeding farms: LLC "Parasol" in Perm and LLC "Nejnesypovskoe" в Uinskiy districts. These farms provide source material from apiaries of one's and other regions country. Breeding farms of reproduction bees Middle Russian breed organized in Altai region (LLC "Taigian honey", Zalesovskiy district), Udmurt Republic (LLC "Russian", Mojaginskiy district), Vologda (LLC "Bees", Vologodskiy district, Kirov (LLC "Legir" Kirovskiy district) and (LLC "Vyatskiy Beekeeper", Tujainskiy district) regions, engaged reproduction bee queens and colonies mainly for apiaries of one's regions.

Carpathian breed (*Apis mellifera carpathica*). Carpathian bees adapted to the poor honey plants and unstable weather with unfavorable climatic conditions because they formed in the Carpathian Mountains with a cool climate and high humidity. Ruttner (1992) and other researchers believe that the bees belong to the population of the "Krainskaya" breed. An important role in the study of biological and economic characteristics of the Carpathian bees was made by the researchers of the Department of beekeeping of the Moscow agricultural Academy named after K.A. Timiryazev, and the Ukrainian station of beekeeping under the leadership named after G.A. Avetisyan.

The bees are gray. They are smaller than Middle Russian bees but bigger than gray mountain Caucasian bees. The proboscis length is 6.4–6.7 mm, the width of the third tergite is 4.5–5.0 mm, the cubital index is 45–50%, and the tarsal index is 53–58%, positive discoid shift, and the curved shape of the back border of the wax plate of the fifth sternite. The body weight of the virgin queen is not less than 185 mg, fertile queen is 210 mg, drone is 235 mg. One of the positive qualities of the bees is their peacefulness when working with them. These bees are more peaceful than other breeds and continue to work on the honey frame removed from the nest. Honey capping is predominantly light. These bees are prone to "quiet" shift and cohabitation of queens. They accept introduced queens well.

They are well oriented in space, so they adapt to the management in the pavilions and platforms better than other breeds. Bees find new sources of honey; they quickly adopt new fields that are more productive. With honey flow starting, bees put nectar first in-store part of the nest and then other parts on the nest. Mobilization activity for the use of honey bees begins already at 8% sugar in nectar. They protect their nests from other bees well, but they are prone to theft. There are some disadvantages to the bees of this breed. For example, the bees clean the nests badly and weakly propolize nests. This leads to a significant defeat of wax moth. The hives with a vertical extension of the nest are preferable for the management of bee colonies (Gubin 1983).

The winter hardiness of the Carpathian bees is higher than gray Mountain Caucasian, but lower than Middle Russian bees. These bees in the autumn significantly reduce the cultivation of brood, releasing of offsprings, and have the opportunity to prepare for the winter properly. In the autumn-winter period, they economize feedstocks. Their resistance to *Nosema* disease, European foulbrood, and other diseases is medium. The spring development of the Carpathian bees starts relatively early and intensively, allowing them to become more efficient in using the early honey flow and pollinate. Egg production of queen bees in the period of intensive growth is 1,200–1,800 pcs, and the maximum is 2,000 pcs per day. Due to the high egg-laying capacity of the bee, colonies are building up to the main honey-gathering force—up to 5 kg. These bees are weak-swarming species.

However, the absence of honey gathering after the end of spring development in the swarm state can move up to 30% of bee colonies. During this period, bees hold 8–15 queen cells. However, from the swarm to the working state they switch relatively easier than Middle Russian bees. Carpathian bees are productively used and have a weak honey yield in the first half of summer, as well as an average honey flow in the middle of the season. They are more effective than the Middle Russian bees pollinate legumes. Honey productivity of the bee colony is 30–50 kg. The beeswax is higher than the starting population. There are 510,000 bee colonies of Carpathian breed. These bees are recommended for breeding in the 28 regions of the Northwestern, Central, Volga, North Caucasus, and southern Federal districts. There are five base breeding farms: PBF "Maikopskoe" located in the Republic of Adygeya, LLC "Pchelopitomnik "Kislovodskiy" of Stavropol region, LLC "Pcheloprom" of Karachai-Cherkess Republic, LLC "Temp", and integrated agricultural production centre "Zolotaya rozga" of North Ossetia-Alania. Based on Carpathian breed, two breed types are developed: "Maikopskiy" and "Moskovskiy".

Breed type "Maikopskiy" of the Carpathian breed. Breed type "Maikopskiy" (patent No. 3885 dated 13.05.2008) was bred by researchers of PBF "Maikopskoe" located in the Republic of Adygeya on the basis of purebred material imported from Zakarpattia (Malkova and Vasilenko 2008). Bees are gray; have a body weight of 103 mg, 6.5–6.9 mm proboscis length, 4.6–7.0 mm is the width of the third tergite, 47.9% cubital index, 52% tarsal index, positive discoidal shift and the curved shape of the back border of the wax plate of the fifth sternite. The body weight of the virgin queen is not less than 190 mg, fertile queen is 215 mg, drone is 240 mg. These bees are peaceful and continue to work on the honey frame removed from the nest. Honey capping is predominantly light. The bees are guided in space well and like to find new sources of honey, and quickly adopt new fields that are more productive. These bees are prone to "quiet" change and cohabitation of queens. They protect their nests from other bees well. They have good winter hardiness and resistance to diseases. Daily egg production during intensive development is 1,900–2,100 pcs. No more than 5% of bee colonies come in the swarm on the apiary. They productively use various types of honey yield, effectively pollinating legumes. Honey productivity of the bee colony is 40–70 kg. The beeswax is good. There are 44,000 bee colonies of breed type "Maikopskiy" of the Carpathian breed. These bees are recommended for breeding in the regions of the North-Western, Central, Volga, North Caucasus, and southern Federal districts. There is a base breeding farm in PBF "Maikopskoe" located in the Republic of Adygeya.

Breed type "Moskovskiy" of the Carpathian breed. Breed type "Moskovskiy" (patent No. 9086 dated 17.05.2017) was bred by scientists of LLC Pchelopitomnik "Kislovodskiy" of Stavropol region and RSAU-MTAA named after K.A. Timiryazev on the basis of purebred material imported from Zakarpattia (Zolina and Mannapov 2017). Bees are gray, have a body weight of 108 mg, 6.4–7.1 mm proboscis length, 4.4–4.8 mm is the width of the third tergite, 33–45% cubital index, 52–55% tarsal index, a positive discoidal shift and the curved shape of the back border of the wax plate of the fifth sternite. The body weight of the virgin queen is not less than 190 mg, fertile queen is 215 mg, drone is 250 mg. These bees are peaceful and continue to work on the honey frame after being removed from the nest. Honey capping is predominantly light. The bees are guided in space well and mobilized to bring nectar containing 8% sugar. These bees are prone to "quiet" change and cohabitation of queens. They protect their nests from other bees well and they are not prone to theft. They have good winter hardiness; the waste for the winter-spring period is not more than 5 percent. They are more resistant to disease than the starting population. They are characterized by early spring development. Daily egg production of queen bees during intensive development is 2,300 pcs. No more than 5% of bee colonies come in the swarm on the apiary. They productively use various types of honey yield. They pollinate legumes effectively. Honey productivity on the apiary of the bee colony is 45–65 kg. The beeswax is good. There are 7,000 bee colonies of breed type "Moskovskiy" of the Carpathian breed. These bees are recommended for breeding in the regions of the Northwestern, Central, Volga, North Caucasus, and southern Federal districts. There are base breeding farm in LLC Pchelopitomnik "Kislovodskiy" of Stavropol region.

Breed Gray Mountain Caucasian (*Apis mellifera caucasica* Gorb.). Gray Mountain Caucasian is the oldest representative of the species *Apis mellifera* L. For the first time in 1916, the bees were described by K.A. Gorbachev. The area of natural settlement of these bees is mountainous regions of the Caucasus and Transcaucasia. Among populations of this breed are the most well known Abkhazian, Mingrelian,

Gurian, Kartli, Imereti, Kakheti, and Kabatapendias breed types, which differ from each other in biological and economical characters (Lekishvili and Khidesheli 1967).

Bees are gray, have a body weight of 108 mg, 6.7–7.2 mm proboscis length, 4.4–4.9 mm is the width of the third tergite, 50–55% cubital index, 55–60% tarsal index, a negative discoidal shift and the straight shape of the back border of the wax plate of the fifth sternite. The body weight of the virgin queen is not less than 180 mg, fertile queen is 205 mg, drone is 220 mg. These bees are peaceful and continue to work on the honey frame when removed from the nest. At this moment, beekeepers can even see the queen bee, which lays eggs in the cells. After inspecting the nest, flight activity of bees is reduced by only 16 percent. Honey capping is dark. The bees are guided in space well. Bees of Gray Mountain Caucasian like to find new sources of honey, and quickly adopt new fields that are more productive. They protect their nests from other bees well and they are prone to theft. These bees are prone to "quiet" shift and cohabitation of queens. They accept introduced queens badly. The bees are characterized by the ability to fly for nectar in unfavorable weather conditions (fog, light rain). These bees are able to live in small colonies, therefore, during the reproduction of queen bees, it is possible to form a nucleus with the weight up to 0.1 kg. With honey flow starting, bees put nectar first in-store part of the nest, and then in the honey chamber. They build wax bridges between the cells, which greatly complicates the disassembly of the nest. The bees clean their nests well. Mobilization activity begins at 10% sugar in nectar. Gray Mountain Caucasian bees are most adapted to the content in the hives of the horizontal system and poorly master the nest in the vertical direction.

In the natural area, where there is a possibility of cleaning hive entrance blocks in winter, these bees winter normally. In the Central, Eastern, and Northern regions of the country in winter hardiness, they are inferior to the Middle Russian bees, especially in the absence of normal environmental conditions (poor quality food, moisture, temperature changes). The bees fare especially badly in winter in the presence of honeydew in feedstocks. In winter, Gray Mountain Caucasian bees keep the relatively low concentration of carbon dioxide (2%) and become very excited when there is any deviation from normal winter conditions (high temperature, etc.). They are much stronger than the other breeds affected by *Nosema* disease, European foulbrood. These bees have relatively early spring growth. In the result during weak and average honey gathering growing brood of Gray Mountain Caucasian bees and the collection of feed in the nests are limited. Egg production of queen bees is relatively low and in the period of intensive development of bee colonies to the honey, harvest does not exceed 1,500 pcs per day. Gray Mountain Caucasian bees are low-swarming bees. Usually, in swarm state, no more than 5% of bee colonies come in the apiary. At the same time, they hold from five to 20 queen cells. Being in the swarm state, bees do not significantly reduce cultivation of brood, construction of new cells, bringing of nectar and pollen. These bees quite easily come from the swarm to the working state in the elimination of the laid queen cells, the expansion of the nest, and the appearance in nature of honey of more than 1 kg per day. The Gray Mountain Caucasian bee is able to efficiently use any weak and intermittent honey flow from berries, fruits, herbs, etc. These bees are more productive than the Middle

Russian bees pollinate legumes, including clover, alfalfa. Honey productivity of the bee colony is 25–40 kg. The beeswax is low (Krivtsov et al. 2009a). There are 360,000 bee colonies of Gray Mountain Caucasian breed. These bees are recommended for breeding in the regions of the North Caucasus and Southern Federal Districts. There are three base breeding farms: Krasnopolyanskaya Experimental station for Beekeeping of Krasnodar region, GUP "Nectar" of the Republic of Ingushetia, LLC "Azamat" located in North Ossetia-Alania (Kostoev 2009).

Breed type "Krasnopolyanskiy" of Gray Mountain Caucasian Breed. Scientists of Krasnopolyanskoy Experimental station for Beekeeping located in the Krasnodar region line breed Gray Mountain Caucasian bees. In 1963, the source material from different regions of Georgia (places of the natural habitat of this breed) was brought to this farm. Based on imported Mingrelian, Kartali, and Abkhazian populations, several lines of Gray Mountain Caucasian bees were separated. Bee colonies of line No. 6 differed by earlier spring development. Line No. 20 was characterized by improved winter hardiness of bees. The bees of line No. 25 had increased egg-laying capacity compared with the initial population. Line No. 34 bees had the longest proboscis and increased honey productivity. The bee colonies of line No. 36 were characterized by improved productive qualities. Various variants of interline crosses are received and tested and the best of them are introduced in production. The next stage was the breed type "Krasnopolyanskiy" of Gray Mountain Caucasian (patent No. 4111 dated 23.06.2008) based on different populations of Gray Mountain Caucasian breed (Krivtsov et al. 2008). Bees are gray, have a body weight of 93 mg, 6.9–7.1 mm proboscis length, 4.6–4.9 mm width of the third tergite, 52.4% cubital index, 55.4% tarsal index, a negative discoidal shift and the curved shape of the back border of the wax plate of the fifth sternite. The body weight of the virgin queen is not less than 185 mg, fertile queen is 210 mg, drone is 240 mg. These bees are peaceful and continue to work on the honey frame removed from the nest. Honey capping is dark. Bees like to find new sources of honey, and quickly adopt new more productive fields. They protect their nests from other bees well and they are prone to theft. These bees are prone to "quiet" shift and cohabitation of queens (Lyubimov and Savushkina 2010). They have good winter hardiness, but bees are not resistant to diseases. Their spring development starts earlier than starting population. Daily egg production in the period of intensive growth of bee colonies to the honey harvest does not exceed 1,500 pcs. These bees are low-swarming. Usually, in swarm state, no more than 5% of bee colonies come in the apiary. These bees use weak honey yield often-intermittent honey collection. Honey productivity of the bee colony is 25–45 kg. The bees intensively propolize their nests, and therefore beekeepers receive some propolis. These bees effectively pollinate legumes, including clover, alfalfa (Krivtsov et al. 2009b). There are 80,000 bee colonies of breed type "Krasnopolyanskiy" of Gray Mountain Caucasian Breed. These bees are recommended for breeding in the regions of the Central, North Caucasus, and Southern Federal Districts. There is base breeding farm Krasnopolyanskaya Experimental station for beekeeping located in Krasnodar region.

Far-Eastern bees (*Apis mellifera* Far-Eastern). Far-Eastern bees originated from the haphazard crossing of Ukrainian steppe, Middle Russian, Gray Mountain

Caucasian, and Italian bees imported from Ukraine and Central regions of Russia into the Russian Far East (Kodes and Popova 2010). Due to long-term breeding of imported bees of different origins in the specific conditions of natural selection and geographical isolation, a population of Far Eastern bees was formed. Comparative tests of Gray Mountain Caucasian, Muddle Russian, Ukrainian, Italian, and Far Eastern bees showed that imported breeds were inferior to local ones in terms of productive qualities. Therefore, breeding in the Primorskiy territory of the Far Eastern bees was decided. With further development based on the selected source material for a set of features, using methods of purebreed breeding, Far-Eastern bees were bred (patent No. 9421 dated in 13.12.2016).

Bees are gray, have a body weight of 107 mg, 6.6–6.8 mm proboscis length, 5.0–5.2 mm is the width of the third tergite, 42.1–45.4% cubital index, 56–57.7% tarsal index, a positive discoidal shift and the curved shape of the back border of the wax plate of the fifth sternite (Sharov 2018). The body weight of the virgin queen is not less than 180 mg, fertile queen is 205 mg, drone is 240 mg. These bees are more peaceful then Middle Russian bees. During the inspection of the honey frames, bees are movable but do not go away. Honey capping is light. In comparison with Middle Russian bees, Far-Eastern bees like to find new sources of honey stronger, and quickly adopt new more productive fields, but inferior to Gray Mountain Caucasian Bees. They protect their nests from other bees well. They do not have a "quiet" shift and cohabitation of queens. They accept introduced queens badly. They keep clean nests and in artificial nests clog up faster than other bees clean and protect theirs well from wax moth. Bees prefer to put the brought nectar first in the honey chamber, and then in the brood part of the nest. Far-Eastern bees are characterized by high efficiency, intensively flying for feed at low temperature and high humidity. These bees have the same winter hardiness like Middle Russian bees. They have lower resistance than Middle Russian bees, but higher than the southern subspecies of bees. Spring development of bee colonies begins relatively early. The daily egg production of queen bees during the period of growth to the honey yield is low and is 1,100–1,600 pcs.

The far-eastern breed is moderately swarming bees. Usually, in swarm state there are fewer bee colonies than Middle Russian bees. Usually, bee colonies move from the swarm impulse in working state with removal of queen cells, expansion of the nest, and starting good honey yield. Far-Eastern bees are most effective in using the late-summer honey collection from various species of lime, bringing up to 30 kg of nectar per day. Honey productivity is 50–100 kg per bee colony (Sharov 2015). Wax productivity is high. They propolize the nest weakly. In recent years, interest in Far-Eastern bees as source material for breeding bees resistant to varroosis has increased significantly. The choice of these bees is because they exist with the *Varroa* mite, successfully surviving for 50 years. The results of tests of Far-Eastern bees imported to a number of US states during experimental infection with *Varroa* mite showed that the population growth rate of this parasite was 2–6 times lower compared with local breeds. This served as the basis for the organization of reproduction of queen bees of similar origin and their implementation to beekeepers of this country. In order to preserve the purity of the Far-Eastern bees in the territory of the Far Eastern

Federal district, the import of other breeds is prohibited. The number of bee colonies is 5,500. There is a base breeding farm—FSBI "Federal scientific center of the Far East Agrobiology named after Chaika" located in Primorskiy Krai.

Destruction the biodiversity of honey bees by pesticides and honey bee protection activity in Russia

The mass death of flight bees in Russia in 2015 is associated primarily with pesticide poisoning. The greatest number of poisonings is associated with the use of insecticides that are highly toxic to bees, in particular, fipronil, and dimethoate for the treatment of rapeseed. In addition, there are facts of treatment with prohibited pesticides (Klochko and Blinov 2016). Such highly toxic pesticides, such as DDT and HCCH, which were banned decades ago due to their high environmental sustainability, are still circulating in agrocenoses. It was found that from soil to flower there is an increase in the content of these pesticides and in bees and honey, they are found in decreasing amounts (Budnikova et al. 2018a,b). In Ukraine, there are reported cases of the presence of banned substances in some drugs, while in the composition, only harmless components are indicated. The number of unregistered drugs among miticides is continuously increasing (Pyaskovskiy et al. 2018). The conditions for conducting beekeeping at a model area in the Kaliningrad region from 1991 to 2014 deteriorated, including the use of pesticides (Gayeva 2015). The protection of bees from poisoning consists in the prior notification of beekeepers about upcoming treatments, limiting the flight of bees, establishing a border protection zone for bees, and processing within certain periods and under established weather conditions. The following classes of pesticide hazard for bees are established:

The first hazard class includes highly hazardous pesticides that kill more than 20% of bees. The border protection zone for bees is at least 4 km; the flight limit is 4–5 days. The second hazard class includes medium hazard pesticides, causing the death of 5–20% of bees. The border protection zone for bees is at least 3–4 km; the flight limit is 2–3 days. The third class of hazard includes low-hazard pesticides, causing the death of 1–5% of bees. The border protection zone for bees is at least 4 km; the flight limit is 1–2 days. The fourth class of hazard includes practically non-hazardous pesticides. Border protection zone for bees is 1–2 km; flight limit is 6–12 hours (Nazarova 2009).

However, these requirements for the protection of bees are practically difficult to fulfill in conditions of a high proportion of personal farms with cultivated areas and private beekeepers with a small number of bee colonies. Pesticides have to be used in accordance with the Instruction for the Prevention of Bees Poisoning with Pesticides. The system of relationships between the plant protection service and beekeepers is often not respected, which leads to the cases of treatment fields, gardens, and forests during an active flight of bees without prior notification of beekeepers. The farm administration is obliged to inform the population, beekeepers located at a distance of 7 km or less, and the veterinary service about the upcoming treatment, at least two days before processing (Klochko et al. 2018). In the case of poisoning of bees with highly toxic pesticides, they die both in the field and on the territory of the apiary and in the hive. A long course of intoxication in case of poisoning with sublethal doses is observed both in summer and in winter when bees eat the pesticide-contaminated feed (Solovyeva 2012). Eating fodder contaminated with pesticides leads to a weakening of the protective functions of the honey bee's body and the activation of weakly pathogenic or conditionally pathogenic microorganisms (Rudenko and Rudenko 2015). The hazard class of pesticides for bees cannot be established only by the results of laboratory experiments, because pesticides circulate in the environment through the accumulation and release by plants. Therefore, field experiments are needed to determine the hazard of pesticides to bees. The technique of such experiments using stationary isolators was developed at the Scientific Research Institute of Beekeeping (Solovyeva 2012). Insecticides and miticides from the classes of chlorine and organophosphorus substances among the old drugs are the most toxic for bees, but the danger of organophosphorus compounds is lower due to the rapid biodegradation in the field (Nazarova 2009). Neonicotinoids have much higher toxicity.

It is assumed that the phenomenon of the bee colony collapse disorder is associated with the use of a new class of pesticides, called neonicotinoids. It has been established that these substances can cause behavioral disorders in bees, such as disorientation, reduced ability to search for food, impaired learning, and memory (Komlatskiy 2018). Neonicotinoids are able to persist in plant tissues for months and even more than a year, and in the soil up to several years. Thus, there is a danger of poisoning bees not only when visiting plants treated in the current season, but also when visiting plants grown in areas treated in previous years, as well as when collecting water from surface sources contaminated by pesticides as a result of their desorption from the soil (Klochko and Blinov 2015). A hypothesis was put forward on the connection of the bee colony collapse disorder with the use of pesticides derived from biogenic amines, such as neonicotinoids and formamidines. The author suggests that the chronic effects of these pesticides cause a violation of acetylcholinergic and octopaminergic transmission in bees. These disorders lead to a weakening of olfactory learning and memory of bees, and thus they cannot return to the hive (Farooqui 2013). Fipronil at a dose of 0.1 ng/bee caused the death of all bees within a week. Upon contact with thiamethoxam at a dose of 0.1 ng/bee, the bees also deteriorated their olfactory memory. The only effect of acetamiprid at a dose of 0.1 μg/bee was an increase in water sensitivity (Aliouane et al. 2009). The synergistic effect of *Nosema* and imidacloprid on the suppression of glucose oxidase activity in bees which reduces their resistance to other infections was shown (Alaux et al. 2010a).

Insecticides cause more or less pronounced effects on the nerve functions that can lead to prevailing behavioral or physiological disorders. The mechanism of insecticidal action involves exposure with various affinities to several molecular targets, which can lead to different effects. The final effect of insecticides strongly depends on circadian rhythms, exposure, age, or stage of the honey bees development, and the season (Belzunces et al. 2012). Residual amounts of pesticides in Spain were found in 42% of the analyzed samples of the spring pollen and 31% of the autumn pollen. Residual amounts of fipronil were found in 3.7% of spring pollen samples, but were not found in autumn pollen samples (Bernal et al. 2010). Imidacloprid toxicity has been shown for bees at sublethal doses of the order of 1–20 μg/kg or less. The presence of residual amounts of imidacloprid in the soil, an increase in its content in plants during flowering, and admission to pollen are shown (Bonmatin et al. 2005). The remains of 19 pesticides were found in a three-year study of the fields of France, from 2002 to 2005, which covered 125 bee colonies, randomly selected by 5 in one apiary. Coumaphos (925 μg/kg) and fluvalinate (487 μg/kg) were detected at the highest concentrations. Fipronil and its metabolites were detected in 16 samples. Imidacloprid and six-chloronicotinic acid were detected in 69% of the samples. The concentration of imidacloprid was in 11 samples from 1.1 to 5.7 μg/kg. 6-chloronicotinic acid was detected in 28 samples in an amount of 0.6 to 9.3 μg/kg. Statistical analysis has not shown any differences between sampling sites, with the exception of fipronil (Chauzat et al. 2006, Chauzat et al. 2009). In particular, bees have twice as little glutathione-S-transferases, cytochrome-p450-dependent monooxygenases and carboxylesterases, which provide for xenobiotic metabolism. The high sensitivity of bees to insecticides may be related to the deficiency of these enzymes (Claudianos et al. 2006). It was shown that the guttation liquid of corn leaves, which were grown from seeds treated with neonicotinoids, contains more than 10 mg/l of pesticide, up to a maximum of 100 mg/l of thiamethoxam and clothianidin, and 200 mg/l of imidacloprid. These concentrations are close to the concentrations of working solutions of pesticides used in the form of an aerosol and sometimes even exceed it, which leads to the death of bees within a few minutes while collecting guttation fluid (Girolami et al. 2009).

In addition, bees can die under the influence of dust when sowing seeds, which were treated with neonicotinoids (Girolami et al. 2012). The non-lethal effect of thiamethoxam causes a high mortality rate of bees by disrupting their return to the hive, on a scale dangerous for the death of the whole colony (Henry et al. 2012a). In laboratory studies of commercially available and potentially valuable neonicotinoids, it was found that nitro derivatives were the most toxic to bees, with average lethal doses of 18 ng/bee in Imidacloprid, 22 ng/bee in clothianidin, 30 in thiamethoxam, and 75 in dinotefuran, and 138 in nitenpyram. Cyan-substituted neonicotinoids exhibit much lower toxicity, amounting to 7.1 μg/bee for acetamiprid and 14.6 for thiacloprid. Piperonyl butoxide increased the toxicity of acetamiprid to bees 6 times, and thiacloprid 154 times, triflumizole 244, and 1141 times respectively, and propiconazole 105 and 559 times, respectively. They affect the toxicity of imidacloprid minimally, increasing it by less than two times. The N-demethylated metabolite acetamiprid, 6-chloro-3-pyridylmethanol and six-chloronicotinic acid did not cause bee mortality at a dose of 50 μg/bee. These results indicate the importance of the *Cytochrome P450* system in the detoxification of acetamiprid and thiacloprid, and hence their low toxicity to bees. The bees did not die during forced contact with alfalfa treated with acetamiprid and its synergist triflumizole at the maximum recommended doses (Iwasa et al. 2004). The death of bees in Italy as a result of sowing corn seeds treated with neonicotinoids due to dust on the plants surrounding the crops was recorded (Marzaro et al. 2011). Growing brood at a low temperature of 33°C did not affect larval mortality when exposed to dimethoate, but the mortality of adults increased significantly. The LD_{50} for larvae was 28 times higher at 33°C than at 35°C, which can be explained by differences in metabolism and, consequently, by a slower intake of poison. Thus, growing brood at a temperature that is different from the optimum can be one of the factors affecting the safety of colonies (Medrzycki et al. 2010). Coumaphos was present in the highest concentration, about 1 mg/kg in propolis taken from all living colonies. Imidacloprid was found at a concentration of 377 μg/kg and 60 μg/kg in propolis. In the combs of dead colonies, that is higher than the levels causing disorientation in bees. Fipronil is found in bees at concentrations of 150 and 170 mg/kg. These concentrations are toxic in themselves.

Other pesticides can impair bees and reduce their productivity (Pareja et al. 2011). A study of the exposure effect of imidacloprid sublethal doses on three generations of brood found that this effect increases the pathogenicity of *Nosema*. Imidacloprid concentrations were significantly lower than those considered harmful to bees (Pettis et al. 2012). The toxicity of imidacloprid was studied in the contact and oral routes of intake for two subspecies of *Apis mellifera*, such as *Apis mellifera mellifera* and *Apis mellifera caucasica*. Oral toxicity showed no difference, while contact LD_{50} was 24 ng/bee for *Apis mellifera mellifera* and 14 ng/bee for *Apis mellifera caucasica* (Suchail et al. 2000). A wide range of behavioral disorders caused by sublethal doses of pesticides, especially insecticides, is described. These disorders range from loss of olfactory sensitivity to the death of flight bees because of a violation of the hive search (Thompson 2003). If *Apis mellifera carnica* bees receive high-quality pollen as feed in the first days of life, they will be more resistant to pesticides than those that receive poor-quality pollen or its substitute. This is a result of the slowing down of enzymatic inactivation of pesticides under protein deficiency conditions (Wahl and Ulm 1983). The negative impact of imidacloprid on the development of the *Bombus terrestris* bumble bee colony and the emergence of the young queen was shown (Whitehorn et al. 2012). The effect of coumaphos on the cultivation of queen bee larvae in a colony without queen bee was studied, and it was found that at a concentration of 1,000 mg/kg coumaphos in the material of the queen cup, all the larvae were removed ten days later. At a concentration of 100 mg/kg, more than half of the larvae were discarded. The bee queen from larvae exposed to 100 mg/kg coumaphos had a significantly lower weight (Pettis et al. 2004). Coumaphos and tau-fluvalinate are used as miticides for treating bee colonies due to their low toxicity to bees. They are lipophilic compounds and are capable of accumulating in waxes with a tendency to cumulate with repeated treatments. It has been established that the toxicity of tau-fluvalinate for bees previously treated with coumaphos increases significantly. The toxicity of coumaphos after treatment with tau-fluvalinate increases moderately. Synergism can be explained by the competition of these miticides for enzymes of the *Cytochrome P450* system (Johnson et al. 2009). Imidacloprid is a sublethal dose of 0.02 ppm (two μl of the solution was applied to the queen bee abdomen), which reduces the viability of the sperm in the spermatheca of the queen bee by 50 percent. Coumaphos at a dose of 100 ppm reduces sperm viability by about a third (Chaimanee et al. 2016). The effects of imidacloprid and coumaphos on the olfactory memory and training of bees were studied. Imidacloprid has a slight effect on the sense of smell of bees, whereas coumaphos caused a moderate deterioration. An unexpected effect is the absence of additive side effects with the combined use of both drugs, and there is even an improvement in learning and memory (Williamson et al. 2012). Amitraz and fluvalinate used for the tick-borne treatment of bee colonies, even at therapeutic doses, worsen the biological characteristics of the bee colony.

Their action is enhanced when the dosage is exceeded, and the acaricidal activity remains almost unchanged (Ilyasov and Shareeva 2014).

Preservation of bee breeds and populations on the territory of the Russian Federation is traditionally engaged in nature reserves, where they live in natural conditions. For this purpose, the reserve "Shulgan-Tash" of the Republic of Bashkortostan, "Visherskiy" located in the Perm region, National park "Orel Polesie" in the Orel region, the reserve in the Baltasinskiy, Mamadyshskiy, and Sabinskiy districts of the Republic of Tatarstan and others exist.

The policy directive about state reserve of regional importance for the protection of native breeds and populations of bees in Russia was developed. It includes general provisions, the purpose, the profile, and the order of education, the mode of organization, protection, and monitoring of compliance with the regime. The reserve is organized on the territory of a radius of not less than 25 km, the conditions of the honey collection, which are typical for the region. In the Central part of the territory occupied by the reserve, an apiary with at least 200 bee colonies is placed, which serves as a base for work. The main method of working with bees in the reserve should be purebreed breeding by the type of closed or panmictic population, which allows one to keep gene concentrations in a high equilibrium state. The reserve supplies raw material to other farms for further breeding work.

Along with nurseries and reserves, the preservation of the gene pool of a certain zoned breed is one of the main tasks of breeding farms for the cultivation of bees. These organizations have a sufficient number of breeding colonies of laboratory-reared honey bee subspecies, use of purebreed breeding, practice in selection, and reproduction and sale of certified breeding production farms in the area of their breeding. According to the state of breeding work, the number of sold breeding products, the productivity of bee colonies, and veterinary well-being in the country, more than 20 breeding farms engaged in breeding of zoned breeds of bees are certified.

It is important when choosing the initial material for the conservation and breeding improvement of bees to control their purebreed. For assignment of bees, traditionally morphometric characters of individuals and the behavior of bee colonies are used. However, these features vary significantly under the influence of environmental conditions. Along with the classical methods of morphometry in Federal beekeeping, a research centre in cooperation with the Institute of General Genetics and Institute of animal husbandry is researching the use of molecular genetic methods of identification of bee breeds. The molecular genetic characteristics of the allele pool of the Middle Russian, Gray Mountain Caucasian, and Carpathian bee breeds using mitochondrial DNA and nuclear DNA microsatellites are given (Borodachev et al. 2015).

For conducting breeding work with bees, normative documents are prepared: regulations on the state natural reserve on preservation of a gene pool of native breed, population of a honey bee, rules of reference of the farms which are engaged in cultivation of bees, to breeding, a technique of carrying out tests for distinctiveness, uniformity and stability of selection achievements of a honey bee, the national standard for a queen bee, and the interstate standard for a bee colony.

Morphology measurements for protection biodiversity of honey bees in Russia

Data on dimensions of exterior features of bees are necessary for studying their systematics and determining the pedigree affiliation in the process of selection work. *Apis mellifera* has natural distribution in Africa and Eurasia, and this species has been introduced to all other continents except Antarctica, and is intensively used in honey production and pollination of entomophilous crops. Honey bees have significant geographic diversity, which is the result of adaptation to local environmental conditions. This geographic diversity disappears under the influence of anthropogenic factors. The uncontrolled import of queen bees, as well as the uncontrolled trade in queen bees, package bees, and bee colonies within the country, threatens regional subspecies and ecotypes, contributing to hybridization. Of no less concern is the deliberate replacement of local subspecies in some regions by imported bees with more preferred features and greater commercial gain. Kozhevnikov (1900b) reports that as early as the 1950s, German beekeepers were interested in the Italian bee-"this devotion even had an epidemic character". In Russia, at the end of the 19th century, there was a passion for the Caucasian bee. According to Alpatov (1927), the transportation of southern races to the north led to large shifts in the natural distribution of races, but in Russia, this process had fewer consequences. At the end of the 20th century, the situation with the import of races has changed dramatically. The disadvantage of these processes due to the financial component is the tendency to reduce the genetic diversity of honey bee populations, which can affect their ability to adapt to new threats. However, large populations of dark forest bees remained in Norway, Sweden, Denmark (Ruottinen et al. 2014), and Ireland (Hassett et al. 2018). The preservation of local bees in these countries is carried out with state support by scientific institutions and beekeeping organizations. Non-hybridized populations of *A. m. mellifera* are also preserved in Russia (Krasnoyarsk region and Tomsk region) (Ostroverkhova et al. 2015, 2016, 2019a). Protected populations of *A. m. mellifera* exist in several European countries (Denmark, Netherlands, Colonsay Island (Scotland), France, Belgium, Norway, and Switzerland) (Pinto et al. 2014). There are national breeding programs for *A. m. carnica* and *A. m. mellifera* bees in Norway. Mating takes place at isolated points, around stationary apiaries there is an area of about 3,500 km^2, where beekeepers are allowed to keep only *A. m. carnica* or *A. m. mellifera* (Bouga et al. 2011). It is necessary to create some protected areas around apiaries engaged in breeding purebred material in Russia, that is, to isolate them from the ingress of genes from other subspecies of *Apis mellifera* and do it with the support of the state, taking into account the experience of European countries (for example, Norway).

Creating a base of reference samples to identify the species and populations that will be used for breeding is required. At the same time, the standard should reflect the natural variability of honey bees and should not be hybridized. Recognized reference samples are of paramount importance not only for determining new ecotypes, but also for determining the location of unknown samples in the existing structure. However, there is a lack of comparability between the used reference samples, which have only general information about the sampling site, without specifying the origin of the samples or the method for their checking (Meixner et al. 2013). Currently, there is an urgent need (both for Russia and in Europe) to develop and adopt a common standard for defining the notion of a "reference sample" and its design rules with organizing the storage of standards and data on them in free access (Meixner et al. 2013). The databank of measurements and reference samples of the Institute of Apiculture (Oberursel, Germany) (Ruttner 1988, Meixner et al. 2013) is not yet freely available, but it is possible to get it when implementing joint projects (Kandemir et al. 2011, Nawrocka et al. 2018). There is also diversity within the subspecies, dividing them into separate populations, but at the same time, these populations do not have an official name. The concept of a population can include any regional natural type of bee that is different enough to give it a name. According to some authors, the scientific description and recognition of the diversity of honey bees in Europe cannot be considered complete, since vast territories, mainly in the eastern part of the continent, have not yet been systematically studied. The same authors summarized the current state of knowledge about the subspecies of the honey bee and population variability in Europe (Meixner et al. 2013).

Identification of honey bee subspecies and races based on morphology. If the origin of the analyzed honey bee samples is known, then several attributes are sufficient to assign a sample to a particular race, which will significantly reduce the time spent on analysis. For example, features obtained by various mathematical operations with measured features are suitable for simplified separation of groups, which is useful for beekeepers. In most European countries, their own independent requirements are used to identify subspecies with a decreased number of features, and in some, laboratories undergo official certification to identify races (see Bouga et al. 2011). Currently, according to Meixner et al. (2013), there is no complete catalog of morphometric changes of honey bees in Europe. There is no single approved system for the identification of European bees, and there is a need to save labor and time costs. Automated geometric analysis of the wing may be the most effective method of identification in the future, but the reliability of separation has not yet been established using only GM of all European subspecies and known ecotypes. It may be necessary to include additional CM features in such an alleged pan-European system that has not yet been developed (Meixner et al. 2013).

The identification of a potential maternal colony as free of African genes is a special case of sample identification, when special software based on only a few morphometric features of honey bees of African and European origin has been used or developed to achieve a quick result (Rinderer et al. 1986, 1987, FABIS (rapid bee recognition system for Africanization) 2019). So Rinderer et al. (1986) used the method of computer measurement according to Daly et al. (1982) and SAS software (SAS Institute Inc., N. Carolina, USA), and in 1987 also SPSS software (SPSS Inc., Illinois, USA) for the separation of African and European bees. However, since the aim of the method is to distinguish between "African" and "European" bees, the clarity of separation is rather low and not suitable for further separation of subspecies (Meixner et al. 2013). At Carl Hayden research centre of beekeeping at the USDA-ARS laboratory, FABIS is used to preliminarily determine "Africanization" and then complete morphometric analysis if it is detected (http://www.ars.usda.gov/Research/docs.htm?docid = 11053).

The monograph of Ruttner (1988) contains the most comprehensive and well-presented database on morphometric characters for most of the honey bee subspecies, based on the use of numerical taxonomy using the features of "the classical morphometry" and discriminant analysis (DA), which is complemented by a number of subsequent studies by various authors. However, since the data is obtained by different methods, the results often do not match. There is a need to combine databases of subspecies and populations obtained using various methods into a single database, which will be a guide for future research projects on identification of subspecies and populations; to transfer reference data into free access and unify existing methods to ensure consistency of reference data and to connect various laboratories having independent reference samples in their storage in a single network.

Discriminant analysis based on a complete set of features is the main research method for determining the distribution of the studied samples among the reference groups. DA process is carried out in two stages. Stage one—the maximum possible set of reference samples is used to confirm they are correctly redistributed into groups. Entering the measured features one by one, you can define a group of features confirming the separation of reference groups, and for further work, only these features should be taken into account. Stage two—the inclusion of the studied sample in the analysis and the determination of the probability of falling into each of the reference groups. However, DA is not suitable for finding new varieties or ecotypes, since the choice is made between a limited and a certain number of reference groups. New types cannot be attributed to any of them, and hybrids cannot be separated from the new types (Meixner et al. 2013).

A sampling of honey bees for morphology measurements

The size of any sample depends on the objectives of the study. In the case of a determination of race affiliation, it should include samples of interest, for example, from the apiary under study or the line of bees. Due to differences between colonies, samples are taken from at least three colonies of bees from an apiary or a line. If it is necessary to investigate the entire population occupying a specific area in order to detect regional differences, then samples are taken from at least five colonies per region, so that sampling errors remain within acceptable limits (Radloff and Hepburn 1998, Radloff et al. 2003, Radloff et al. 2010). Sample size is the number of individuals in the sample. For the analysis of the exterior, samples of individuals (bees, virgin and laying queens, drones) are taken. Bees are selected at the age of 2–3 days (Goetze 1964) or kept for winter. It is necessary to select such a number of individuals that will provide analysis of the sample using the method or a combination of methods (this number can be calculated with the use of the coefficient of variation, Cv). European authors recommend sampling of 30–40 workers from a colony (Meixner et al. 2013).

When selecting material for research, it is necessary to take into account the seasonal variability of bees. So Mikhailov (1927a) studied the seasonal variability of exterior features (the proboscis length, the length and width of the right forewing, the number of hamuli (hooks) on the right hindwing, the sum of lengths of tergites 3 and 4) and found that the number of hamuli (hooks) was the feature with the least variability (did not give significant differences between generations), and the length of the right wing gave the greatest variability. He also found that during the season all measured parameters change in a varying degree. Mikhailov (1927a) recommends collecting honey bees in the same season, preferring the second half of summer (hatching bees from the frame) or autumn-winter. He also states that on the question of seasonal variability of bees' exterior characteristics, the study should be repeated, i.e., he considers this question not fully investigated yet.

We recommend taking into account that the sample will be used only for morphological analysis, as well as the necessary material supply (Sample Bank) for various additional purposes (including a possible exchange of material with other researchers); one should take 60–100 bees. Moreover, in the case of especially valuable purebred material, one should take 150–200 individuals from the colony. Virgin and laying queens should be taken in the quantity necessary for scientific purposes (given their limited number). Given that the coefficients of variation on the exterior features of drones have higher rates, the number of drones in the sample can be doubled compared with the number of bees. One sample should contain individuals selected from only one bee colony (Meixner et al. 2013). The sampling is carried out from: (a) the generation of bees kept for winter, starting from the third decade of September after feeding with sugar syrup with no brood from the last frame (it is necessary to exclude the queen from getting into the sample). At the same time, this sample can be used to determine varroosis after treatment with miticides. These terms are adopted in FSBSI "Federal Scientific Center of Beekeeping" and are approximate for the central part of the country and can vary depending on the geographical location of the area and weather and climatic conditions; (b) if it is required to select during the period of the brood presence, it should be done from the brood zone (Meixner et al. 2013); (c) individuals aged 2–3 days (Goetze 1964) can be obtained by placing a comb frame with outgoing bees into a net probationary ward for 2–3 days; (d) a sampling of drones is more complicated because they are only available seasonally and are more prone to wander than working individuals (Jay 1969, Currie and Jay 1988, Neumann et al. 2000). As one of the variants for the controlled sampling of drones from a particular bee colony, a drones' cell can be used, which is placed in an insulator of perforated zinc (or perform tractor) before the drones exit. You can then mark the drones with a marker for marking the queens (if more mature drones are required) and remove the insulator ward or select 2–3 days old drones; (e) virgin queens are taken from the cells 2–3 days after leaving the queen cell.

Sampling can be carried out in any container, from which they can be easily shaken out into a container with water heated to 100°C (if you are not supposed to carry them alive). Aluminum cups or small plastic jars with a wide neck with a lid are suitable for this purpose. Samples of living bees, queens, drones should be fixed, scalding them with drinking water heated to 100°C or exposing them to diethyl ether so that they throw out proboscis. Otherwise, the proboscis will remain bent and cannot be measured (Alpatov 1945, 1948, Bilash and Krivtsov 1983, Meixner et al. 2013). After fixation, each sample must be packaged separately into a container (for example, a well-washed plastic bottle with a wide neck, a plastic jar, a glass jar, or any other suitable container) or a gauze fardel (made of medicinal cotton gauze or gauze medical bandage), which are put into a total container. For large volumes, manual installation is recommended for seaming vials, which actually implies packing samples in these vials, and they can already be packaged in any suitable container. If samples from different colonies are placed in a common container, each sample should have a label made of thick writing paper, on which only the number of the bee colony from which the sample was taken, and the date of the sampling, are clearly written with a carbon pencil. It is impossible to make inscriptions using other means of writing since in this case, the inscriptions will dissolve in the preservative solution. On the label of the general capacity or in the accompanying documentation, is the following are indicated—the numbers of bee colonies, the date of sampling, the name of the location of the apiary (settlement, district, region, country), the geographical coordinates of the apiary, the origin of the queen and where it was obtained, if there are documents confirming the pedigree bee colonies, and copies of these documents are attached. Fardels with samples are poured from the top with a 70% (v.) solution of ethyl alcohol (C_2H_5OH), where chitin remains sufficiently soft for preparation (can be stored in 95% ethanol), carefully sealed, and stored until preparation. If long-term storage is supposed, as in the case of bee sampling, it is better to store in a glass container rather than plastic, since the alcohol gradually evaporates through the walls of plastic bottles. It is better to use glass jars of various sizes with a neck for a screw cap. If the morphometric analysis is conducted by a third-party laboratory, the bee samples are sent in any accessible way that does not prohibit the transport of alcohol-containing liquids or their delivery.

The main honey bee morphology measurements used in Russia

Subspecies, ecotypes (races) often only slightly differ in the mean values of several measurements; therefore, additional statistical methods are necessary to separate groups. The concept of numerical taxonomy was introduced into the systematics of the honey bee by DuPraw (1964, 1965a,b), and further developed by Ruttner et al. (1978). The fundamentals of the methodology for studying and measuring individual chitinous parts of bees were laid by Kozhevnikov (1900a). In his monograph "Materials on the natural history of the bee (*Apis mellifera* L.), Kozhevnikov (1900a, 1905) carefully analyzed the available data in the entomological and apicultural literature on genus *Apis* and *Apis mellifera* L., in particular, and having conducted many different morpho-anatomical investigations, made a number of conclusions concerning the taxonomy of the genus, morphology, and anatomy of bees. Further development of the work of Kozhevnikov concerning morphometry was obtained by his students Alpatov and Mikhailov. In studies on morphometry, various types of features have been used since the works of Kozhevnikov (1900a). There are, for example, signs of coloring, linear, and angular measurements, as well as a separate group of calculated features, including indices and sums. In recent years, they have been supplemented with feature "Wing shape" used in "geometric morphometry". Meixner et al. (2013) give the most frequently used features. The main set of

36 features described by Ruttner (1988) contains the features recognized in the "classic morphometry" of most countries. The groups of features, used by different researchers, vary considerably in their numbers from the "core set" described by Ruttner. Sometimes other features are introduced, but their choice is largely arbitrary. Since the comparisons, studies based on features with different compositions are difficult. When studying some unknown variations, it is recommended to use 25 features suggested by Meixner et al. (2013), in combination with the analysis of the wing shape (19 landmarks).

These minimum measurements will provide: (a) an extensive base for comparing the variability under study with some known distributions of features; (b) an accurate and reliable count of the characteristics representing the numerical description of the morphological variability of bees (Meixner et al. 2013). The forewing of the bee is in first place in the number of measurements and indices that can be obtained from this organ. However, there are discrepancies in the measurement and analysis of the morphometric characteristics of the wing, whose venation scheme is of particular interest in the systematics of insects as a whole. Currently, various approaches are being used for the morphological analysis of the front wing.

Classic and geometric morphometry of honey bees. The main problem associated with an increase in the number of main measurable features is the lack of backward compatibility. That is, the results of studies based on geometric morphometry (GM) cannot be compared with reference data obtained using the wing classic morphometry (CM), accumulated in previous works by various authors (including descriptions of reference subtypes in a monograph by Ruttner 1988). The classic morphometry (CM) of the wing includes variability in wing size (length and width) (Alpatov 1927) and venation: the lengths of the "A" and "B" segments of the medial vein of the third radio-medial (3rd cubital) cell and various indices (Alpatov 1935, Goetze 1964) and angles. For the first time, measuring the angles was introduced into bee morphometry by DuPraw (1964), who proposed measuring 17 angles between the wing vein joints, and Ruttner reduced this number to 11 angles (Ruttner et al. 1978, Ruttner 1988).

Recently, instead of morphometric methods characteristic of a bee wing, a number of studies use geometric morphometry (GM), based on the theory of the form. The GM uses the coordinates of landmarks located at the intersections of the wing veins of bees (Baylac et al. 2008, Tofilski 2004, 2008). Kozhevnikov (1900a) studied the venation of the wings of bees, i.e., the shape of individual cells and the place of discharge of individual veins, as a possible species trait. GM appeared as a "technical tool" for analyzing geometric problems formulated by Thompson d'Arcy in 1917 (1992). Its main advantage is geometric images into which numerical data can be embodied (Pavlinov and Mikeshina 2002).

The GM procedure begins with the fact that the objects under study are combined by their centroids at the point of intersection of the space axes. Then, their centroid sizes are reduced to unity, due to which their alignment with the standard is achieved: thus, the "dimensional factor" is excluded from the subsequent analysis. The alignment of each instance is due to its isometric change, in which the difference between the centroids calculated for all labels (Procrustes method) or a certain pair of labels (baseline method) is minimized between it and the standard. After this, an isometric "rotation" of the aligned objects relative to the standard occurs in such a way as to minimize the total difference in the coordinate values for all the landmarks (cited in Pavlinov 2001). More information about the GM can be found in works in Russian (Pavlinov 2001, Pavlinov and Mikeshina 2002, Vasilev et al. 2018) and English (Bookstein 1991, Rohlf and Marcus 1993, Rohlf 1993, Bookstein 1996, Rohlf 1996, Zelditch et al. 2004a,b.) The data, namely the coordinates of the landmarks, are obtained using "on-screen digitizers", for example, with the help of two programs tpsUtil and tpsDig2 (or tpsDig) (Rohlf, UK) sequentially from a previously digitized image of the object (Vasilev et al. 2018). This and other software related to the GM, articles on the topic of GM and others can be found on the website http://life.bio.sunysb.edu/morph/.

The obtained data is then analyzed using various methods of multivariate statistical analysis (principal component analysis (PCA), canonical analysis (CA) and discriminant analysis, and others (Vasilev et al. 2018). For this purpose, various software are used, such as MorphoJ (designer Klingenberg, UK, site http://www.flywings.org.uk/MorphoJ_page.htm) (Klingenberg 2011), IMP (designer Sheets, USA site http://www3.canisius.edu/~sheets/) (Vasilev et al. 2018), and others. The GM method is used to separate the subspecies of honey bees and their hybrids in Brazil (Francoy et al. 2006, 2008), Poland (Tofilski 2004, 2008), and other countries, as well as to analyze the distribution of *Apis mellifera* subspecies along the evolutionary lines (branches) and the relationships between them (Miguel et al. 2011, Kandemir et al. 2011). Some authors, Tofilski (2008), established a slight superiority of GM over CM, others, Miguel et al. (2011) indicated that GM is more appropriate than CM to identify the Africanization process, and Kandemir et al. (2011) consider GM a reliable tool for the separation of bee subspecies that have advantages over CM. Various types of software have been developed for the automatic identification of subspecies of honey bees using GM: DrawWing (Tofilski 2004, 2007, 2008), ApiClass online system (Baylac et al. 2008).

DAWINO method is applied on a commercial basis in the Czech Republic. The abbreviation "DAWINO" stands for "discriminant analysis with a numerical output". In essence, this is a nice name of the ordinary discriminant analysis performed using Beewings 1.20 software licensed by the Research Institute of Bees in Dol. A brief description of this method can be found at http://www.beedol.cz (the English version) and it is mentioned in articles (Cermák and Kaspar 2000, Satta et al. 2004, Floris et al. 2007, Bouga et al. 2011, Meixner et al. 2013). This method and/or data obtained using it was used in Norway, Switzerland, Slovakia, and Macedonia. Now, this software is no longer available at the market, and the authors do not engage. However, judging by the information provided on their website, this method combines obtaining landmarks coordinates using GM software and calculating KM parameters on their basis: 17 angles (all angles are DuPraw 1964), 7 linear measurements, 5 indices, and one area. All these angles and other parameters have metric units of measure for features and can be combined with other morphological features of the body in further analysis. According to the reviews of Macedonian colleagues in the course of their work, some inaccuracies were found with 9 features that were excluded from further analysis, and with the remaining 21 features this software works fine.

Currently, there is a certain inconsistency regarding the sequence and number of marking points of various authors when analyzing the shape of the wing. Some authors use only one wing cell for GM (Francoy et al. 2006). Others, in their various works, change the number of marked points, the number of points varies from 18 to 20, or some of them begin marking with "0" and others with "1". There are several schemes for the sequence of marking points on the image. Some authors do not indicate which front wing they took: right or left. Others take the left front and the third ones choose the right front (Tofilski 2008, http://drawwing.org, Nawrocka et al. 2018, Miguel et al. 2011, Baylac et al. 2008, ApiClass, DAWINO (http://www. beedol.cz) Cermák and Kaspar 2019, Kandemir et al. 2011). To exclude the further parallel development of incompatible databases in the morphometry of honey bees, Meixner et al. (2013) made a number of proposals for the standardization of wing measurements. These suggestions are as follows: (1) to store all future data in the form of landmark coordinates (instead of the format of derived features, such as angles) to facilitate the exchange of data between different studies and research groups; (2) to use the landmarks layout illustrated in the ApiClass example (http:// apiclass.mnhn.fr or in the article by Meixner et al. 2013) and shown in the following Fig. 7.1. From these coordinates, used in most geometric studies, it is possible to calculate all 30 features of the DAWINO method, including the angles suggested by DuPraw (1964).

Preserving the coordinates of the landmarks instead of the calculated features will allow one to keep the data available when developing analysis methods. Calculating the metric parameters by landmark coordinates is possible using the IMP software (Integrated Morphometrics Package) (Vasilev et al. 2018). As noted above, due to the lack of backward compatibility, the coordinates of the landmarks cannot be obtained from the features of the wing CM, but several attempts have already been made to associate GM with the subspecies features obtained when using CM. For this purpose, the available reference samples of the data bank (Oberursel, Germany) were measured by the GM method (Kandemir et al. 2011, Nawrocka et al. 2018). Meixner et al. (2013) posed the question whether geometrical morphometry should permanently replace classic morphometry, that is, accurate, powerful and effective shape analysis based only on the geometry of the wing should replace the full set of CM features.

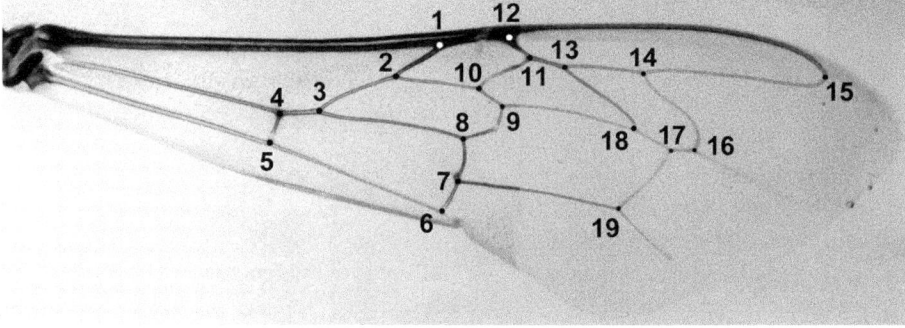

Figure 7.1: Location points used in the system ApiClass.

For phylogenetics, GM is more suitable in comparison with the features of CM and is a method comparable to a certain extent with molecular methods. So Miguel et al. (2011) compared the data obtained using GM and microsatellite analysis and established their comparability. The same authors believe that GM is more suitable than analysis of mitochondrial DNA or CM when screening and identifying the process of Africanization. In confirmation of previous authors, Oleksa and Tofilski (2015), when comparing the three methods for identifying subspecies based on the use of microsatellite locus, *COI-COII* mitotypes, and GM of the forewing, found that the allocation of colonies to the same groups is the same in microsatellite analysis and GM. The authors believe that GM can be used to detect hybrids between *Apis mellifera mellifera* and *Apis mellifera carnica* (Oleksa and Tofilski 2015).

GM should be used to study the phylogenetic relationships between populations, ecotypes, subspecies, where the use of CM can lead to improper conclusions. A group of authors based on previous studies and existing standards created free software for identifying subspecies and evolutionary lines of bees (Nawrocka et al. 2018). Meixner et al. (2013) consider that in order to take into account the variability of honey bees as a numerical record of the morphology of subspecies and ecotypes, it is necessary to use CM with a large set of features to demonstrate the currently existing features of subspecies or ecotypes, in addition to the question of their phylogenetics. GM of venation can replace the metric features of a wing even within the framework of CM, but so far, no attempt has been made to combine these methods (Meixner et al. 2013).

Methods of morphology measurements of honey bees are used in Russian Federal Beekeeping Research Center

When measured using MBS-9 stereomicroscopes (or similar), separate chitinous parts of the bee are immersed in glycerol, then excess glycerol is removed, and the parts are placed on a glass slide successively as they are prepared. As well, the styling is performed on the principle of the same-type parts of the chitinous skeleton on a separate glass, while maintaining the sequence of the prepared bees. After all the bees have been prepared the glass is covered with the second one, then with a glass rod from the top, it is shedded with glycerin drop by drop so that all parts of the bee are in glycerol. Preparation of individual parts of the chitinous skeleton of a bee to obtain an image through a flatbed scanner is carried out similarly to preparation when measured through MBS-9 with some differences. After filling the glass, it is quickly turned over, and the side with the object is put on the scanner glass. Then it is scanned. When using automatic measurement systems (geometric morphometry), such as DrawWing (http://drawwing.org/node/4) and ApiClass (http://apiclass.mnhn.fr), samples are prepared according to the guidelines on the sites.

Today, a wide range of measurement methods are available (including a stereo microscope with a reticle scale, real-time video measurement, photographing, or scanning and analysis in commercial and free image analysis programs). The choice of measurement method does not matter, only under one condition—measurement devices must be carefully calibrated and checked for the presence of optical distortion (in the case of cheap optics). Calibration is an important basis for the correct representation of all measurements related to size and it cannot be ignored. In order to obtain compatible data in different laboratories, it is always necessary to carry out measurements or analysis in the same way for all laboratories. This is important because personal habits or "local techniques" in different laboratories can introduce distortions that can cause problems and lead to incompatibility of data sets. Before starting work, inexperienced researchers should seek advice and use cross-validation, measuring the same samples. It is also necessary to conduct a crosscheck between the laboratories involved in determining the species of bees by the morphometric method, especially when mastering new methods (Meixner et al. 2013).

Measurements are made under a stereoscopic microscope MBS-9 (or other similar) using an eyepiece with diopter reset with a removable scale. To determine the linear dimensions of an object, it is necessary to calculate the number of graduations of the scale, which fit into the measured area of the object, and multiply this number by the number indicated in the conversion table corresponding to the increase in the microscope at which the measurement is made (according to the passport of MBS-9). The length of the wing is measured with magnification x8.21, and all other features are measured with magnification x16.35. Linear measurements made in divisions of the eyepiece scale are then converted into millimeters, and indices are expressed as a percentage.

When describing features, the use of words "length" and "width" is due to different meanings when used in zoology ("length" in relation to the axis of the body) and in the spoken language ("length" in relation to the proportions of the object). Therefore, when the meaning is not entirely clear, the terms "longitudinal" and "transverse" are used. When describing measurements of the wing, points are taken at intersections of imaginary midlines veins (Ruttner 1988). In these recommendations, each measurement has a serial number in the form of an Arabic numeral; measurement points (if their placement is required) are indicated by Latin letters. A detailed description is given for those features that have been used so far at FSBSI "Federal Beekeeping Research Center" (Russia). For other features, recommended by Meixner et al. (2013), the measurement technique can be studied according to Ruttner et al. (1978) and Ruttner (1988). The features found on paired organs (wings, legs) or on abdominal segments (tergites, sternites) are measured similarly to those given.

Feature 1—"Length of the proboscis". The method of measurement was proposed by Alpatov (1927*) (here and thereafter the year is given by the source, where the measurement of this feature is shown for the first time in the figure). He received the total length of the proboscis by adding up four consecutive measurements. Currently, if the proboscis is expanded, one common measurement is used (Fig. 7.2).

Feature 2—"Length of the right forewing" (Alpatov 1927). At present, one measure is taken of the total length of the wing from the highest point on its base to the opposite edge over the longest distance (Fig. 7.3). Feature 3—"Width of the right front wing" (Alpatov 1927) is measured perpendicular to the length of the wing in its widest part (Fig. 7.3). Features 4 and 5—the lengths of the segments of the "media" (M) vein–"YZ" (A) and "XY" (B) (Goetze 1964, Ruttner 1988) are measured as the distance between points: X–at the intersection of the "media" (M) vein with

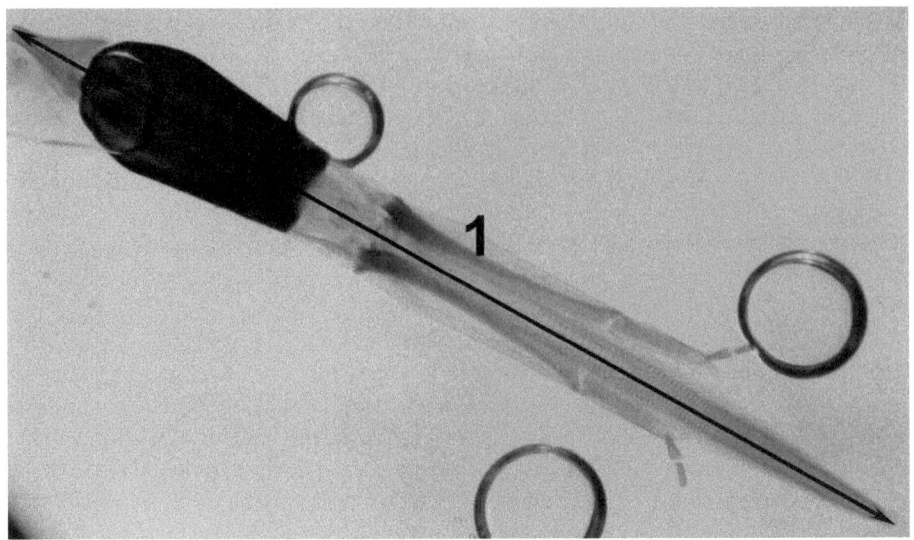

Figure 7.2: Feature 1- "Length of the proboscis".

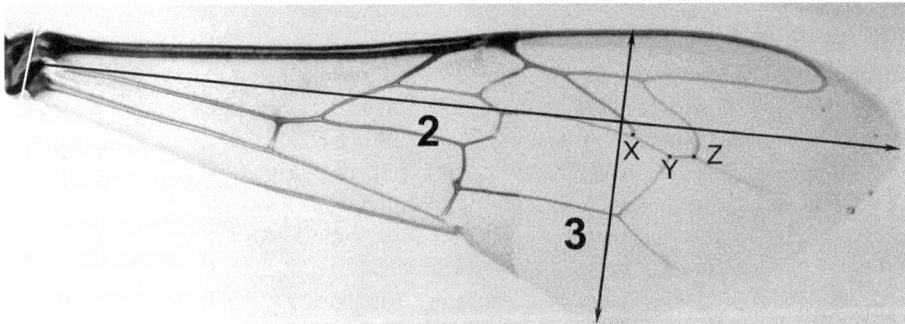

Figure 7.3: Features 2-5- "Right forewing".

the "2 radial sector-media" (2rs-m) vein: Y–the "media" (M) vein with the "2nd media-cubitus" (2m-cu) vein and Z–the "media" (M) vein with the "3rd radial sector-media" (3rs-m) vein (the names of the veins, cells, and their abbreviations are given by Gauld et al. 1988) (Fig. 7.3).

Feature 6—"The discoid shift" (Goetz 1964) is determined by drawing a line through two maximum points from each other "marginal cell" (MC) points ("C" and "D") and building a perpendicular to this line, which must pass through the center of vein intersection "radial sector" (Rs) and veins of the "3rd radial sector-media" (3rs-m) ("E"). Sign "0" is placed if this perpendicular passes through the center of the intersection of veins "cubitus" (Cu) and "2 media-cubitus" (2m-cu) ("F"), respectively, " + " if to the left of this intersection, and "-" if to the right. Using the measurement of a digital image allows you to go from "-"/" + " to a numerical characteristic, i.e., to measure the angle of the discoid shift (Fig. 7.4).

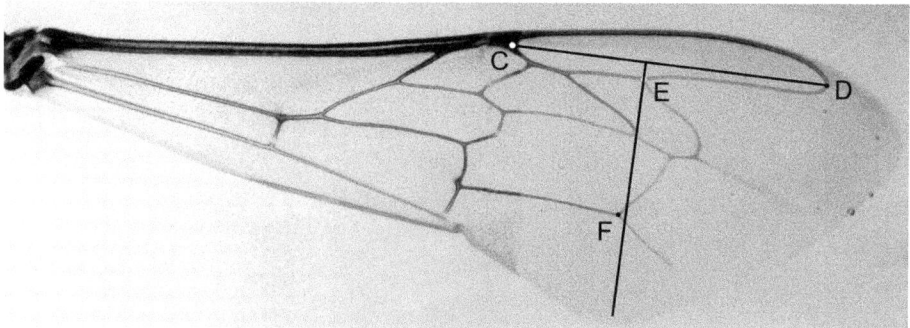

Figure 7.4: Feature 6- "The discoid shift" (Goetze 1964).

Feature 7—"Length of tergite 3–longitudinal distance" (Alpatov 1927) is accepted to take along the axis of the bee's body, in connection with which it turns out to be less than the width. Lengths of tergites 2 and 4 can be measured (Fig. 7.5).

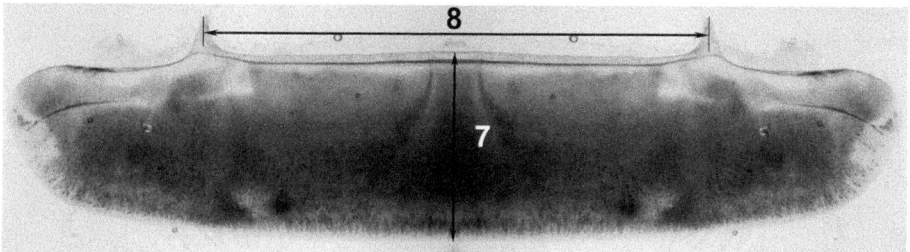

Figure 7.5: Features 7- 8 on the third tergite.

Feature 8—"The conventional (transverse) width of the third tergite" (Bilash and Krivtsov 1983) is measured between the centers of the tergite protrusions, called apodeme (Fig. 7.5). Feature 9—"Length of sternite 3–longitudinal distance" (Alpatov 1927) (Fig. 7.6). Feature 10—"The conventional (transverse) width of the third sternite" (Bilash and Krivtsov 1983) is measured between the centers of protrusions, called apodeme (Fig. 7.6). Feature 11—"Length of the wax mirror on the third sternite" (longitudinal distance) is the shortest distance (Ruttner 1978) (Fig. 7.6). Feature 12—"Width of the wax mirror on the third sternite" (Alpatov 1927) (transverse distance) is the largest diameter of the plate. In this case, the thickness of the plate edge should not be taken into account—the reference points must be taken on its inner side in order to determine the "clean" dimensions of the mirror (Fig. 7.6). Feature 13—"Distance between wax mirrors on the third sternite" (Ruttner et al. 1978) (Fig. 7.6).

Feature 14—"Length of the first segment of the right hind tarsus" (Alpatov 1927) (Fig. 7.7). Feature 15—"Width of the first segment of the right hind tarsus" (Alpatov 1927). When measuring this feature, bumps located on the lateral edges are included in the measurement (Fig. 7.7). Calculated indexes are features 16, 17, and 18. Feature 16—"Cubital index"—currently there are two methods for its calculation: (a) according to Alpatov (1935), it is determined by the ratio of the length of the smaller

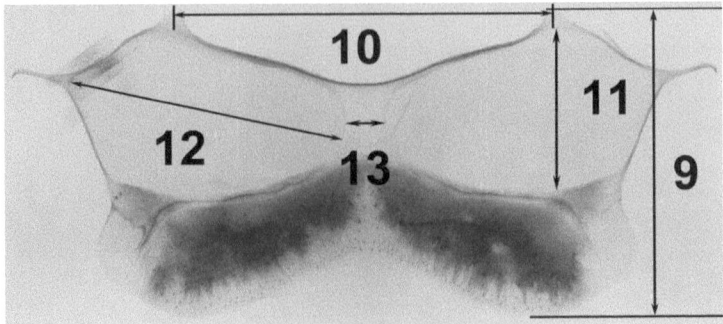

Figure 7.6: Measurements 9-13 on the third sternite.

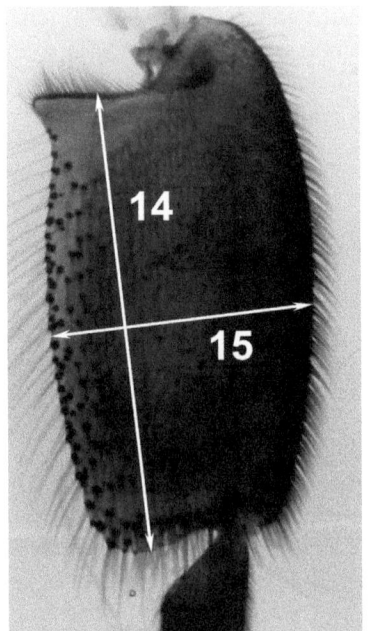

Figure 7.7: Features 14-15 on the first segment of the right hind tarsus.

segment "YZ" to the length of the larger "XY" and is expressed as a percentage; (b) according to Goetze (1964) it is determined by the ratio of the length of the larger segment "XY" to the length of the smaller "YZ". Feature 17—"Tarsal index" (according to Alpatov 1948) ("wide-leg" index). This sign is determined by the ratio of the width of the first segment of the right rear tarsus to its length and is expressed as a percentage. Feature 18—"Linear load index (LLI)". It was proposed by Ragim-Zade (1975) to characterize the potential ability of bees to collect food, expressed as the ratio of the wing length to the total length of 3 and 4 tergites (currently not used).

The countable and qualitative feature is feature 19—"The hamuli (hooks) number on the hind wing" (Fig. 7.8). Kozhevnikov (1900a) noted that feature "the

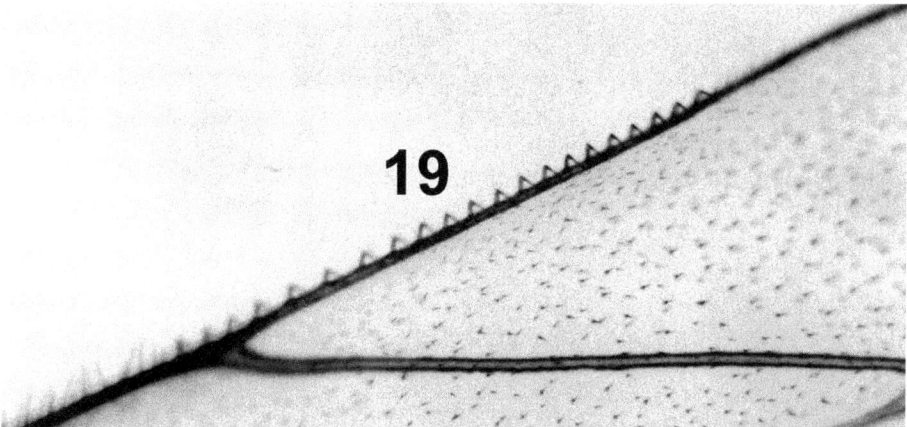

Figure 7.8: Features 19-"The hamuli number on the hind wing".

hamuli (hooks) number on the hind wing" does not have a specific value; it does not give a diagnosis but can be used for a species characteristic (except *A. florea* F.).

Kozhevnikov (1900a) also noted that this attribute has no significance for the race separation. Goetze (1964) points out that if species of genus *Apis* have differences in the hamuli number that can be related to the size of their bodies, then among subspecies and races of *Apis mellifera* it is more difficult to derive this pattern and this feature does not guarantee any accurate racial characteristics. Ruttner (1988) considers this feature unsuitable for taxonomic separation and excludes it from the group of main features used. Feature 20—"The color class according to Goetze" (Goetze 1964) is determined on the second and third tergites by the area filled with the yellow pigment of its surface using a nine-grade scale. Ruttner et al. (1978) gives a 10-grade scale and assesses the color of 2–4 tergites. In our opinion, the assessment of this feature is rather subjective and difficult, especially for an inexperienced employee.

Measurement on a computer using software and obtaining a digitized image of the parts of honey bees

The first researchers measured manually, then they began to use software (Daly et al. 1982), which significantly increased the accuracy (according to Meixner et al. 2013). The systems used to obtain a digital image can be divided into the following groups. (1) Systems based on stands of microscopes, macroscopes, stereo microscopes, and other similar devices and using various digital video (except for eyepiece) and cameras (mirror and mirrorless) without an objective lens as image capture devices. Cameras are mounted on microscopes, etc., by means of various adapters. Microscopes, etc., must be checked for optical distortions before use. (2) Scanners: (a) slide scanners provide an image with high sharpness and good contrast, since they have a built-in lens and, compared with previous systems, are relatively cheaper, (b) flatbed scanners are not recommended for some automatic systems (http://apiclass.mnhn.fr) because of poor contrast and low image sharpness, but due to their availability, it is possible to use them if manual alignment of dots is assumed. For scanning, it is necessary to use a flatbed scanner that has a CCD type sensor, since this type of sensor provides a better scan quality, i.e., a greater depth of field than scanners that have a CIS type sensor. (3) Digital mirror and mirrorless cameras mounted on special stands, equipped with macro lenses and ring or macro flashes. In these systems, the surface of preparation should be parallel to the plane of the matrix. It is not recommended to use digital cameras of the "bridge" type due to optical deformations given by the lenses used in them.

When using a flatbed scanner, scanning is carried out almost immediately after the preparation is ready, since glycerin absorbs moisture from the air, dilutes and flows out, resulting in deterioration of the quality of the image and difficulty in measuring the preparations. Scanning is best done through a transparent film substrate (this will prevent the scanner glass from being scratched and speed up scanning, eliminating the need to wipe the scanner glass). First, a film is placed on the scanner glass and the preparation on the slide is placed on it. From above, everything is covered with a piece of paper. Additionally, one can use a small load in order to press the glass better and ensure good visibility of the preparation. Scanning is carried out with a resolution sufficient to measure objects on the monitor screen, but not less than 1200 dpi (the higher the scanning resolution is, the better image quality can be obtained, but the more time will be spent). The measurement on a computer is carried out with the help of the software designed for image analysis, which allows one to do the required measurements and carry out measurements in the right units and with sufficient accuracy (that is, providing output data with a certain number of decimal places). The grouping of this software is presented in the diagram (Fig. 7.9). Note to Fig. 7.9: Beemorph- Russell Talbot, UK; CooRecorder, CBeeWing—Cybis Electronic and Data AB, Sweden; AltamiStudio—Altami, Russia, ScopePhoto 3.0—ScopeTek, China).

The software can be obtained free of charge: by downloading from the Internet (free software) or as a bonus when buying some equipment (for example, an eyepiece camera). The software can also be purchased separately. When measuring, the measurements described above are used. When getting CM indicators, before the measurement it is necessary to calibrate the program using a digital image of the scale, which has the necessary accuracy divisions and is obtained under the same conditions as the image of the chitinous parts of bees. The scale must be verified by the appropriate government agency. For calibration instructions, see the appropriate software help. In GM, the tpsDig Rohlf program (Vasilev et al. 2018) is used to place landmarks manually. Along with this, software operating in semi-automatic (Quezada-Euán et al. 2003, Steinhage et al. 1997) and automatic (Steinhage et al. 2007, DrawWing-Tofilski 2004, ApiClass-Baylac et al. 2008) measurement modes are created and used. These modes have reduced the time of analysis and improved the accuracy of identification of bee species.

Mathematical processing of morphology results. In mathematical processing of the data obtained on a computer, depending on the level of its complexity, different software is used (MS Excel, Microsoft, USA, Statistica, StatSoft, USA, etc.). Only measurable primary characteristics are used for analysis, not sums, or indices. When measuring on MBS-9, the data is recorded in a special notebook, and then entered into

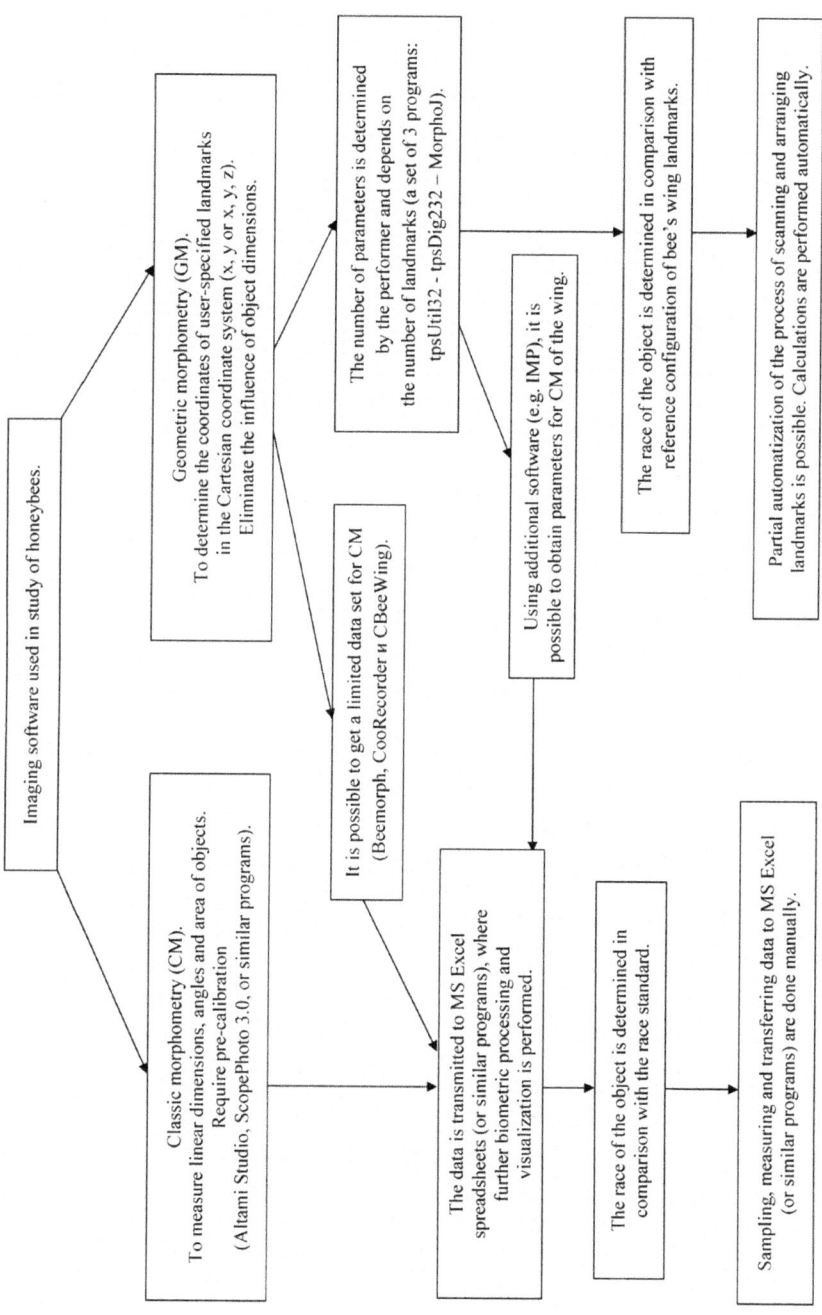

Figure 7.9: The softwares used in the study of honey bee morphometry.

an appropriate computer program, for example, MS Excel or a similar one, having the possibility of mathematical data processing. When measured on a computer, data is immediately transferred to similar software. Previously, the data was processed using a calculator, according to the formulas given in the relevant literature (Plokhinskiy 1969, 1970, and others). As there are several schools on biometric processing, it is necessary to check the obtained data of mathematical processing for compliance with the results obtained in the calculations on the calculator using the formulas set forth by Plokhinskiy.

For more complex methods of multivariate statistical analysis (principal component method (PCA), canonical and discriminant analysis methods, and others), Statistica software or something like it are used. Some elements of mathematical analysis are also present in the GM software (for example, in MorphoJ and others). Two main methods (principal component method (PCA) and grouping k-means clustering) are used to determine whether the samples under investigation represent one or several groups. If these two methods lead to a clear definition of the group, then the data obtained is further processed by a discriminant analysis (DA), which determines the significance of group differences and the accuracy with which the samples are redistributed into groups.

Attention should be paid to the fact that gaps in the sample structure may give the impression that there are two or more different groups that are checked statistically, but the difference will disappear with a more uniform sampling range (Radloff and Hepburn 1998). At the same time, true groups are characterized by sudden morphological changes depending on their geographical origin, which can be confirmed by geographic areas and attitudes to physiographic (Meixner et al. 2013). More information about the methods of analysis and the software used for this can be found in the relevant literature (Vasilev et al. 2018, Zelditch et al. 2004a,b).

Conclusions

Twelve breeds and honey bee populations were included in the state register of breeding achievements of the Russian Federation. It has set detailed biological and economic characteristics of bee colonies of Middle Russian, Bashkirian, Carpathian, Gray Mountain Caucasian, Far-Eastern breeds and breed types: "Priokskiy", "Orlovskiy", "Tatarskiy", "Burzyanskaya Bortevaya" of Middle Russian, "Maikopskiy", "Moskovskiy" of Carpathian, and "Krasnopolyanskiy" of Gray Mountain Caucasian. A certain place is given to the protection of the gene pool of breeds and populations of honey bees in keeping them in nurseries and reserves, where they are in natural habitat, as well as in breeding farms for breeding bees of a certain origin. Currently, more than 20 breeding farms are engaged in the breeding of zoned breeds and the sale of breeding products to apiaries in the regions of zoning of these bees have been certified. Annually, these farms deliver to 200,000 queen bees and bee colonies. However, the achieved level of production of breeding material, primarily of the Middle Russian bees, does not meet the demand of consumers for these products. Therefore, in the coming years, along with the preservation of the gene pool, it is planned to organize new breeding farms to increase the production of purebred queen bees and bee colonies of zoned breeds and types of bees.

Preservation of breeding resources of bees, their breeding improvement, and rational use will increase the productivity of bee colonies to 20%, and significantly increase the yield of entomophilic crops due to their pollination activities. The development and adoption of a common standard for the definition of the concept of a "reference sample" and the rules for its design will ensure their comparability, and organize the storage of standards and data on them in free access, on mutually beneficial conditions, to all researchers. Provision and storage of the samples under study and the measurement using a unified methodology will provide an opportunity to obtain comparable results. Obviously, pesticide poisoning is one of the important causes of the death of bee colonies. This is due to the ability to cause not only the rapid death of bees, but also the accumulation of contaminated feed reserves, wax pollution, which leads to a long course of poisoning. Pesticides make bees vulnerable to other pathogens and reduce colony strength. Particular attention should be paid to the new class of pesticides, named neonicotinoids. These substances cause the collapse of bee colonies. Collapse occurs not only because of the rapid death of bees, but also because of specific behavioral disorders leading to the death or weakening of the colony.

CHAPTER 8

Breeding Better and Healthy Honey Bees is the Only Way to Save A Native Biodiversity

Gregorc, A.

Introduction

Honey bee colonies in an apiary usually have different genotype, and therefore morphological, economic, and ethological characteristics vary from colony to colony. A stock of bees, with a combination of traits that characterize a particular group of bees, is considered to be a breed. When a specific breed exhibits aggressive behavior or it is inclined to swarm, it may lose a great extent of its economic value. Furthermore, morphological, behavioral, genetic, and economic characteristics of individual workers or honey bee colonies are considered and evaluated in order to perform the effective selection. Specifically, the most important breed traits are disease resistance, honey production, gentleness. In different breeding programs, other traits are also included: brood viability rate, aggressiveness, disease resistance, pest resistance, defensive behavior, swarming tendency, life span, hygienic behavior, pollen collection, comb building, propolis collection, royal jelly production. Differences in observed characteristics are also due to cross-breeding as a result of mating with a variety of drones. Swarming is considered an important negative characteristic in selection and queen rearing process. Swarming behavior should be controlled in order to produce honey. It is important to know that several selective traits have an effect on colony strength and subsequently on honey production. As an example, in inbreed, brood viability of honey bee colonies is reduced, and therefore colony strength is reduced with significantly affected honey production. Colony strength would also be increased with increasing workers' life span and many other traits, including developing disease resistance, hygienic behavior.

University of Maribor, Faculty of Agriculture and Life Sciences, Pivola 10, 2311 Hoče, Slovenija; ales.gregorc@um.si

Beekeepers establish a breeding program to evaluate the performance of various honey bee races. Honey bee race is referred to as specific geographically determined bee characteristics as a result of natural selection in their homeland. Native bees are thus adapted to their original environment, and further selection needs to be performed to adjust beekeepers' requirements. Pure race present natural resources and material for economic breeding. Genetic variability and preserving the autochthonous managed honey bee populations are also an important objective of performing the quality breeding program. The quality selection and the quality of reared queens ensure economic production in the beekeeping industry. Selection objectives and criteria are important tools in breeding bees for developing productive beekeeping using native honey bee strains.

Popular races are autochthonous distributed in geographic areas and other strains of bees or crossbreeds are also a priority in beekeeping practice. We need to expose disease resistance as a selective trait, which correlates with the honey bees' "hygienic behavior" and contributes to reducing the diseases. This has direct implication in potentially reducing the use of medicines in honey bee colonies and a better perspective for production of safer honey bee products.

Tolerance and resistance characteristics

When we select and rear honey bees, two professional terms need to be considered and potentially introduced into breeding practice. They are tolerance and resistance, which refer to different mechanisms to enable bees in reducing the effects of the parasite *Varroa* (*Varroa destructor*) or any pathogenic or another physiological factor, especially with repeated exposure. The body thus becomes less responsive and develops the ability to overcome effects without disease appearance (Dorland 1990). In a honey bee apiary, the tolerance is the ability of bees to live in association with *Varroa* mites, or other pathogenic and/or nonpathogenic agents. These external factors have no negative effects on survival, colony population size reduction, or other aspects related to honey bee fitness. When *Varroa* transfer secondary infections, bees may be developing resistance to all those pathogenic organisms, and the final effect is developing tolerance to *Varroa* infestation. In bees and in the process of breeding, lack of sensitivity to insecticide or acaricides is also important, with emphasis and is a result of continued exposure or even genetic change. An important experience is also developing drug or pathogen resistance that arises by natural or artificial selection.

The ability of organisms to remain unaffected or slightly affected is thus considered resistance. Resistant bees are able to maintain low levels of *Varroa* infestation or other pathogens due to known or unknown bee characteristics. Absolute resistant bees, *Varroa* or other pathogens would not infest or infect individual bees or an entire colony. In respect of *Varroa*, this is not realistic, but some other bacterial or viral agents may not affect bees, and lastly, individual bees or an entire colony can be more or less resistant. Low infestations or infections that remain below levels in resistant bees would not seriously compromise colony health. There are also complex relationships between individual workers and entire colony that need to be respected in honey bee selection and breeding. Individual bees may reflect their relative resistance to other bees in a colony, and finally reflect development to a more or less resistant colony.

Honey bee colony defense systems

Individual bees evolved physiological, immunological, and behavioral defense systems toward pathogens and parasites. The colony-level response, as opposed to the immune responses within an individual bee, presents coordinated behavioral cooperation among individual bees in a nest. The collective defense against parasites is termed "social immunity" (Cremer and Sixt 2009). We can observe continuous simple interaction between two adult honey bees in a colony. When hundreds or thousands of individuals within a honey bee colony interact, the social immune responses would have a defensive effect against pathogen or parasite on a colony level. At that stage, particular attention needs to be implemented to how this information could be incorporated into beekeeping practices or selective programs to improve bee vitality and health.

Grooming behavior of honey bees

The ability to remove foreign particles and pollen from the body is an important behavior for an individual bee when grooming herself, or one bee may groom another bee: autogrooming or allogrooming behavior. Autogrooming has a significant impact as a defense mechanism for genetic resistance to the tracheal mite, *Acarapis woodi* (Danka and Villa 2000), the causative agent of Acariosis. During allogrooming, adult bees remove foreign particles or parasites from each other. This behavior was described as an important mechanism of defense against phoretic *Varroa* by the original host, *Apis cerana* (Büchler et al. 1992), and by Africanized *A. mellifera* in the tropics (Moretto et al. 1993). One of the possible mechanisms of resistance to *Varroa* in Africanized bees is "auto- and allogrooming" behavior, where bees brush particles from themselves or brush bees from their nestmates. It has long been known that different strains of bee differ in their resistance to *Varroa* and even more, resistance levels can be developed through the selection process performed at the beekeeping operation. It is demonstrated that Africanized bees, which are important hybrids of *Apis mellifera scutellata*, nowadays breed in Brazil and appear to have more resistance than European strains. Another example of grooming behavior efficacy is as a defense mechanism to *Varroa* in *Apis cerana*. Grooming behavior may potentially be included in honey bee breeding programs for queens rearing and reproduction, and increase the ability of honey bees to resist *Varroa*. As grooming activity involves workers licking and chewing, secondary increased incidence of physical contact between workers can potentially result in viruses transmission. Therefore, during selection process, tests need to be performed in order to identify potentially increased incidence of virus infection.

Hygienic behavior of honey bees

Disease resistance is known to correlate with the "hygienic behavior" of worker bees. This is a genetically controlled collective response by adult workers to recognize dead brood and then remove the infected or damaged brood (larvae and pupae) (Gramacho 1999). Hygienic behavior was originally defined as the ability of honey bees to detect and remove brood infected by *Paenibacillus larvae* from the nest. Hygienic bees may uncap comb cells containing dead, sick, or damaged brood, and remove this brood from the colony (Woodrow and Holst 1942, Rothenbuhler 1964). It was found that hygienic behavior is a significant defense mechanism against brood diseases, including chalkbrood disease (Gilliam et al. 1988), *Varroa* mites parasitism (Boecking and Spivak 1999, Spivak and Reuter 2001, Wilson-Rich et al. 2009).

Hygienic behavior of honey bees is also responsible for an important mechanism of defense against *Varroa*. Bees detect and remove *Varroa*-infested pupae. *Apis cerana* worker bees, in their colonies, regularly detect *Varroa*-infested pupae. Workers may either make a hole in the wax capping, or may remove the pupa and subsequently release the mite confined in the brood cell (Boecking and Spivak 1999). When the mites are released from the cell-capping opening, they become exposed to allogrooming between adult workers. This resistance mechanism is well known in *Apis cerana*, and may be activated in heavily parasitized host *Apis mellifera*. Successful reproduction of mites in *A. cerana* colonies is limited to drone brood pupae. It is possible that colony-level resistance of *A. cerana* to *Varroa* will largely reflect the seasonal production of drones, and thus limit the opportunity for mite reproduction (Fries et al. 1994, Rosenkranz et al. 2010). In queen breeding operations, monitoring the hygienic behavior capacity of honey bee colonies as a mechanism of defense against *Varroa* is very important. It needs to be a regular practice for selecting mother queens prior to breeding and reproduction.

Rapid population growth of *Varroa* in colonies is due to *Varroa*'s ability to reproduce in both drone and worker brood of *A. mellifera*. The removal of mite-infested worker brood through hygienic behavior would be a highly desirable trait in honey bee selection against varroosis. Killing any mite offspring during the reproductive cycle and reducing *Varroa* reproductive success would have a negative cumulative effect on the *Varroa* population dynamic in colonies. It was experimentally confirmed that workers in selected colonies for hygienic behavior were able to remove up to 60% of the experimentally infested pupae (Spivak and Boecking 2001).

Detection of parasitized and diseased brood is based on workers' olfactory stimuli (Masterman et al. 2001, Spivak et al. 2003). Volatile compounds associated with chalkbrood-infected larvae identified in a low concentration specifically associated with mite-infested pupae induced a response in very sensitive bees. Bees had positive electroantennogram response to one volatile substance, phenethyl acetate, isolated from infected larvae. A strong hygienic response was achieved in very low concentrations by colonies selected for rapid-hygienic behavior (Swanson et al. 2009).

It is likely that the olfactory stimuli that hygienic bees use to detect *Varroa*-infested pupae are associated with the bee's wound response to mite feeding, although the mite's offspring or feces accumulation may also be important. A line of bees, named VSH for *Varroa* sensitive hygiene (Harris 2007), that displayed hygienic behavior was also found. VSH colonies were able to detect and remove mite-infested pupae (Harbo and Harris 2005), and they removed more infested pupae compared with colonies from the hygienic line bred based on the freeze-killed brood assay (Ibrahim and Spivak 2006). It was subsequently found that they remove the mite-infested brood only after the mite has initiated oviposition, indicating the stimulus must reach a critical intensity (Harris 2007). It was also found that the bees of VSH line do not always remove the mite-infested pupae, and when the comb cell capping is opened, the mite is allowed to escape (Harris 2008). There is considerable potential for understanding mechanisms of mite resistance through continued studies of the VSH line. In beekeeping practice, specific phenotypic characteristics have been observed, and through selection activities, more tolerant honey bee lines to *Varroa* can be reproduced.

Colony test for the hygienic ability

Early detection of diseased brood is critical for developing resistance. In experimentally developing resistance, the bees must be able to detect dead or damaged brood and remove it from comb cells before the pathogen reaches the infectious stage within the bee host. This characteristic is important in colonies selected for rapid-hygienic behavior, and demonstrates resistance to American foulbrood and chalkbrood in the field (Spivak and Reuter 2001). Colonies with enhanced hygienic behavior have been shown to be significantly less diseased (Spivak and Reuter 1998). Differences in the ability of colonies to recognize and remove dead larvae and pupae after infection by the bacterium *Paenibacillus larvae*, the causative agent of American foulbrood, also exist between colonies. Therefore, every beekeeper has the option to follow characteristics of his colonies, perform the selection, and subsequently rear and reproduce productive queens. The importance of selecting hygienic colonies lies in the fact that diseased, non-hygienic colonies produce less honey than healthy, hygienic colonies. It was found that non-hygienic colonies often had symptoms of naturally occurring chalkbrood disease (*Ascosphaera Apis*) (Spivak and Reuter 2001). Hygienic behavior can be increased in a population after queen selection without mating control from 66.25% to 84.56% in five years (Palacio et al. 2000). Successful selection in a designed honey bee colony population is an option to reduce the incidence of diseases in an operation. Consequentially, the use of chemicals in selected honey bee colonies to control diseases may be significantly reduced (Gonçalves and Gramacho 2003). The test for the hygienic behavior presented in Fig. 8.1.

An important aspect of introduction grooming behavior characteristic through the selection process is to increase removing diseased brood in selected breed colonies. Through handling, removing or ingesting, diseased particles are facilitated when the pathogen has reached the transmissible stage. This may happen also in colonies consisting of slow-hygienic bees with lower olfactory sensitivity to diseased brood. Therefore, only in colonies with highly infected brood, the stimulus level is high enough that bees detect malformations and begin to remove decomposing brood particles. The dynamics of hygienic behavior and other group-level social activities are the main tasks of so-named colony "social immunity" (Cremer and Sixt 2009) performance. Hygienic behavior has a direct effect on reducing clinical symptoms of economically most important honey bee disease. These group-level dynamics

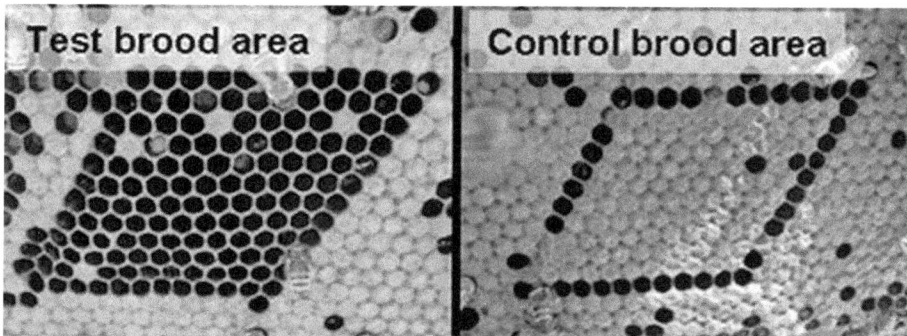

Figure 8.1: Test and control areas 24 hours after the perforation of cell cappings and killing pupae (left photo). Pin test after piercing 100 cells containing young pupae. Majority of cells are opened, pupae removed, and comb cells are completely cleaned out. Control brood area with solid cappings (right photo).

can thus be observed to be employed for selection in every beekeeping operation, and is therefore accessible for every beekeeper. We suppose that both mechanism grooming behavior and/or hygienic behavior needs to be properly incorporated into the breeding program.

Necrophoric behavior

The removal of dead adults from the nest is also named necrophoric behavior to represent collective behavior and favors colony health by reducing contact with potential pathogens. *Necrophory* is an effective and important activity of workers, because most studies have focused on the behavior in relation to genetic determinants of task specialization. Some workers removing dead bees or debris from the hive bottom may have an impact on the reduction of dead bees and pathogens. Beekeepers in their breeding program may also include the appearance of that behavior into their program.

Honey bees have developed natural defense systems that are relevant to their health in managed colonies. These defenses can be better enabled by both management and breeding decisions by the beekeeping industry. It is therefore important to recognize possible variations between honey bee populations or even between colonies in an apiary and to explore them for breeding objectives.

Evaluation of the defensive behavior

Defensive behavior has been demonstrated under genetic control and is a complex trait involving individual worker behavior and a coordinated colony response (Guzmán-Novoa and Page 1994). Common sources of attack stimulus for honey bees include alarm pheromone, vibrations, carbon dioxide, hair, and dark colors (Crane 1990). The beekeeper usually recognizes this phenomenon at both levels, when individual worker bee shows aggressiveness or defensive behavior as a response to external disturbance. Aggressive honey bee colonies should not be selected as queen mother or drone colonies for reproduction and mating. Defensive behavior is thus an important selection criterion in genetic improvement programs for all autochthonous honey bee races.

Special emphasis on aggressiveness is given in geographical regions with Africanized honey bees (*Apis mellifera* L.) (Sheppard et al. 1991). The venom of Africanized honey bee is not more painful or voluminous than venom from workers of pure race honey bee. Africanized honey bees are much more sensitive to alarm pheromone in comparison with temperate honey bees, and they produce a larger quantity of alarm pheromone than temperate honey bees (Pirk et al. 2011). The odors to coordinate colony defense are released from workers' sting glands and glands located in the head. The stinging response in Africanized honey bees is also more intensive and only a minor disturbance, such as a slight motion, vibration, or odor can induce a vigorous response. They normally respond as "group attacks" and follow moving intruders for up to a kilometer (Winston 1992). Africanized honey bees respond 2.4 times faster to alarm pheromones and about 30 times as fast to a moving target (Collins 1985). Prompt defense response is most likely a result of adaptation to life in tropical climates, where there is a higher rate of predator attacks on colonies. Temperate honey bees normally attack only to defend their colony, but will also attack if they are seriously disturbed outside the nest. It was found that high aggression is negatively correlated with ectoparasitic mite presence, and increased aggression is decoupled from negative health outcomes (Rittschof et al. 2015).

Any honey bee characters wanting to be selected in the genetic breeding program need to be quantified and evaluated. A reliable approach to such evaluation is detecting differences in defensive behavior between colonies at the apiary. It is even more important to evaluate defensive behavior between different bee strains or ecotypes using several measurements (Andere et al. 2000). For selecting less defensive honey bees, the behavior may be measured under controlled experimental conditions in the laboratory (Collins and Blum 1982). The metabolic test is employed to quantify defensive behavior based on the measurement of oxygen consumption in honey bees after stimulation with an alarm pheromone (Moritz 1985). This laboratory test allows us to perform a testing objective independent of any environmental influence. Aggressive colonies should not be selected for reproduction.

Impact of *Varroa* on colonies health and survival of honey bees

Varroa as the external parasite has physical and pathological effects on an individual bee or on the whole colony level. *Varroa* attack both adults and developing bees. Parasitized brood is injured with reduced larval protein content, and subsequently bee body weight is reduced and organ development is affected, and finally, worker or drone life is shortened (Bowen-Walker and Gunn 2001). Emerging bees may be deformed with missing legs or wings and together with DWV, microbes, and reduced immune competency, the survival of adult bees is significantly affected in untreated colonies. Highly infested colonies that are not examined for mites and effectively treated may die or contribute to increased winter mortality or queenlessness. *Varroa* mites are also a vector in transmitting a number of viruses from infected to healthy bees. Some viruses demonstrate typical clinical symptoms in contrast to other viruses that can be detected only using molecular diagnostic methods. Viruses associated with *Varroa* mites in colonies are: DWV, ABPV, Chronic Bee Paralysis Virus (CBPV), Slow Bee Paralysis Virus (SPV), Black Queen Cell Virus (BQCV), Kashmir Bee Virus (KBV), Sacbrood Virus (SBV) (Allen and Ball 1996, Martin 1998, 2001, Tentcheva et al. 2004, Martin et al. 2010). *Varroa* parasitization together with viruses as secondary infections influence weakened bees' immune systems and contribute to increased risk of colonies mortalities. It was also shown that highly pathogenic DWV and ABPV associated with *Varroa* in highly infested honey bee colonies contribute to winter colonies losses (Berthoud et al. 2010). Proper *Varroa* diagnosis and timing of the treatment of colonies together with sufficient efficacy is imperative for beekeepers to preserve their selected honey bee stock.

Good beekeeping practice in control diseases of honey bees

Honey bee diseases, their diagnosis, and control is an important part of good beekeeping practice. Regardless if beekeepers own one, hundreds, or thousands of honey bee colonies, they need to develop beekeeping management and also a bee disease management program. Beekeepers must learn and get a basic knowledge of bee diseases, their clinical symptoms, prevention, and control methods. To manage activities regarding bee diseases, they need to examine their colonies regularly. Whenever a beekeeper enters an apiary, he needs to check the activities of the colonies, including flight activity at the entrance of the hive. Spring flight activity includes observation of pollen gathering, the intensity of flying, and presence of dead or moribund bees at the entrance. We usually compare those activities between colonies at the same apiary. When outside temperature allows, we may inspect a few colonies, especially colonies that appear abnormal at the hive entrance. We may quickly observe the size of early spring cluster or evaluate a number of combs populated in the hive. In "backload" type hives, we may see dead bees on the hive bottom board. During colony examination, we may also sample dead bees or collect whole debris from the hive bottom. Collected material from hive bottom in early spring may be examined in the laboratory for the presence of *Varroa* or even spores that cause American foulbrood (*Paenibacillus larvae*). Robbing can also be detected at the hive entrance by observing the vigorous activity. A part of "good beekeeping practice" is also making notes for each apiary or even colony of findings at the hive entrance, checking for the appearance of dead bees, looking for immature stages of bees, chalky bee pupae, or other abnormal findings.

Feeding bees to keep colonies healthy

In the past decade, beekeepers have faced unprecedented honey bee colony losses, averaging 30% annually (Rennich et al. 2012). This decline is attributed to the combined effects of colony stressors, including colony collapse disorder, pesticide exposure, management practices, pests, and pathogens (Oldroyd 2007, Rennich et al. 2012). Specifically, *Nosema* sp. infection reduces a colony's adult honey bee population size, brood production, and honey yield (Botias et al. 2013). Beekeepers commonly supplement colonies with commercially available pollen and nectar substitutes. Pollen patties or high-fructose corn syrup are used to stimulate colony growth and to combat losses resulting from multiple stressors.

Adequate nutrition is integral to a properly functioning colony (Haydak 1970). Pollen increases brood fecundity, worker longevity, and honey yields (Mattila and Otis 2006). Furthermore, the degree of the increase in fitness is dependent on the nutritional quality and source of the pollen (Li et al. 2012, Pirk et al. 2009). Recent findings even suggest that the quality of the pollen diet may influence the strength of an individual honey bee's immune response (DeGrandi-Hoffman et al. 2010, Rinderer et al. 1974). Therefore, satisfying the nutritional needs of a colony is critically important for overall colony health and success. Pollen substitutes also vary in their influence on overall colony health (DeGrandi-Hoffman et al. 2008). There are numerous products available on the market, yet little is known about the contribution of these pollen substitutes to brood stimulation, colony growth, and pathogen resistance.

In addition to providing pollen substitutes, nectar supplementation is common in the beekeeping industry (LeBlanc et al. 2009, Ruiz-Matute et al. 2010). High-fructose corn syrup is a common nectar supplement in the United States. High temperatures are used to manufacture high-fructose corn syrup from cornstarch. As a result, a carbohydrate breakdown product, 5-Hydroxymethylfurfural (L'5-(hydroxymethyl) furan-2-carbaldehyde, HMF) is found in the final product.

Hydroxymethylfurfural in honey bee feeds

Honey is a highly concentrated mixture of mainly two dissolved sugars, fructose, and glucose, plus at least 22 other composite sugars (White and Doner 1980). Additionally, there are about 70 other compounds, including proteins, vitamins, minerals, organic acids, aromatic compounds, and various derivatives of chlorophyll (Lipp 1994). Many more honey components may remain undiscovered. Potentially fraudulent honey may be identified by detecting specific breakdown products (metabolites), such as hydroxymethylfurfural or 5-hydroxymethyl-2-furaldehyde ($C_6H_6O_3$) (HMF). HMF can be selectively produced from keto-hexose, notably from D-fructose (Shalumova and Tanski 2010) and other acidic media containing dissolved monosaccharides (Teixido et al. 2006). Normally, HMF is present in honey in trace amounts (Basumallick and Rohrer 2001). The rate of HMF formation in foods depends on environmental temperature, the type of sugar, pH, and the concentration of divalent cations in the medium (Lee and Nagy 1990). Excessive heating or inappropriate storage conditions can increase HMF levels. HMF is recognized as a marker of quality deterioration for a wide range of foods containing carbohydrates (Morales 2009). Inappropriate heat processing of honey affects honey fermentation and reduces honey quality (Tosi et al. 2001). In fresh honey, HMF can occur at concentrations as high as 15 mg HMF/kg honey, but it normally occurs at levels between 0.06–0.2 mg HMF/kg (Basumallick and Rohrer 2001).

The Codex Alimentarius of the World Health Organization (WHO) and the European Union (EU Directive 110/2001) have defined a maximum HMF quality level in heat-treated honey (40 mg HMF/kg). Above that level, honey quality begins to deteriorate. HMF concentration increases above 20°C. Temperatures inside a hive normally exceed 20°C (~28–30°C), and in summer can reach as high as 40°C or more, where the concentration of HMF can reach 10 mg/kg of honey (Ribeiro et al. 2012). That is approximately one-third of the HMF concentration known to be harmless to bees (30 mg/kg, Jachimowicz and El Sherbiny 1975). Although these summer levels of HMF are considered nontoxic to bees, a few studies actually confirm a safe level of HMF in honey bee colonies (Bailey 1966). Low concentrations of HMF < 10–15 mg/kg in honey does pose a little risk to honey bees, but toxic concentrations of HMF seems to induce lethal intestinal tract ulceration (Bailey 1966). About 150 mg HMF/kg of commercially acid-hydrolyzed inverted sugar syrup can cause 50% bee mortality within 16 days (Jachimowicz and El Sherbiny 1975). HMF concentration in inverted syrup for feeding bees may not exceed 20 mg/kg, as it is in most honey (Kammerer 1989).

There is no standard limit value of HMF in bee nourishment. Approximately 250 ppm HMF in honey bee diet is considered toxic (LeBlanc et al. 2009). High concentrations of HMF in honey stores could represent a factor in early deaths of bees and the extinction of bee colonies (van der Zee and Pisa 2010). It is therefore important to understand the potential adverse effects of high HMF doses on honey bees, especially when beekeeper use a variety of supplements for feeding of bees.

Examination of adult bees and brood

The colony size and its strength, together with the behavior of adult bees, is an important indication for the health status of colonies. Symptoms of the majority of honey bee diseases or even disorders induced by no-pathogenic agents (pesticides) are frequently similar. In any suspicious clinical disorder, we need to confirm causative agent by performing specific laboratory diagnostic procedures.

Colony strength evaluation is an important procedure that helps to establish a standard procedure for the clinical examination of colonies prior to collection of adult bees or brood samples. Colony status is normally established by estimating the adult honey bee population and, in some cases, the amount of brood in a hive. Normal colony strength can vary with time of the year and management practice by the beekeeper. When examining large numbers of colonies in different apiaries, consistent procedures and definitions for inspection need to be used in order to ensure consistency and repeatability of examination and evaluation. Important factors need to be considered during colonies examination—presence and quality of the queen, including queen age; the size of the worker population, indicated with numbers of combs covered with bees; the amount of brood. Queen activity within last 3 days is indicated by finding eggs in comb cells, and quality of brood is indicated by finding a solid, good-sized brood pattern. Examiner needs to be careful not to injure the queen during the inspection. The total area contains healthy brood in any stage of development, including eggs or larvae in open cells and capped brood. Each side of the frame with brood or without brood needs to be examined and evaluated. Healthy brood has a single egg in the bottom of the cell, glistening white larvae, and smooth capping. Brood should not be scattered. That is a typical symptom for infectious brood diseases and an indicator for other malformations (inbreeding, intensive nectar flow).

At the time of the examination of the colonies, we need to consider findings in several colonies at the same location in order to know better any possible differences in diseased or affected colonies by external influences. For clinical examination, all hives must be accessible.

A bee inspector or veterinarian generally performs clinical examination on colonies before transportation. Alternatively, adult bees or brood may also be sampled from the hives in the apiary that are randomly or systematically selected. When suspicious colonies for any disease is found, it needs to be examined very carefully. A bee inspector or veterinarian needs to use proper personal protective clothing, including protective wear, glows, gumboots, or plastic tie boots. There is also a need to use a clipboard with the data sheet for recording findings, toothpicks/matchsticks, jars/bags for collected samples, a marker to label samples, and a camera. An experienced bee inspector or veterinarian may perform a rapid clinical examination, which is not disruptive to the colony, and the examiner will not crush the queen. Dead colonies found during colonies examination need be recorded and any infectious disease needs to be excluded. Colonies found with symptoms of AFB must be marked, and the beekeeper or regulatory agency will make sure they are abated, as required by law.

When examining the colony for brood disease, we need to find combs with brood. If there is no brood present, find combs where brood was previously present, and some remaining brood can still be found in comb cells. If the colony is vital with fertile queen, uniform egg laying patterns from the center of the comb to the outer edges can be found. The same pattern is seen also for larvae and pupae. Brood capping is normally solid and uniformly brown and convex. A mixture of capped and uncapped brood gives a scattered appearance and may be a symptom of brood disease. Every beekeeper needs to recognize and distinguish between normal brood pattern and that of the diseased brood.

During the examination, some abnormalities may be observed in brood appearance and structure of brood in the comb. Comb cells with abnormal cappings need to be more carefully examined for changes in brood or their remains. Larvae and pupae could be decaying and that decaying material may be dry, seen as scales laying lengthwise on the bottom of comb cells. Most important contagious brood diseases are American (AFB) and European foulbrood (EFB). Some other diseases or non-contagious conditions may sometimes show features similar to AFB or EFB, and those symptoms are considered important for differential diagnosis. Therefore, differential diagnostic procedure can be used by professionals, concomitantly, or alternately with protocols, guidelines, or other diagnostic procedures to establish the correct diagnosis.

American foulbrood (AFB) is an infectious disease of honey bee larvae and pupae. The causative agent is spore-forming bacteria, *Paenibacillus larvae* specific to honey bees. It is highly contagious for young larvae and will weaken and, in most cases, kill a honey bee colony. Spores can be found on old combs from infected colonies on beekeeping equipment. Therefore, the destruction of the comb from diseased colonies and equipment is required to prevent the spread of AFB to neighboring colonies. Early detection of the disease is important to reduce spreading it to healthy bee colonies inside and between apiaries. This is especially important in queen rearing and other specialized apiaries.

Infection development. The spores of this microorganism are only visible under a high-powered microscope with a magnification of 1000. AFB spores enter per os into the larval gut, they swallow and within 24–48 hours, the spores germinate and develop into vegetative "rods". Vegetative bacteria invade the hemolymph and body tissues, killing the infected larva before pupation, usually immediately after the brood cell is capped (Djukic et al. 2014). When the vegetative rods spread the whole larvae, the decaying larval material may have a mucus-like consistency, detected by "ropey test". The dead, decaying larva will adhere to the tip of the stick, the adhesive material stretching for up to 2.5 cm before breaking and snapping back (Shimanuki and Knox 2000). The consistency of the decaying brood at the first stage is soft and in a few weeks, the dead brood becomes dry and forms scale. At that stage, "ropey test" cannot be used, and the vegetative bacteria transform into spores. In the remains of a single infected and decaying honey bee larva, approximately 2,500 million highly resistant and contagious spores may be found.

The caps of brood comb cells with dead older sealed larvae or young that are upright in the cells are usually darker than the caps of the healthy brood. Diseased larvae have concave capping that are often punctured. During the disease development within the colony, a scattered brood with an irregular pattern of sealed and unsealed brood comb cells. Considering disease development in the colony, it is important that the bees that handle diseased brood are on average, 15–18 days old. Therefore, they are older than typical nurse bees feeding young brood, so the likelihood of the same bees returning to feed larvae and potentially transmitting spores into larval food is reduced (Arathi et al. 2000). In the colony, effects of the preventive mechanism of temporal polyethism may help reduce the movement of pathogens from carriers, adult hive bees, to the most susceptible larvae in comb cells.

Clinical symptoms typical for AFB can be easily distinguished from the normal, solid pattern of healthy brood cells observed in brood combs (Shimanuki and Knox 2000). Early detection of AFB is important in order to proceed to proper diagnosis and destruction of infected colonies and disinfected contaminated equipment.

Diagnosis. After finding typical clinical symptoms of AFB, the pathogen bacteria may be identified using a microscopic examination of dead larvae smear or by cultivation on selective culture media. Final confirmation may be performed by applying biochemical, serological tests, or performing the Polymerase Chain Reaction (PCR). PCR is very sensitive, and specific result for pathogen confirmation can be obtained (OIE Manual of Diagnostics 2004). There are also commercial "AFB diagnosing kits" available, based on serological evidence of the pathogen agent, convenient for use in the field also for beekeepers or bee inspectors. In clinically indifferent cases where the scattered brood is observed, colonies are weakened and other non-specific symptoms are present, misinterpretations may occur.

Control. Effective beekeeper inspection services and their help to beekeepers in examination of colonies is very important in finding potential outbreaks. Early diagnosis of AFB contributes to minimizing economic damage to apiaries. Diseased honey bee colonies are killed and hive materials belonging to the colonies are disinfected or destroyed by burning. The bees are usually killed by poisonous gas, such as the burning of sulfur powder. In some countries, diseased colonies are instantly killed by pouring a small amount of gasoline. Dead bees with combs, the honey, old hives, and the contaminated equipment are the main carriers of spores and should be destroyed by burning. Combs, without brood, can be preserved if an examination of wax samples in the laboratory does not reveal *Paenibacillus* spores. Well-preserved hives can be sterilized by briefly scorching with a torch. Plastic parts should be cleaned and brushed with 3 to 5 percent sodium hydroxide. During spring season, strong colonies are getting diseased, and the killing of the bees can be avoided. In that case, treatment may include performing artificial swarm by shaking bees into a decontaminated hive with new combs. After establishing artificial swarms, the bee entrance is closed and they are placed for two days in a dark quite cool room and then exposed on a bee yard and fed. In some countries, beekeepers who destroy their AFB-infected colonies receive compensation, either directly from the government or from beekeepers' organizations, or specialized insurance agencies.

Early destruction of colonies is better than having AFB spread so that a large numbers of colonies must be destroyed. It is suggested for beekeepers selling honey bees or used beekeeping equipment to have a legal permit. Through the process of third-party validation, it can be demonstrated that the material has been inspected, and determined healthy. In some geographical areas, beekeeping operations are in close proximity to each other and therefore honey bee colonies are exposed to diseases at other colonies within a 3 to 8 km radius. It is important that beekeepers in newly established AFB outbreaks take immediate steps to examine honey bee colonies, diagnose the disease, and eliminate AFB. Where the AFB is found, restricting the movement of potentially infected material is important.

Using hygienic bees that detect and remove diseased brood from the colony is an alternative to AFB disease treatment (Spivak and Downey 1998). In order to breed hygienic strains of honey bees in the breeding program, a test for selection of hygienic bees needs to be included.

Extension activities and prevention. Beekeepers need to be educated and be familiar with the clinical symptoms of AFB as the disease is notifiable. Furthermore, beekeepers need to monitor the health of their colonies regularly by inspection of the brood nest during the brood season. When there is no brood in continental climatic conditions, during winter, old brood combs in a storeroom need to be examined. Specifically, bee inspectors need to instruct beekeepers to examine colonies with brood in early spring, during the main honey-producing season, and in autumn when hives are prepared for winter. They also need to help beekeepers during the season when specific operations are performed, such as re-queening, establishing new nucleus colonies. Special attention concerning examination of colonies should be performed in queen rearing operation. In these operations they manage nurse colonies, queen bank colonies, and queen mother or drone mother colonies.

In general, there is no cure for AFB. Beekeepers need to take steps to prevent an infection from establishing itself in a beekeeping operation. During suspicious colonies examination, it is essential that adequate protective clothing, including a bee veil, is worn, and techniques for safe handling of bees are understood before opening hives and collecting samples. Preventive beekeeping practices need to be introduced in every operation. This includes regular sterilization of beekeeping equipment during colonies examination and when moving between bee yards. Hive tools should be scraped free of wax and propolis and disinfected by heating at high temperatures. If a colony shows symptoms of AFB, the hive tools should be heat sterilized before being used on another colony. It is advised to use disposable gloves; otherwise, gloves need to be scrubbed with soapy water. Soap will not destroy spores on gloves or hands, but it helps remove spores from gloves or hands by vigorous hand washing. There is also a need to clean and disinfect other tools, such as a smoker, beekeeping brush. In AFB affected apiary, the beekeeper needs to prevent swarming, as swarms may be contaminated with AFB spores, and avoid management activities that encourage robbing behavior in bee yards.

European foulbrood (EFB). The appearance of European foulbrood disease is not confined to Europe alone, as the disease is found in all continents where *Apis mellifera* colonies are kept. It is important to know that *A. cerana* colonies are also subject to EFB infection in India. EFB is generally considered less virulent than AFB, although higher incidence and losses in commercial colonies have been recorded in some European countries, including the UK, France, Switzerland. Incidences of the disease are correlated with climatic conditions and nutritional stress factors in the colony. EFB is highly contagious but infection may remain without visible signs for a long period. Cooler wet weather and poor nutrition will promote the incidence of this disease. EFB is characterized by dead and dying larvae, which can appear curled upwards, brown or yellow, melted, and rubbery. Sudden outbreaks of the disease usually appear in early spring. It can be a result of a change of seasonal conditions and other stress-related factors, such as nutritional deficiencies, shifting the bees, domination of the colony by older workers.

The causative agent is Gram-positive, not spores-forming bacteria, *Melissococcus plutonius*. Honey bee larvae are infected per os and in the larval gut, the bacterium competes for food. After infection, the larva dies before the cell is capped. Some infected larvae may survive and develop into an adult bee. Infected larvae die when they are four to five days old.

Diagnosis. On the basis of the clinical signs, diagnosis is not always reliable, because AFB and a number of non-disease conditions and viral diseases have similar symptoms: scattered pattern of sealed and unsealed brood, colony weakening. Laboratory examination is obvious in order to confirm EFB diagnosis, by carrying out a microbiological test using selective culture media (OIE Manual of Diagnostics 2016, Bailey and Ball 1991). Biochemical tests or the PCR test can be applied in order to confirm the causative agent.

Control. In a weak infection, hygiene behavior is often sufficient to overcome the disease. A young vigorous queen will always do better than an older queen will, and using selected disease-resistant breeding stock to increase resistance to the disease will contribute to disease control. Maintaining hive hygiene and regular replacement of brood nest combs will contribute to reducing the concentration of the causative agent over the brood area. Requeening on a regular base can strengthen the colony by giving it a better, reproductive queen, thus increasing number of workers to remove infected larvae from the hive.

The chalkbrood disease. Chalkbrood, a fungal disease, causes serious damage to beekeeping in temperate America, Europe, Japan. However, in Asia, it is rarely considered to be a serious honey bee disease. Chalkbrood is caused by the fungus *Ascosphaera Apis*, spores that form during sexual reproduction. Infection by spores of the fungus is usually observed in larvae that are three to four days old. Infection occurs either per os or through the body surface, percutaneous. The disease is usually most prevalent during the spring with intensive fungal growth, and cool and humid, poorly ventilated beehives (Gilliam et al. 1978). Besides environmental conditions, the appearance of the disease is also due to biotic factors, such as differences in fungal strains and the genetic background of the bees. Various strains of *Ascosphaera Apis* as a causative agent show a difference in the level of virulence (Glinski 1982). Genetic background of bees is a very important factor affecting the incidence and severity of the disease. There are also other factors, including general colony health status, and stress conditions affecting the colony from the environment. During the selection process, it is imperative to improve resistance to infectious diseases in a variety of honey bee stocks (Gilliam et al. 1983, Spivak and Downey 1998). Chalkbrood appears when pre-existing stressors, both biotic and abiotic, are present and affect honey bee colony.

The chalkbrood disease is easy to identify by finding typical mummies in comb cells at the hive bottom board or at the hive entrance. Infected brood can be easily removed from the comb cells by tapping the comb. We have many preventive biotechnical measures to prevent disease outbreak. Most important measures are to maintain strong, vital colonies, and to use stocks that show resistance to brood diseases. It is therefore important that in every beekeeping operation, beekeepers follow the incidence of chalkbrood and select colonies for breeding and queen rearing that do not show clinical symptoms of chalkbrood.

Nosemosis, Paralysis, and Septicemia are the primary diseases of adult bees. Nosemosis is supposed to be a disease that affects wintering colonies, causing serious damage in continental climatic regions with the long winter season. The problem of poor honey bee health and mortality of colonies is caused by microsporidia, *Nosema cerana*. The pathogen spread from its original host, the Asian honey bee (*Apis cerana*), to the European honey bee, *A. mellifera*. *N. ceranae* induce several effects on individual bees or at a colony level. Most often, effects are demonstrated in reduced bee health. The immune system is affected by energetic stress, and finally resulted in increased mortality of colonies (Antúnez et al. 2009, Mayack and Naug 2009, Seitz et al. 2015).

In *Nosema*-infected colonies, the normal bee polyethism is changed in terms of the changed progression of bee tasks during the first 21 days of development (Lecocq et al. 2016). *Nosema cerana* infection induces that infected nurse workers are prematurely shifted to foraging activities (Wang and Moeller 1970). Healthy adults typically begin foraging when they are 22 days old. Nurse-aged bees *Nosema*-infected are twice as likely to engage in precocious foraging than non-infected bees (Goblirsch et al. 2013). Newly emerged adult bees that are infected with *N. ceranae* spores and observed for 14 days exhibited traits of older bees. This behavior may be an adaptive mechanism for the colony and limits the spread of infection inside the hive (Evans and Spivak 2010).

Factors affecting resistance to *Nosema* infection include the total number of bees in the colony and the size of the hindgut of individual bees. From beekeeping practice, it is supposed that Italian bees tend to be slightly susceptible to *Nosema* and resistant to paralysis and septicemia. Brother Adam in his breeding activities indicates that he has found no obvious resistance to *Nosema*, except possibly in the Egyptian bee (*Apis mellifera lamarckii*). Caucasians tend to be very susceptible to *Nosema*, though selected strains exhibit some resistance. Several researchers have noted that the eastern honey bee (*Apis cerana*) seems to be almost immune to *Nosema*.

According to the breeding programs in place in some countries, a sample of worker bees from "breeder" colonies is collected and examined for the presence of *Nosema* spp. spores microscopically. The number of *Nosema* spores, per examined bee, is then calculated, and colonies with the lowest spore load are then selected for further queen rearing. Colony productivity and brood production depend on the quality of the queen and its age (Tarpy et al. 2000).

Tracheal mites (*Acarapis woodi*) usually cause a moderate level of infestation and result in an infested colony with poor wintering ability. If more than about 30 percent of the workers are infested going into winter, the colony will probably die. Resistance appears to be based on behavioral and anatomical differences. Bees with the highest level of resistance are currently from England, where bee populations were decimated in the early 1920s. As the highly susceptible bees were killed, only the resistant colonies survived. The net result is that bees of English origin have a high level of genetic tolerance to tracheal mites. The typical pattern seen when a colony dies from tracheal mites is a colony with a handful of dead bees and almost all the honey stored for wintering still in the hive. During flying days in winter, bees flew out and died. A typical clinical sign of *Acarapis woodi* infestation developing during the winter period is plenty of bees on the ground in front of a hive with bees crawling slowly away, wings disjointed. After losing most of the infested adult bees, the few remaining start rearing brood in the forthcoming spring. Diseased colonies are also exposed to starvation during early springtime, and therefore have a minimal chance to survive.

Resistance to brood diseases

Resistance to brood diseases has been found for the following—American foulbrood, European foulbrood, Sacbrood, and Chalkbrood. There are several other brood diseases caused by viral, bacterial, and fungal agents, but none has as much effect as the first four. As expected based on their parasite and pathogen pressures, honey bees have developed both individual and group mechanisms to combat the disease. Grooming, nest hygiene, and other behavioral traits can reduce the impacts of pathogenic bacteria, fungi, and parasitic mites. For example, "hygienic behavior" first described for honey bees (Rothenbuhler 1964) is now a classic example of a social defense, whereby workers identify and remove infected larvae from among the healthy brood (Spivak and Reuter 2001). Resistance seems to center around hive cleanliness and brood nutrition, with emphasis on hygienic behavior, which has a tendency to uncap and remove diseased brood. Carniolan honey bees have a high average level of resistance to brood diseases. Italians, on the other hand, show resistance to varying degrees, and respond readily to genetic selection.

During research and performing breeding activities in queen rearing apiaries, strains of honey bees that exhibit some tolerance to AFB through genetically inherited traits have been identified. These traits are not sufficient to manage an AFB outbreak, and do not offer sufficient protection for AFB outbreak when spores infect a critical amount of brood in a colony. Selection procedures in breeding stations consider characteristics mainly to improve resistance by introducing better hygienic behavior.

Behavioral resistance of honey bees to *Varroa destructor* is very important, and includes grooming and hygienic behavior. Both systems are very important because they also act as a mechanism against other diseases of the honey bee colony. We will discuss specific characteristics important for *Varroa* research and developing breeding activities for any pure honey bee subspecies or race. Brood attraction, capped period, reproduction, and infertility of *Varroa* are important selection options to develop resistance to *V. destructor* infestation.

Brood attraction and duration of capped brood

Varroa mites enter into the drone brood rather than into the worker brood, and thus the reproductive success in drone cells is much higher than in worker brood (Martin 1995, Donze et al. 1996). Drone brood is approximately 12 times more invaded by *Varroa* than worker brood (Boot et al. 1995a). Higher attractiveness of drone brood can partly be explained by a 2–3 times longer development period, and by the larger surface of drone cells compared with worker cells. It is expected that drone brood encounter 3.4–5.1 times higher infestation rate than workers brood (Boot et al. 1995b). Drone larvae secrete longer and stronger emission of the kairomone signals in comparison with worker larvae, and induce preferences of *Varroa* to drone brood (Trouiller et al. 1992). Other factors that attract *Varroa* to worker brood are also indicated—cuticular hydrocarbons of eight-day-old worker larvae (Rickli et al. 1994) and broods from different stock origins show different levels of attractiveness to the *Varroa* (Guzmán-Novoa et al. 1996). Reproductive success of the *Varroa* in the drone brood is thus much higher than in worker brood, and is probably result of a natural selection process for the higher preference of mites to invade drone brood (Fuchs 1992). It is also evident that high mite infestation levels or a small amount of drone brood in the colony results in a relatively high number of mites per drone brood cell, indicated as high drone brood infestation. It is, therefore, a relevant diagnostic or biotechnical *Varroa* control method to apply drone-trapping comb. Those methods can be applied in a beekeeping operation in order to monitor relative infestation of colonies. Prior to selection, mother queen or drone mother colonies need to be tested to ensure that healthy drones and queens are ready for mating.

A shorter period of the capped brood has an effect on reduced reproduction of the *Varroa* (Büchler and Drescher 1990). *Varroa* reproduces only in capped brood, and immature mites in brood die at the time of emerging bees. It is estimated that the reduction of the duration capped period by 1 hour results in an 8.7% reduction of the colony infestation during an observation time of 18 months. The shorter capped period can be partly explained in a lower reproduction rate and successive lower infestation of Africanized bees in comparison with European bees (Camazine 1986). It was also shown that shorter brood capped period of selected colonies does not affect honey production (Siuda et al. 1996). The duration of the capped period is also influenced by the behavior toward sealed brood cells, and not by the pre-capping care the brood receives during larval development (Bienefeld 1996). Taken together, the capped period duration has a genetic basis, and therefore there is a possibility of selection of successful honey bees with a shorter capped period. Selection of drones rather than queens would be more effective due to higher variability in the capped period of drones than queens (Moritz and Jordan 1992). Described selection approach can be introduced into breeding and queen rearing program.

Reproduction and infertility of *Varroa*

Varroa mites reproduce more in drone brood than in worker brood, because some physiological factors of non-reproductive mites are observed only in worker cells (Martin and Cook 1996). A female *Varroa* has on average 1.5–2 reproductive cycles in its life (Fries and Rosenkranz 1996), with a range of 0–7 cycles (De Ruijter 1987). It was found that *Varroa* ovipositioning after entering into brood cell might be stimulated by prior feeding on adult bees or the bee larvae, respectively, and shorter feeding period has a slightly reduced fertility potency in female mites (Beetsma and Zonnenveld 1992, Rosenkranz 1990). Multiply-infested cells also have an impact on reduced *Varroa* reproduction as a result of the existence of chemical factors in female *Varroa* and subsequently, the number of daughters per mite decreases in multiply-infested cells (Eguaras et al. 1994, Nazzi and Milani 1996). It is also evident that *Varroa* has lower reproduction potential in tropical Brazil. That bee expressed approximately two times greater proportion of non-reproductive mites in comparison with honey bees in a temperate climate in Europe (Ritter and De Jong 1984, Moretto et al. 1991).

It is evident that the association between *Varroa* and *A. mellifera* in some regions had a potential biological ability that specific honey bee stocks have developed. Over a long period of association of a host and its pest with the help of performing breeding activities, there is an opportunity for the development of resistance mechanisms of the host to its pest. There is some evidence in beekeeping operations indicating variability between honey bee colonies in resistance of honey bees to *Varroa*. In addition to resistance traits of individual larvae, *Varroa* Sensitive Hygienic behavior (VSH) of workers has been demonstrated as a useful indicator for developing resistance of honey bees to *Varroa* mites in breeding stocks. This behavior can be included in breeding activities, and selected honey bees can detect infested capped brood and destroy *Varroa*. VSH bees have been developed and are now successfully used in beekeeping operations in the U.S. and are available for individual beekeepers (Harbo and Harris 1999, Ward et al. 2008).

Breeding better bees

Selection model in beekeeping includes numerous parameters, such as the initial population of mites, duration of brood developmental, and the capped period of workers and drones. *Varroa* phoretic period, *Varroa* preference to drone brood, *Varroa* infertility level, the number of *Varroa* reproductive cycles, and winter mortality of *Varroa* mites are also considered for selection (Fries et al. 1994). Beekeepers or breeding organization can also incorporate additional variables associated with *Varroa* parasitism, such as monitoring mortality of bees, natural *Varroa* mortality, and *Varroa* population growth in the host colony into their breeding program.

Considering a variety of selective characteristics, at least three breeding programs within North America have been performed in order to produce *Varroa* resistance stocks, incorporating a variety of imported honey bee races also (Ibrahim et al. 2007). In European beekeeping conditions, the rich diversity of natural honey bee races (subspecies) and local varieties (ecotypes) offers enormous genetic resources for selection of honey bees with *Varroa* resistance. Nowadays, the most important autochthonous subspecies for selection programs are employed—*A. m. carnica*, *A. m. ligustica*, *A. m. macedonica*, and others widely spread throughout Europe. Furthermore, in addition to selection on traits affecting the reproductive rate of *Varroa*, which is considered the most important trait, selection on grooming and hygienic behavior would contribute to a reduction in *Varroa* population. Furthermore, *Varroa* resistant stock need to be continuously maintained with sufficient gene pool, and controlled matings in queen propagation are required for achieving progress in developing better honey bee colonies. These honey bee colonies normally require fewer acaricides treatments for *Varroa* control, and they demonstrate longer survival without *Varroa* control, and simultaneously they retain the commercial qualities desired by beekeepers. It is also evident from the point of view of beekeeping that slow mite population growth during rearing season is the ultimate indicator for developing *Varroa* resistant stock. This very important characteristic may be used in the selection process on mite resistance in circumstances when behavioral and physiological mechanisms are not considered or stay unknown. Diagnostic methods to establish the level of infestations in colonies are well developed, simple, and standardized, and normally easy to use for beekeepers. It is therefore important to incorporate *Varroa* diagnostic methods into beekeeping practice and to incorporate them into testing of large-scale colonies for desired selection programs for progress. Performance tests and calculating breeding values for honey production, swarming behavior, gentle temper, and strength of colonies are currently the main criteria considered in a variety of breeding programs in European countries. Developing and propagation of honey bee colonies resistant to *Varroa* is also getting attention in breeding activities for individual beekeepers or for the breeding organization. This is an important aspect in reducing chemical treatments applications against *Varroa* and potential accumulation of residues in honey bee products and developing Varroa mites resistant to acaricides. It can be concluded that the disease resistance of honey bee colonies, the survival of the colonies in local climatic conditions, and adaptation to local environmental and management conditions are not enough to be included in current breeding programs.

When applying a breeding program to develop *Varroa* resistant stock, we need to consider secondary viral infections. Acute paralysis virus (APV) and deformed wing virus (DWV) associated with *Varroa* have become more pathogenic to honey bee colonies (Sumpter and Martin 2004). Every honey bee colony is also continuously preyed on by hornets, wasps, and members of closely-related genera, such as spiders. All these predators prey on honey bees by waiting near the hive entrance and grabbing a bee on its way in or out of the hive. Colonies that aggressively guard and defend the hive will be resistant, but tend to sting beekeepers more often. Guard bees normally stand near the hive entrance, and challenge intruders and soldier bees, and tend to sting more than younger house bees. It is supposed that only resistant colonies show strong hive defensive behavior expressed by guard and soldier bees, and that they also fly outside to forage and collect nectar more often. Soldier bees are therefore also responsible for gathering more honey. The large variation in the percentage of soldier bees in different colonies allows us to differentiate colonies in this characteristic, and give an option for selection. A higher percentage of soldier bees in some colonies is an indicator for production of a larger amount of honey, and therefore may be an important selection criteria. Therefore, a breeding program that includes selection to increase the percentage and quantity of active foragers is also commercially important.

In many countries, individual beekeepers initiated selection and breeding activities in their apiaries using their own stock as a basic preselection on a large population, and breeding queens for their own use. Governmental or non-governmental institutions often support beekeepers' activities, and they establish specialized bee-breeding organizations for coordination breeding and queen rearing activities. They also tend to preserve pure honey bee races and to stimulate commercial queen rearing activities. Different selection tools have been incorporated into beekeeping practice to achieve breeding objectives of a certain honey bee population.

Basic selection and raising queen honey bees

In our beekeeping operations, it is often found that some honey bee colonies produce more honey than others, some colonies are more gentile, and some colonies are calmer sitting on a brood comb than others. After performing specific tests, it is possible to find that some colonies are more hygienic, have a good laying pattern, and are even better producers of honey. It is also observed that some colonies are even better in surviving long winters in continental climatic conditions than other colonies of the same apiary. These differences between colonies are observed in an apiary at one location with the same conditions and types of flora. Often these production differences are a result of variations in the strain of bee and the quality of the queen in the individual colonies, and have a genetic background. Observations about the performance of colonies conducted by individual beekeeper are very important in order to perform basic pre-selection program, and for queen rearing as a next step. In that way, every beekeeper can contribute to his/her own gene pool variation, and therefore in the whole beekeeping sector, there is sufficient variability in the honey bee population that can be preserved.

The individual beekeeper may have different reasons to decide for selection in honey bee apiary and objective for queen rearing. Beekeepers are firstly, by queen rearing, self-sustainable by making up for winter losses and to increase hive number in an apiary. Genetic diversity within the apiary is a challenge for the beekeeper and diversity in a certain geographic area is a challenge for the beekeeping sector or breeding organization. There is also an economic interest of beekeepers in massive rearing quality queens for business. Therefore, in order to improve the stock quality, beekeepers can change the strain (or type) of a bee in colonies for selection, by removing the queen and replacing her with a queen of the desired strain. This procedure is known as requeening a colony. Requeening is also used to replace queens that are old, or have reduced egg-laying capacity.

Every beekeeper can get laying queens from a queen breeder, or they may be reared by any beekeeper who has a good understanding of bee behavior, bee handling, and beekeeping. There are some practical tips for beekeepers and queen breeders about diseases and mite resistance. Hygienic behavior test is modest to conduct using honey bee colonies in the selective apiary and is helpful to avoid AFB, EFB, chalkbrood, and other brood diseases, as well as *Varroa* infestation. Testing for hygienic behavior in our breeder queens is the ultimate approach to rear quality queens. It is also supposed that tracheal mite or *Varroa* resistance development is an option to improve the quality of honey bee queens. Reducing the amount of *Varroa* control treatments, and subsequently breeding queens, would be a good approach to develop resistance in the bee population. Therefore, every beekeeper can produce even better queens than a queen breeder. Crucial aspects are conducting selection processes and following proper queen rearing technology. There are several steps that need to be considered—selecting queen mother and drone mother colonies, grafting larvae, organizing mating station, deciding the type of mating nuclei used in rearing, and deciding the duration of allowing the mated queen to lay eggs in mating nuclei. A commercial queen producer, for instance, often uses "baby nuclei", where the queen can lay eggs only a few days after mating. There are several options to improve the queen-rearing technique, as small-scale queen breeders can spend more time in rearing quality queens and simultaneously testing their performance.

"Bee breeding" is, therefore, a process of selecting queen mother and drone mother colonies with desired characteristics that are used for queen propagation by "queen rearing" techniques. Bee breeding is a continuous process in every modern beekeeping operation, with the aim to produce their own quality queens.

Queen rearing techniques

Under three different colonies conditions, honey bees may raise natural queen cells. Swarm cells are built when the colony is preparing to swarm, a natural colony reproduction. Populous colonies normally develop several swarm queen cells simultaneously during springtime. The majority of swarm queen cells can be found at the bottom of combs. When colonies are preparing to swarm, eggs are laid in queen cell cups built by the workers. The situation is similar in the case of supersedure, where the colony raises only a small number of queens to replace an old or deteriorating queen. Queens raised under swarm or supersedure circumstances are usually well developed, and emerged queens have good egg laying potential and are considered high quality. As the larvae are destined to become queens, they are fed the appropriate diet for queen larvae from the time of hatching. Swarm cells look similar to supersedure cells, but there are usually more of them. Supersedure queen cells appear where the colony is trying to raise a new queen. Supersedure can occur in spring and summer. The current queen is still alive but is getting lees reproductive, old, or damaged. In some way, it is no longer up to the task of laying a large numbers of eggs. It occasionally happens that the colony builds up supersedure cells soon or early next spring after a newly mated queen has been obtained from commercial queen producer. The cutting out of these cells will usually result in the colony settling down. If the colony is strong, it would be possible to produce another queen from that supersedure queen cell. There are usually 1–3 supersedure queen cells. The supersedure condition in a colony can be triggered within the hive by the beekeeper in order to raise large numbers of quality queens in well-controlled conditions.

When the queen is missing or dead, emergency cells are built on existing worker larvae in the comb cell. Some worker cells containing larvae are modified to become queen cells. These larvae may have been fed worker royal jelly for some time and the queen raised may have a reduced egg laying capacity. Obtaining queens from an emergency queen cell is usually not a priority of beekeepers.

Artificial queen rearing

Queens are raised from the same diploid, fertilized eggs as are workers bees. Newly hatched, a female larva is either queen or worker caste. The food that newly hatched larvae receive is royal jelly produced by nurse bees. During the first three days of larval development, there are small differences in the composition of royal jelly. The larva grows rapidly, apparently developing precursors of both castes. On the third day after hatching, a major difference in diet occurs when pollen is included in the diet of female larvae destined to become a worker. It is therefore a short period of three days to develop worker or queen bee, depending on the diet quality. After this, the caste formation is only partly reversible, resulting in intercastes with mixtures of phenotypes of both castes. When grafting three-day old larvae, some phenotypic or physically unrecognizable responses may occur in developing intercastes. It is therefore judicious to rear queens from the youngest larvae.

For quality queen rearing, it is important to graft larvae under the age of 24 hours. Due to grafting at this age, the queen larva will be not exposed to the worker diet. When greater numbers of queens are needed and when it is necessary to control characteristics of colonies in the apiary by a selection of stock, it is essential to rear queens by grafting larvae from selected colonies. This method used for queen rearing is the most efficient of all methods of controlled queen rearing.

Queen rearing practice

An abundant supply of nectar and good quality pollens are needed for rearing quality, reproductive queens. High-quality drones from selected drone mother colonies for mating with the newly emerged virgin queens are essential at the mating station. Strong and well-established starter and cell raising colonies ensure queen cell development. A queen mother colony is selected for obtaining larva for grafting. Queen and drone mother colonies need to demonstrate ideal morphological, ethological, and economic characteristics. More specifically, selective criteria include race characteristics, gentle temperament, disease resistance, low swarming tendency, and excellent honey production.

For contemporary queen rearing for massive production or the individual beekeeper's own use, queens need to be well organized and well prepared. Before starting rearing procedures, a starter colony for the initial stage of raising queen cells, and cell building colony for the development of queen cells need to be established. Queen rearing starts with grafting honey bee larvae, and later there is a transfer of the mature queen cells to honey bee nucleus colonies for the mating stage. Grafting requires the use of queen cell cups, made of beeswax or plastic, into which young larvae are transferred. Young larvae should be supplied with plenty of royal jelly in their worker comb cells in the carefully selected queen mother colony. Grafting can be performed quickly and without time-consuming searching for suitable larvae by confining breeder queen to a single usable comb inserted into a cage made with queen excluder walls. The next day, a comb with eggs ±24 hours old is transferred to the feeding compartment. Three days later, larvae suitable for grafting are developed. Grafted larvae into queen cup cells are transferred into the starter colony.

Starter colony

It can be made up by dequeening a colony or by establishing a new strong colony with brood, bees, pollen, and honey. They need not occupy a standard hive. It can be established in five or seven-frame nucleus. For starters, the queenless colony must be strong with nurse bees and have abundant supplies of honey and pollen. An empty box with bottom board and the lid are populated with four combs of unsealed brood with adult bees and a comb of unsealed honey, and pollen with bees is positioned on each side of the brood. Empty combs may be placed in the box to fill the gaps. Queenless cell builder colony are maintained by weekly additions of sealed and emerging bees, or by shaken bees from several colonies. A good practice is to give supplementary feed to the starter colony routinely. Queenless colonies accept inserted larvae and start queen cells more rapidly than a queenright starter colony would. The state of queenlessness will stimulate the nurse bees in the starter colony to feed and produce more brood food for young queen larvae. A graft of 15 to 30 queen cup cells is inserted in a gap prepared in advance between the youngest larvae and the pollen comb in starter colony.

The cell builder colony

For cell builder colony, a strong colony fully occupying at least double-story hive should be selected. It must have a young vigorous queen, and its population should be maintained by adding sealed or emerging brood, or by bees shaken from brood nests of other colonies. Queen excluder between two hive chambers serves to confine the queen to the bottom box. This brood chamber should contain a balanced amount of brood and empty drawn comb for the queen to lay eggs. Two combs of very young larvae should be placed in the center of the super. There should be a gap of one comb frame between the two brood combs in the super just wide enough to take the comb frame with accepted queen cells (or just grafted larvae), that will be inserted later, approximately 24 hours after establishing the cell builder colony.

The remaining space in the upper hive compartment is filled with combs of honey and pollen. Combs of unsealed honey and pollen are placed alongside the frames of unsealed larvae. This will simulate a natural brood nest. Nurse bees will be recruited to feed the unsealed larvae in the super. The bees in the super will recognize that the new brood nest is not being occupied by their queen, and begin to respond to the natural supersedure instinct. This condition with the presence of freshly started cells (or grafted larvae) transferred into the cell builder colony will induce acceptance and start larvae rearing instantly. A light nectar flow and abundant pollen provide the best field conditions for rearing queen cells. When supplementary feeding of the cell building colony becomes necessary, the sugar syrup offered will simulate needed nectar. The larvae in queen cells are raised above the queen excluder into the upper body.

The "ripe" queen cells are taken from cell builder colony on the tenth day following grafting. Bar with queen cells and plenty of bees on them need to be gently brushed and cells distributed to hives or mating nuclei for mating. Queen cells may be inserted into an incubator with controlled humidity and temperature for virgin queens emerging. If they are kept longer in the cell builder colony, also called a finisher, any queen emerging early will destroy all other queen cells with queen pupae just before emerging. Queen pupae in queen cells are very sensitive to shaking or vigorously brushed bees from them, and therefore manipulation with queen cells need to be very careful.

Grafting in beekeeping

Grafting is the process of transferring young larvae form queen mother colony into prepared queen cup. It is an important task in queen rearing procedure. These young larvae are transparent, almost a little larger than an egg. Grafting needs to be conducted in a suitable place with the necessary ambient temperature of approximately 25°C, and humidity of approximately 50% RH, and suitable lighting. Light, temperature, and humidity must be adjusted to offer the best possible conditions, which ensure optimal larvae survival during grafting. Grafting may, therefore, be done indoors under suitable artificial controlled conditions, and in the open, if atmospheric conditions are suitable, without exposure of larvae to direct sunlight, and the work is performed as quickly as possible.

The type of grafting tool for larvae transfer has no special impact on larvae survival rate. It is important that larvae be picked up without damage. The grafting tool needs to follow the curve of the bottom of the cell and has the advantage of taking a little royal jelly with the larvae. Grafting tool with its apex needs to be inserted under the back of the tiny floating larvae without touching it.

A variety of grafting tools can be used by individual small scale beekeepers. A shaped matchstick or small paintbrush is satisfactory, as the larvae stick well to the brush. A sharpened quill is still used by some queen breeders, or the commercial "Chinese" grafting tool made from goose quill is very popular, and is used by many queen breeders nowadays. Grafting equipment should be cleaned and disinfected regularly in order to avoid any risk of biological contamination.

When everything is ready to graft, remove the frame of larvae from the queen mother colony and carefully brush the bees from it and carry it to the grafting location. The frame with larvae should rest on adjustable support in a position where the light falls into the cells, avoiding direct sunlight. It is very important that both hands are free to hold the bar with cell cups with the index finger pointing at the cell cup to be fixed.

Before grafting, a drop of a mixture of royal jelly and water may be used to wet the queen cup just before the young larva is placed in them. This is important to prevent larvae from drying out during the procedure of grafting and transferring grafted material into a starter colony. It is important to transfer grafted bars of cells into the starter colony as quickly as possible so that nurse bees start feeding the larvae. When all the procedures of grafting are short enough, the acceptance rate of larvae may be over 95 percent.

A queenless starter colony has plenty of nurse bees, sufficient stores of honey and pollen, and active foraging bees. The starter colony is motivated by the emergency impulse, and ready to accept the bars of newly grafted up to 24 hours old larvae. One day after grafting, once the cells have been started off in the starter colony, the bars of cells from the starter colony need to be removed and transferred into the cell builder colony. At that time, the acceptance rate of cells between bars can vary, depending on the care and skill of the operator. During all procedures, feeder needs to be refilled in case there is no pasture in the environment.

At that stage, new bars with grafted larvae may be grafted and given to the same starter colony. Many experienced beekeepers simply omit using a starter colony for accepting grafted larvae. They transfer freshly grafted larvae directly into the cell building colony and obtain successful acceptance percentages.

Queen cells incubator

From the fifth day after grafting and the day before emerging, on the eleventh day, the queen cells can complete their development. During this time larvae can be kept under a controlled environmental condition in an incubator. Constant temperature is at around 35°C, with 75% RH. Newly emerged queens in an incubator are more difficult to accept by workers in a new colony or mating nuclei than queen cell which is about to emerge. This method of emerging queens allows the beekeeper to mark virgin queens before introducing them into mating nuclei.

There is also an option to keep queen cells in contact with the workers by the end of their development. As soon as queen cells are sealed, they can be transferred into small emerging queen cages, where queen cells are separated from each other. Caged queen cells may remain in the cell builder colony or may be transferred into a separate incubator colony, which runs as a normal cell builder colony (finisher).

Both queen larvae and pupae are very sensitive during development, and special care needs to be taken that they are not damaged or cooled. Between the seventh and ninth day after grafting, the young larvae have spun their cocoon and are hanging from the top of the queen cell, and any handling should be avoided. At days 10 or 11 after grafting, when queen cells are ripped, just before transferring them to the colonies, queen cells may be turned upside down and checked for the silhouette of the future queen. Live pupae would move when one end of the bar is gently shaken.

Mating of virgin queens

Ripe queen cells may be placed in the queenless colony or into the mating nucleus the day before they hatch. The alternative is that the newly emerged virgin queen may be inserted into the newly established mating nuclei. Queen cells are best left in the cell building colony until the day before they are due to emerge, and then taken from and distributed to nuclei. It is therefore safe to say that queen cells may be given to nuclei on day 10 after grafting larvae. When a mating nucleus is established and exposed to the mating location, it is essential to take care of neighboring drones. There can be "wild" drones accidentally in the area of mating. If a selection program is being conducted, there is a need to establish an isolated mating station, where controlled mating with selected drones can take place. In that way, required racial, morphological, ethological, and genetic characteristics can be considered results of the selection program conducted by beekeeper operations or breeding organizations in some countries. A complete guarantee of controlled mating can be achieved by conducting insemination.

A virgin queen may take one or more orientation flights prior. Virgins are sexually mature and mating takes place five to six days after the queen's emergence. During the mating season, the drones gather in drone congregation areas, 10–40 m above ground, and they emit a pheromone that can attract other drones, thereby increasing the size of the congregation. These drone congregations can contain as many as 11,000 drones from up to 240 different colonies Virgin queens join the vicinity of the drone congregation area and mate with as many as 20 males mid-air (Koeniger et al. 2005). Mating normally takes place during calm weather and warm days. If queens are of poor quality, it may be due to bad weather that queens stay in the mating nuclei for several days, even though those queens could have repeated their mating flights several times.

Drone population

One of the most important concerns for queen breeding is to obtain a large enough selected drone population. In order to get quality drone selection, drone mother colonies should be tested and developed. Queen breeder usually looks for a unique line of inheritance selected for desired characteristics. The selection process for drone mother colonies is performed geographically separate from selection for queen mother colonies in order to avoid inbreeding.

Selection criteria for queen rearing

Since the beekeeping production depends on the selection of honey bee colonies, selection objectives and criteria used are important tools in bee breeding. The most important measurable colony characteristics are gentleness and the tendency to remain calm on the comb, overwintering ability, spring colony development, swarming tendency, honey production, resistance against diseases.

The implementation of breeding programs of different autochthonous honey bee races and other strains of bees are the beekeeping priority, and practiced in several European and non-European countries. The first step in the selection of several breeding programs is race confirmation using morphological and behavioral observations, genetic analyses, and measurement of colony productivity. Morphological analyses can employ cubital index (Ci) calculation and coloration of abdominal segments as a criterion comprised in a breeding program for some races. The cubital index is known as the morphometrical method for discrimination of *A. mellifera* subspecies. Calculated Cubital index demonstrates the ratio of the lengths of two wing veins on edges of the cubital cell on honey bee wing. Which sample of worker bees analyzed will produce a graph with a single peak is race specific. In a graph, a curve demonstrating two or more distinct peaks is indicative of interracial breeding (Ruttner 1988). Economic goals can be considered through the evaluation of the selection aims. Beekeepers recognize a variety of bee colony characteristics. The characteristics most often included in the breeding program are calmness of the bees on the combs, and weak swarming tendency. The surface area of capped brood in the productive colony is a reflection of the queen's fecundity. It is an important indicator for the capacity of the colony to build up its population and to produce honey and hive products, including pollen, propolis, and royal jelly. Honey bee colonies need to be strong enough for use in commercial pollination services. The brood area in the colony is thus useful information on the brood activity of the colony.

The swarming tendency of a colony used to be considered a desirable characteristic some decades ago. Nowadays, inclination to swarm is not appreciated and such colonies are not selected for breeding. Estimating overall colony strength allows the possibility of ranking colonies for their economic potential, and is a useful method for selecting productive colonies (Pechhacker and Leichtfried 1991). Some characteristics, such as entleness, the tendency to remain calm on the comb, and swarming behavior are considered, and the brood area on a central brood comb are scored according to a four-point system. The international standard for testing and selecting honey bees was first established as technical recommendations for methods to evaluate the performance of bee colonies at the Apimondia symposium "Controlled mating and selection of the honey bee" in 1972 (Ruttner 1972). Colony strength is measured in the form of the number of combs occupied in the hive that are occupied by bees. Characteristics of colony performance are assessed several times during the season, and an average score is calculated (Gregorc et al. 2008). Honey yields from test colonies are determined regularly for each colony by weighing combs before and after honey extraction. The difference, in kilograms, is the amount of honey produced per colony.

Conserving genetic diversity

Preserving the genetic diversity of honey bees in different geographical regions is very important, particularly in those regions in which the local beekeeping industry includes breeding activity (Kraus 2005). Breeding autochthonous honey bee races is supported by EU regulations. The main objectives are to restock hives in order to ensure the quality and safety of bee breeding, beekeeping technology, and honey bee products. Council Regulation (EC) No 1804/1999 supports organic beekeeping based on using indigenous honey bee rather than imported stock (Moritz et al. 2005). If the majority of beekeepers in an area participate in the breeding program, it helps preserve the genetic variability of the honey bee population. The intensity of selection effort is thus greatly improved and accelerated.

The quality of the reproductive female, the queen in the colony determines the colony performance. Commercial queen breeding requires the mass production of large numbers of high-quality fertile queens, which are reared in accordance with a particular breeding program. A high-quality queen should demonstrate the desired physical characteristics which can be assessed by her body weight, number of ovarioles, size of the spermatheca, number of spermatozoa, and being diseases and pests free. Colonies led by 1 or 2-year-old queens with a higher number of ovarioles are more productive than colonies headed by older queens with respect to brood production and honey yield (Avetisyan 1961, Woyke 1984). Queen quality and colony production parameters are determined by the relations between the weight of the queen and the number of ovarioles in the queen's ovary (Wen-Cheng and Chong-Yuan 1985, Gilley et al. 2003). There are also indications that the weight of the queen, brood production (Akyol et al. 2008), size of the spermatheca, and number of spermatozoa are indicators of the quality queen (Woyke 1966, Bieńkowska et al. 2008).

Honey bee breeding activities are increasing worldwide, together with the demand for quality queens. Queen breeders are therefore encouraged to follow practical productive selection procedures within specific breeding programs. They also need to follow professional recommendations for quality rearing in order to satisfy increasing market requirements.

Honey Bees in Latin America

Requier, F.

Introduction

The western honey bee (*Apis mellifera* L.) is a native species from Africa, Europe, and Western Asia (Ruttner 1988). This species evolved four evolutionary lineages that present a huge diversity of 31 subspecies (also called geographic races) (Ruttner 1988, Sheppard et al. 1997, Sheppard and Meixner 2003, Meixner et al. 2011, Chen et al. 2016). During the last glacial period, the European population of *A. mellifera* retreated to the South into glacial refuges. Geographical barriers, such as the Alps, Pyrenees, and Balkan Mountains contributed to the reproductive isolation of the populations over the four evolutionary lineages (Ruttner 1988, Requier et al. 2019). When the European glaciers retreated, three evolutionary branches were free to re-expand northwards. The so-called O evolutionary lineage of the European population (e.g., *A. m. caucasica*) extended from Caucasus to the North (*A. m. pomonella*) and to the West (*A. m. anatoliaca*), while the C branch (e.g., *A. m. ligustica*, *A. m. carnica*, *A. m. cecropia*) extended from the Balkan Peninsula to north-eastern coasts of the Mediterranean Sea and Central Europe, and the M branch (e.g., *A. m. mellifera*, *A. m. iberiensis*) extended from the Iberian Peninsula to Western, Northern, and North-Eastern Europe (Ruttner 1988, Requier et al. 2019). The African population was less affected by glacial events and evolved independently through the A branch (e.g., *A. m. scutellata*, *A. m. adansonii*, *A. m. capensis*). These evolutionary lineages and associated 31 subspecies vary in terms of morphology (Ruttner 1988), molecular (Franck et al. 2000b), behavior, and chemistry characteristics (Yusuf et al. 2015).

Honey bees are generalist insect collecting pollen and nectar from diverse flowering plants (Haydak 1970, Requier et al. 2015a,b). They are social insects living in a colony, where larvae, adults, and the queen rely dominantly on nectar

Universidad Nacional de Río Negro, Mitre 630, San Carlos de Bariloche, Río Negro, CP 8400, Argentina; Department of Animal Ecology and Tropical Biology Biocenter, University of Würzburg, Germany; requierf@gmail.com and pollen for food (but also oils and honeydews). Nectar is the main source of energy for adult flights. Pollen is consumed by both adults and larvae (Keller et al. 2005), and is their only source of protein and lipids (Haydak 1970, Requier et al. 2017a). Honey bees create honey reserves from nectar dehydration to feed the in-nest population even during the period without access (or low availability) to floral resources surrounding the colony nest (e.g., rainy days, winter period, dry season). This latter characteristic explains the particular interest of human civilizations in this species. Numerous prehistoric rock paintings are proofs of the honey collection of wild colonies from humans—so-called honey hunting—since thousands of years in Europe and Africa (Pager 1973, Dams 1978). Over the last centuries, beekeeping techniques have been developed, transforming the honey hunting on the wild colony to the trapping of swarms and the breeding of colonies in hives. These domestication techniques greatly improved the amount of honey collection and facilitated the beekeeping practices towards the human-related selection of some subspecies and their genetic hybridization to optimize honey yield and management efficiency. Modern beekeeping is now based on the use of standardizing hives (e.g., Dadant and Langstroth 10-frames hives), that allows year-round intense management with more than several hundred hives per beekeeper, hive movement over large distances, pest and pathogen treatment, and control over reproduction (e.g., swarming control, queen rearing, and artificial insemination).

Similarly to many other domesticated species, the beekeeping and *Apis mellifera* has been spread all over the world with human migratory flows, especially during the colonial period of human history. Honey bee populations therefore have a human-mediated spread beyond their native range, toward a worldwide introduction of *Apis mellifera* (Pirk et al. 2017, Requier et al. 2019). In this chapter the status and trends of honey bee populations that spread beyond their native range is presented, with the specific attention on the case study of Latin America. This chapter describes the characteristic of honey bees in term of genetic and health, and its importance for human well-being in this region. Moreover, particular attention is paid to the taxonomic diversity of other honey-producer bee species that are native to Latin America, and their biotic interactions with the introduced populations of *Apis mellifera*.

The worldwide expansion of *Apis mellifera*

Apis mellifera originates from Africa, Europe, and Western Asia (Ruttner 1988, Whitfield et al. 2006b, Requier et al. 2019), and is particularly well known as a honey producer. Therefore, the single species, *Apis mellifera*, is often cited as "the honey bee" despite the fact that there are at least eight other honey bee species (Mathias et al. 2017). For instance, the giant honey bee *Apis dorsata*, the dwarf honey bee *Apis florea*, and the Bornean honey bee *Apis koschevnikovi* only exist as wild populations (Requier et al. 2019). The preferential domestication of *Apis mellifera* is linked with its productivity. *Apis mellifera* is the most productive species of honey bees explained by its native range (Fig. 9.1). As other *Apis* species live in a mild tropical climate, they have no need to amass large food stores because they can find flowers to meet their needs throughout the year (Ruttner 1988). In contrast,

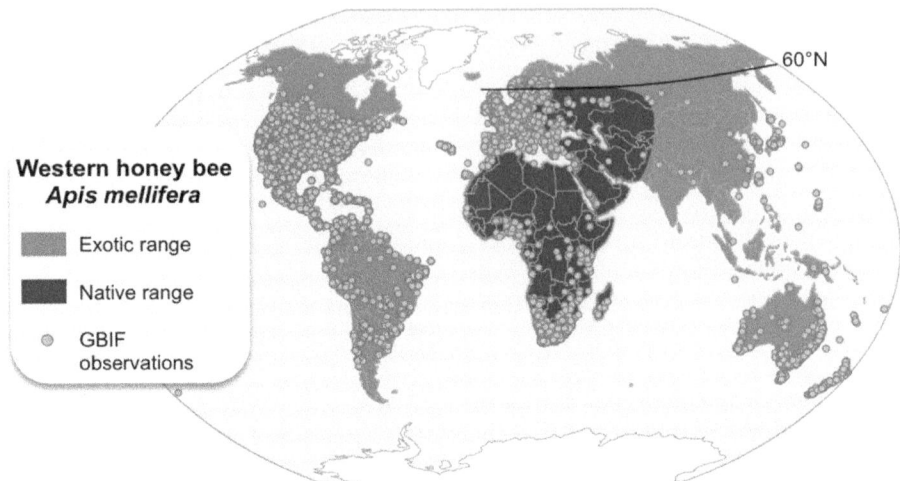

Figure 9.1: The worldwide expansion of the western honey bee *Apis mellifera*. In Europe, the native range of the species is limited around the 60°N (Ruttner 1988). The Eastern limits of the native range include the Middle East, Kyrgyzstan, Western China, and parts of Kazakhstan (Ruttner 1988). The GBIF dataset (in dots, GBIF Occurrence Download 2018a) shows the human-mediated introduction of the species throughout the rest of the world, and are plotted using the maps and rgdal R packages (R Core Team 2018).

Color version at the end of the book

Apis mellifera has evolved over regions with periods of flower limitation (e.g., rainy days, winter period, dry season), leading to the selection of building characteristics of honey reserves. *Apis mellifera*'s great yield capacity has led it to be used by beekeepers and its introduction worldwide (Fig. 9.1). The western honey bee is now the most widely-distributed honey bee species. The large database from the Global Biodiversity Information Facility (GBIF)—an international organisation that focuses on making scientific data on biodiversity available online, such as global distribution data on plants, animals, fungi, and microbes—shows the worldwide distribution of the western honey bee (with 115,816 records from 545 published datasets, using TaxonKey = "*Apis mellifera* Linnaeus, 1758", GBIF Occurrence Download 2018a, Fig. 9.1). This human-mediated spatial expansion covers its introduction in the American continent, Australasia, Asia, and North Eurasia.

The introduction of *Apis mellifera* in Asia comes from interests of honey productions by the use of European subspecies (e.g., *A. m. mellifera*). This introduction has lead to mixed success, given that local beekeepers have concentrated their production on the use of the introduced species despite the native one (e.g., *Apis cerana*) (Li 1998, Moritz et al. 2005). Although this introduction has improved the honey yield and beekeeping productivity, it has potentially affected local populations of native species. Indeed, drones and queens of *A. mellifera* and *A. cerana* mate, but do not produce viable offspring (Ruttner and Maul 1983). In the surrounding landscape of beekeeping activities using the introduced *A. mellifera*, the native *A. cerana* queens could have poor mating success (Li 1998, Moritz et al. 2005). Moreover, the inter-species competition for floral resources could affect the survival probability of the native populations (Paini 2004, Cane and Tepedino 2017, Geslin et al. 2017, Mallinger et al. 2017, Geldmann and González-Varo 2018, Requier et al. 2019). The introduction of *A. mellifera* in Asia and its interaction with *A. cerana* has lead to the transfer of the ectoparasitic mite *Varroa destructor* from the local Asian population to the exotic one (Oldroyd 1999, Anderson and Trueman 2000). Till date, the *Varroa* mite is one of the most important factors weakening the honey bee populations (see below).

Trends in *Apis mellifera* introductions over Latin America

The first introduction period of *Apis mellifera* in Latin America (and further in the Americas) took place in the 17th century for honey production interests (Crane 1984). The introduction includes several European subspecies (e.g., *A. m. carnica, A. m. iberica, A. m. ligustica, A. m. mellifera*). However, in the tropical regions of Latin America (called LA thereafter), introduced European subspecies were far less successful than in North and Central America (Sheppard 1989). The colonies imported from Europe were often only poorly adapted to tropical conditions. Colonies did not gain desired productivity levels (Delgado and del Amo 1984), giving rise to the plausible plan to import tropical African rather than European honey bee subspecies for honey production (Moritz et al. 2005). In 1957, Warwick E. Kerr imported about 170 queens of the African subspecies *Apis mellifera scutellata* from local populations of Tanzania and South Africa, and introduced them in the state of Sao Paolo (Brazil) (Fig. 9.2). This second step of *A. mellifera* introduction in LA was initiated with the aim to breed more productive honey bees to improve beekeeping yield in tropical and subtropical regions. Twenty six swarms of this experimental program escaped into the wild. Moreover, a larger scale program was established to promote the breeding and the distribution of vast numbers of *A. m. scutellata* queens in Brazil, highlighted by post introduction molecular analyses (Hall 1990, Smith 1991a). This has facilitated the widespread expansion of the African honey bee over the Americas within a few decades (Fig. 9.2). It bypassed all attempts to stop its spread through large-scale international control programs and was first detected in the United States near Brownsville in October 1990 (Schneider et al. 2004, Moritz et al. 2005). *Apis mellifera scutellata* is now well established in the southwest of the United States and is frequent in Texas, New Mexico, Nevada, Arizona, southern California, Puerto Rico, and the Virgin Islands, but seems to have markedly slowed its further spread northward in the last decade (Fig. 9.2) (Kono and Kohn 2015, Rangel et al. 2016). Indeed, a large data collection of 618 specimens from 35 countries (using the Barcode of Life Database) (Ratnasingham and Hebert 2007, Barcode of Life Database 2018) clearly shows the current extent of *A. m. scutellata* in the Americas (data covers 1993 to 2017), with probables, extend limits in North America and Patagonia (south of Argentina). This database synthesizes the DNA barcode data that are available worldwide. Here a selection of high-quality sequences of honey bees (*Apis mellifera*) using the barcode index number (BIN) AAA2326 was performed (see more detail on the procedure in Kono and Kohn 2015). Honey bee sequences scored as African subspecies have a cytosine at position 2382 (Crozier and Crozier 1993) within the *COI* gene, all others have thymine at that position.

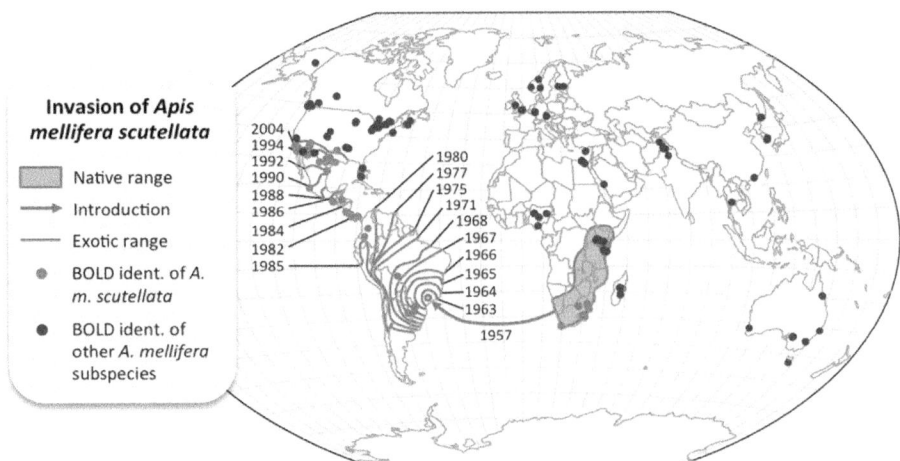

Figure 9.2: Introduction and invasion of the African honey bee *Apis mellifera scutellata* in the Americas. The native range of *A. m. scutellata* covers the southwestern of Africa, from Tanzania to South Africa (Ruttner 1988). The trends of the exotic range expansion are based on former studies (Spivak et al. 1991, Visscher et al. 1997, Moritz et al. 2005, Kono and Kohn 2015, Rangel et al. 2016). The BOLD dataset (in dots, Barcode of Life Database 2018) confirms the human-mediated wide distribution of *A. m. scutellata* in the Americas, and are plotted using the maps and rgdal R packages (R Core Team 2018).

Color version at the end of the book

Controversy has surrounded the terminology used to name *A. m. scutellata* populations in LA (Schneider et al. 2004). Various terminologies were used, such as "Africanized honey bee" and "neotropical African honey bee". Although introgression of European alleles has occurred, African genetic and behavioral characteristics have been largely preserved during the invasion process in LA (Schneider et al. 2004, Moritz et al. 2005). Therefore, the terminology of "African honey bee" was selected here. The invasive populations of the African honey bee have also achieved public notoriety as "killer bees". There was an obvious and measurable increase in stinging behavior associated with the African genotype, and it has had a major impact on beekeeping management, which might make it appropriate to consider the African honey bee in the Americas as an invasive pest (Moritz et al. 2005). However, Brazilian beekeepers, for instance, seem to have adapted their operations to the behavioral changes of the honey bees and often prefer to run their apiaries with African honey bees, given their increase in honey yield (Gonçalves et al. 1991).

The difference between the African honey bee and European honey bees has since been subjected to extensive research (e.g., Spivak et al. 1991, Schneider et al. 2004, Kono and Kohn 2015, Rangel et al. 2016). Beyond its tends toward intensified aggressiveness in LA, the African honey bee populations differ to their European counterpart with (Fletcher 1977a,b, 1978, Winston et al. 1983, Winston 1992, Moritz et al. 2005): (i) more swarms per colony (Winston et al. 1981), (ii) more migratory swarms (Smith 1960), (iii) more male sexuals (Otis et al. 2002), (iv) smaller colonies (Hepburn and Radloff 1998), (v) a shorter generation time (Hepburn and Radloff 1998), (vi) a shorter queen development time (Hepburn and Radloff 1998), (vii) colony usurpations (Danka et al. 1992, Vergara et al. 1993, Schneider et al.

2004), (viii) increased fighting ability (Hepburn and Radloff 1998), (ix) different pollen diet diversity and resource exploitation (Villaneuva and Roubik 2004), (x) seasonal migration (Hepburn and Radloff 1998), and (xi) no excessive honey hoarding necessary for overwintering (Hepburn and Radloff 1998).

Diversity of honey bees in Latin America

Two large scale studies of genome sequence have shown the evolutionary history of the honey bee *Apis mellifera* in LA (Whitfield et al. 2006a, Wallberg et al. 2014). Behind highlighting the origin of the species—*Apis mellifera* comes from Africa-those studies also described the geographical and temporal pattern of diversification in the New World. Interestingly, these studies showed a differential distribution of the initially introduced European subspecies in LA, with a preferential occurrence of *A. m. mellifera* in Brazil, Paraguay, and North Argentina (the province of Misiones), while a hybrid between *A. m. mellifera* and *A. m. carnica* is more distributed toward the south of LA. The second wave of introduction involving the African honey bee *A. m. scutellata* allowed its general establishment through a hybrid *A. m. mellifera-scutellata* above the latitude of 33°11.603'S. A hybrid zone between *A. m. mellifera-carnica* and *A. m. scutellata* occurs in the latitude range of 31°04.993'S and 33°11.603'S. Moreover, these studies pinpoint a similar establishment of hybridization with *A. m. scutellata* in southern Texas over the period 1983–2001. After the invasion of African honey bees, the European mitotype declined to about 30% within 4 years, but remained stable thereafter (Pinto et al. 2004). The presence of "African" mitotypes in European honey bees and vice versa has now been documented in several American regions, which shows that hybridizations between African and European honey bees do occur and can be detected at the population level (Clarke et al. 2002, Moritz et al. 2005). Nevertheless, after 50 years of contact and interbreeding with European honey bees, the African genome has experienced relatively low levels of European introgression (Ratnieks 1991). It seems that over the colonizing history of the African honey bee over Americas, a non-hybridizing "colonizing" front of migratory and reproductive swarms spread north (Ratnieks 1991) and hybridization with European honey bees occurred behind this front (Moritz et al. 2005). The introgression of African honey bees in the originally introduced European populations is therefore established, even if the variable is in term of density over the continent. However, the spread of African honey bees has slowed down since 2000 (Fig. 9.2).

Talking about honey bee diversity in LA, one should also consider all the honey-producer bees. Indeed, LA includes native honey-producer bees, well known as stingless bees (tribe of Meliponini). They are also social bees living in colonies, building nests, and creating honey reserves, but have the particularity to be adapted to tropical and subtropical climate and to be forest-related with the use of resins. Stingless bees also use resin as a source of chemical compounds which they incorporate in their own cuticular profiles (Leonhardt et al. 2009, Leonhardt et al. 2011a). By adding resin-derived compounds to the self-produced compounds already present on their body surfaces, the bees increase their chemical and thus functional diversity (Leonhardt et al. 2011b), and thereby, e.g., improve protection against predators (Leonhardt et al. 2015). Similarly, access to diverse and chemically different resin sources ensures protection against different antagonists (Drescher et al. 2014), indicating that many bees not only need floral diversity, but also a broad spectrum of plant species providing non-floral resources. Stingless bees are very diverse in term of behavior, morphology, and species richness (Velthuis 1997). The large database from the Global Biodiversity Information Facility (GBIF) was used to show this species richness and diversity. Using the taxonomic filter (i.e., the TaxonKey) of "Apidae" (2,207,047 records from 994 published datasets, GBIF Occurrence Download 2018b), and then selecting the records with a genus corresponding to the stingless bees, 163,233 stingless bee records were selected. This large dataset clearly shows the tropical specificity of stingless bees, given that most of the records are centered between the latitudes 24°16.623'N (Central Mexico) and 25°58.931'S (North Argentina) (Fig. 9.3). A total of 269 species of stingless bees have been recorded in GBIF, in LA, with 33 species well known and kept as honey producers (Table 9.1).

Figure 9.3: The GBIF observations of stingless bees species in Latin America. The GBIF dataset (in dots, GBIF Occurrence Download 2018b) shows the spatial distribution of native stingless bees over Latin America, and are plotted using the maps and rgdal R packages (R Core Team 2018). Stingless bees are native honey-producer bees in Latin America. A total of 269 species have been registered in the GBIG dataset, for which 33 species are well known as honey-producers (GBIF Occurrence Download 2018b).

Table 9.1: List of 33 stingless bee species registered in the GBIG dataset, known and managed as a honey producer in Latin America.

Genus	Scientific name	Author name
Cephalotrigona	*CephaloTrigona capitata*	(Smith 1854)
Frieseomelitta	*Frieseomelitta doederleini*	(Friese 1900)
	Frieseomelitta varia	(Lepeletier 1836)
Melipona	*Melipona asilvai*	(Moure 1971)
	Melipona beecheii	(Bennett 1831)
	Melipona bicolor	(Lepeletier 1836)
	Melipona capixaba	(Moure and Camargo 1995)
	Melipona compressipes	(Fabricius 1804)
	Melipona crinita	(Moure and Kerr 1950)
	Melipona eburnea	(Friese 1900)
	Melipona fasciata	(Latreille 1811)
	Melipona fasciculata	(Smith 1854)
	Melipona favosa	(Fabricius 1798)
	Melipona flavolineata	(Friese 1900)
	Melipona fuliginosa	(Lepeletier 1836)
	Melipona marginata	(Lepeletier 1836)
	Melipona panamica	(Cockerell 1919)
	Melipona quadrifasciata	(Lepeletier 1836)
	Melipona rufiventris	(Lepeletier 1836)
	Melipona scutellaris	(Latreille 1811)
	Melipona seminigra	(Friese 1903)
	Melipona subnitida	(Ducke 1911)
	Melipona yucatanica	(Camargo, Moure and Roubik 1988)
Paratrigona	*Paratrigona subnuda*	(Moure 1947)
Partamona	*Partamona helleri*	(Friese 1900)
	Partamona seridoensis	(Pedro and Camargo 2003)
Scaptotrigona	*Scaptotrigona bipunctata*	(Lepeletier 1836)
	Scaptotrigona mexicana	(Guérin-Méneville 1845)
	Scaptotrigona polysticta	(Moure 1950)
	ScaptoTrigona postica	(Latreille 1807)
	Scaptotrigona tubiba	(Smith 1863)
Tetragonisca	*Tetragonisca angustula*	(Latreille 1825)
	Trigona angustula	(Latreille 1825)

Surprisingly, among the list of stingless bee species producing honey, we can see that several species have been discovered very recently (e.g., Melipona capixaba (Moure and Camargo 1995)). This suggests that other species are probably unknown, and that very little is known about stingless bee systematic (but see Velthuis 1997).

Honey bee health

Large-scale monitoring of bee populations (wild and managed) have revealed an ongoing decline in both European wild bee populations, as well as the managed honey bee population in the United States and across Europe (Biesmeijer et al. 2006, Potts et al. 2016). In 2006, the Colony Collapse Disorder syndrome (CCD) was observed in the USA (Cox-Foster et al. 2007, vanEngelsdorp et al. 2007, vanEngelsdorp et al. 2008, Oldroyd 2007, vanEngelsdorp et al. 2009). The Colony Collapse Disorder consists of a sudden colony depopulation of the adult population of a colony, without any evidence of disease of the high load of parasites and pathogens (vanEngelsdorp and Meixner 2010). Not only have bee populations decreased, but also beekeeping productivity (Ellis et al. 2010b, Potts et al. 2010a). In fact, current global honey production is decreasing, independent of the stagnation in beehive livestock numbers (Potts et al. 2010a). Many factors are likely involved in the cause of bee decline. In particular, land-use change (i.e., habitat loss, landscape fragmentation, and decrease of resource diversity and abundance) was suggested to be a major cause for the decline of wild bees in agricultural areas, as it results in the loss of floral-food resource and nesting habitats (Biesmeijer et al. 2006, Roulston and Goodell 2011). The agricultural disturbance of floral resource availability and composition, resulting in the scarcity of particular flowering species and/or monotonous offers of specific crop flowers, affect managed honey bee and wild bee population dynamics and survival (Crone and Williams 2016, Requier et al. 2017a, Kaluza et al. 2018). Pesticides, e.g., neonicotinoids as well as pyrethroids and fungicides, were also highlighted to severely affect the behavior and fitness of honey bees and wild bees (Goulson et al. 2015). Additionally identified critical causes are diseases, pathogens, and parasites (Fürst et al. 2014). Despite highly intensive research efforts, however, no single main driver and/or mechanism could be identified (Goulson et al. 2015). Bee decline is therefore best presumed as a multifactorial cause (Potts et al. 2010b, Goulson et al. 2015, Potts et al. 2016, Henry et al. 2017), with three main groups of stressors involved: lack of flowers, pesticide exposure, and parasites and pathogens, which occur simultaneously or even reinforce each other through synergistic effects (Fig. 9.4). Thereafter, the mechanisms underlying each stress factor are shortly described.

Lack of flowers

While floral resources certainly have an impact on the honey bee longevity, especially through the over-wintering survival, which is totally dependent on the level of reserves stored during the spring and summer seasons, there is little demonstrated evidence of a direct link between floral resources decrease and honey bee colony losses (but see Requier 2013, Requier et al. 2017a). For example, colonies having suffered CCD symptoms in comparison with control (without CCD symptoms) revealed

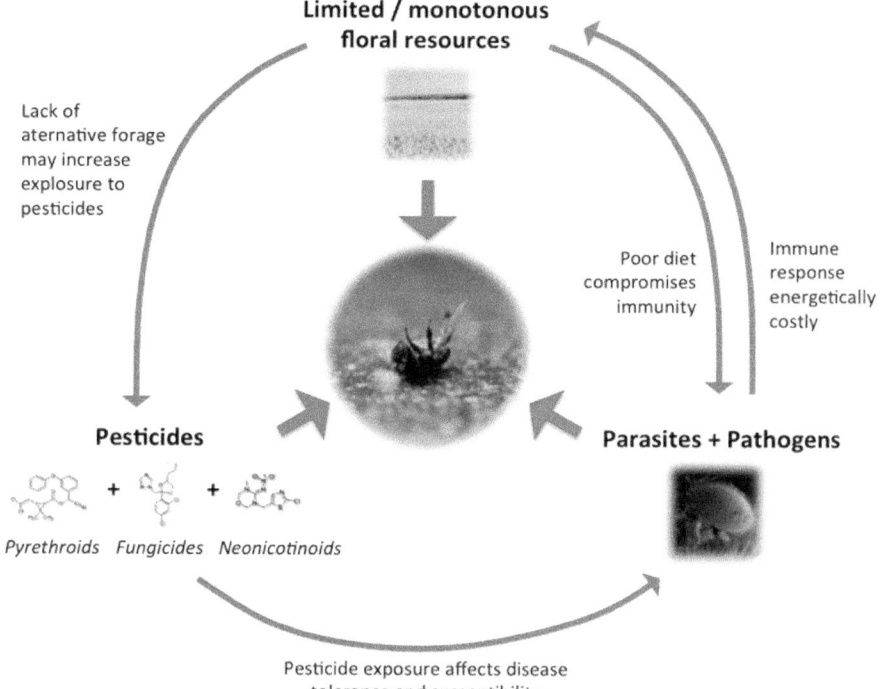

Figure 9.4: The decline of bee pollinators- a consequence of multiple stresses. Redraw from Goulson et al. (2015).

Color version at the end of the book

no differences in individuals' body mass, levels of protein, and fat (vanEngelsdorp et al. 2009). However, life history, growth, and survival of honey bees are closely associated with the availability of resources in their environment (Haydak 1970, Keller et al. 2005, Naug 2009, Brodschneider and Crailsheim 2010, Requier 2013, Requier et al. 2017a).

Only few studies suggest a possible alteration of foraging areas for honey bee by the current agricultural intensification and land-use changes (Naug 2009, Decourtye et al. 2010, Requier 2013, Requier et al. 2015a, 2017a), an assumption supported by beekeepers who mention the lack of food resources (quantity, diversity and quality) as one of the main causes of honey bee decline (VanEngelsdorp et al. 2008). Experimental laboratory studies show effects of the lack in resource quantity (Toth et al. 2005, Mattila and Otis 2006), diversity and quality (Alaux et al. 2010b, Di Pasquale et al. 2013) on honey bee health (Brodschneider and Crailsheim 2010, Decourtye et al. 2010). Deprivation in pollen food resource (pollen shortage) causes a decrease in honey bee population (Keller et al. 2005, Requier 2013, Requier et al. 2017a), and greater susceptibility of individuals to parasites or pathogens (Mayack and Naug 2009, Alaux et al. 2010b, Requier 2013, Requier et al. 2017a). In addition, the amount of pollen brought back to the hive is known to affect the physiological metabolism (Alaux et al. 2011), immunocompetence (Alaux et al. 2010b), and tolerance against pathogens, such as viruses (DeGrandi-Hoffman et al. 2010). In addition, pollen food-shortage reduces the sensitivity to pesticides (Wahl and Ulm 1983), the learning and memory capacities (Mattila and Smith 2008), the regulation of foraging ontogeny and activity (Toth et al. 2005, Feigenbaum and Naug 2010), the social interaction (Schulz et al. 2002), the behavioral development (Schulz et al. 1998), the inspection and feeding of larvae by workers (Huang and Otis 1991), the pheromonal regulation (Fischer and Grozinger 2008), and the alteration of physiological responses (Willard et al. 2011). Based on experimental approaches, it was shown that variability in diversity and quality of the resource act negatively on the lifespan of honey bees (Schmidt et al. 1987, Schmidt et al. 1995), and on the development of the hypopharyngeal glands (Pernal and Currie 2000). Pollen diversity, in turn, ensures the immunocompetence of honey bees (Naug and Gibbs 2009, Alaux et al. 2010b).

Pesticide exposure

Some scientists and many beekeepers believe that pesticides have a leading role in colony-loss (e.g., Oldroyd 2007, Henry et al. 2012a,b, 2014), probably in interaction among pesticides (Prado et al. 2019), and with other stressors (Alaux et al. 2010b, Vidau et al. 2011, Goulson et al. 2015). In cereal agricultural landscapes, honey bees are exposed to a variety of pesticides since they forage extensively on flower-blooming crops, such as oilseed rape (Brassica napus), maize (Zea mays), or sunflower (Helianthus annuus), that are routinely treated against insect pests with insecticides (Mullin et al. 2010, Prado et al. 2019). Systemic pesticides in particular diffuse throughout all plant tissues and eventually contaminate nectar and pollen resources that are used by bees (Rortais et al. 2005, Prado et al. 2019). Pesticide exposure can induce sub-lethal effects, such as disorders in movements, behavioral and spatial orientation of foragers,-, all possibly strongly affecting honey bee colony dynamics. Many symptoms of agro-chemicals treatments are visible on the motor function of honey bees, e.g., uncoordinated movements (review in Desneux et al. 2007). Since 2001, there has been increasing evidence in the negative effects of a new group of insecticides, the neonicotinoids (e.g., imidacloprid and thiamethoxam) on honey bees. First symptoms of poisoning appearing after imidacloprid exposure cause stationary behavior in bees (Medrzycki et al. 2003), but also staggering, tumbling, hyperactivity, and tremors (Suchail et al. 2001), as well as a reduction in the number of flights by foragers (Decourtye et al. 2011). Exposure of the honey bee to thiamethoxam causes behavioral disorientation, which decreases the probability of homing flight to the hive (Henry et al. 2012a, 2014), and a decrease of the adult population size in the colony, which can generate the collapse of the colony (Henry et al. 2012b). Pettis et al. (2012) further highlighted an interaction effect between imidacloprid and pathogens, with a significant increase in the Nosema pathogen infection. Moreover, Di Prisco et al. (2013) showed the occurrence at sub-lethal doses of clothianidin induced viral proliferation. These evidence of interaction converge toward the current hypothesis of multiple drivers affecting the decline of bees (Potts et al. 2010b, Goulson et al. 2015, Potts et al. 2016, Henry et al. 2017) (Fig. 9.4).

Parasites and pathogens

"Parasites and pathogens" is the most frequently cited factor involved in the mechanism underlying honey bee decline, including the ectoparasitic mite *Varroa destructor*, viruses, and bacteria. Special attention has been paid on the *Varroa* factor, and its link with over-wintering mortality of honey bee colonies (Faucon et al. 2002, Dahle 2010, Rinderer et al. 2010, Nguyen et al. 2011). Other studies have then highlighted the interaction between *Varroa destructor* and viruses (Chen and Siede 2007, Berthoud et al. 2010, Dainat and Neumann 2013, Francis et al. 2013), suggesting that the infestation of *Varroa destructor* could facilitate the susceptibility to viruses. The viral pressure mainly includes the Deformed Wing Virus (DWV, transmitted by *Varroa*), Acute Bee Paralysis Virus (ABPV), Chronic Bee Paralysis Virus (CBPV), and Kashmir Bee Virus (KBV). These viruses have a sub-lethal effect on individuals in the colony, causing morphological or behavioral disorders. In interaction with other factors, they reduce the fitness of the colony and increase over-wintering mortality of colonies (Berthoud et al. 2010, Carreck et al. 2010a,b, Martin et al. 2010). The over-wintering mortality can be also increased by the action of bacteria (Evans and Schwarz 2011), or microsporidian *Nosema ceranae* (Cox-Foster et al. 2007, Paxton et al. 2007, Higes et al. 2008, Cornman et al. 2012).

Status of the *Apis mellifera* populations in Latin America

Vandame and Palacio (2010) performed a global review on the status of honey bee populations in Latin America, recently updated by Maggi et al. (2016). Both reviews greatly show the wide board of stressors present in Latin America, very similar to those present in the USA and Europe (i.e., the factors described above). This list of stress factors includes health problems, such as Varroosis, RNA viruses, and Nosemosis, and pesticide exposure. Surprisingly, the lack of flowers was not pinpointed; although a recent study shows that beekeepers suggest the lack of flower as an important factor of colony losses in Argentina (Requier et al. 2018a). Interestingly, Africanized honey bees seem to be less affected by *Varroa* mites than the European subspecies (Medina and Martin 1999, Seeley et al. 2015). Global concerns about the loss of honey bee colonies have motivated monitoring programs of colony losses over the past 10 years. Among the most famous ones, the Bee Informed Partnership (BIP) has developed a national monitoring program in the United States carried out annually since 2007 (Kulhanek et al. 2017), while the consortium COLOSS (Brodschneider et al. 2016) has developed its in Europe (with additional non-European countries). These programs reveal that the loss of managed honey bee colonies can reach up to 25–50% every winter in Europe (Brodschneider et al. 2016) and North America (Kulhanek et al. 2017). However, Latin America lacks published data on the loss rates of honey bee colonies (Requier et al. 2016, 2017b, 2018b, in press), and too little is known about the regional situation. For instance, Vandame and Palacio (2010) concluded that "there are no reports of massive colony losses in Latin America", while Maggi et al. (2016) stated that "several cases of colony losses and colony depopulation were reported by beekeepers throughout the continent, yet no accurate data has been published to date".

Recently, Requier et al. (2018b) have presented validated estimates (i.e., published in an international scientific journal with a peer-review process) of colony losses over Latin America (Fig. 9.5). This data comes from individual research initiatives at a regional or national scale based on questionnaires using a citizen science approach (Requier et al. 2016, 2017b, 2018b, in press). Among data collected across five countries over the last 7 years, annual loss rates varied between 0.1% (95% Confidence Interval 0–0.3%) for Northern Patagonia (Argentina) in 2012–13 and 62% (95% CI 57.9–66.1%) for the multi-annual Brazilian survey (Fig. 9.5). Winter losses (where a thermic winter occurs) appear to be higher than seasonal losses in Argentina, but not in Uruguay. Even if very useful and informative, these data sets result in low participation rates from beekeepers, confirming the difficulty in carrying out monitoring programs of colony losses in Latin America (Maggi et al. 2016, Requier et al. in press). Furthermore, most of these monitoring initiatives have been developed independently from each other, which can further

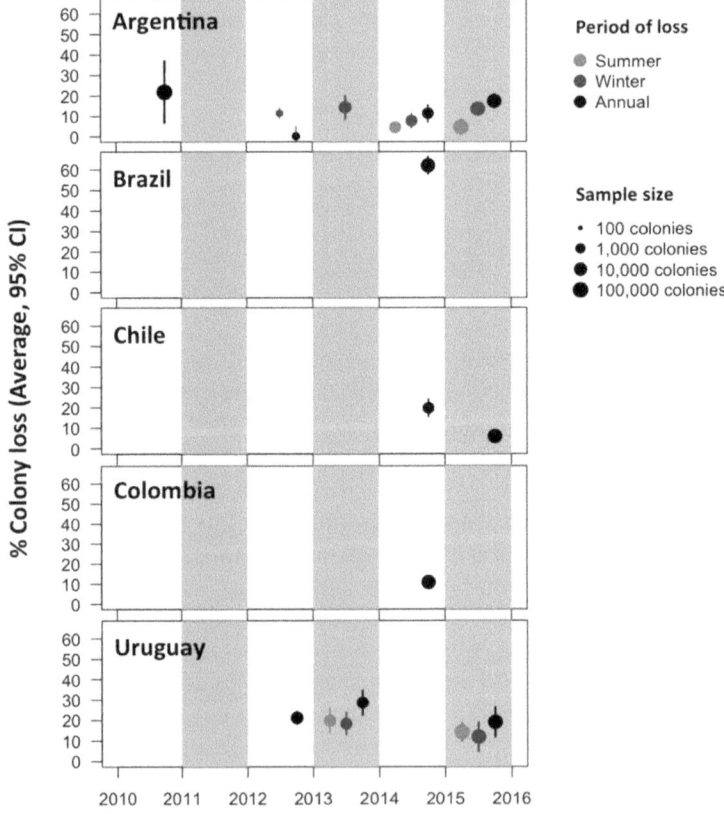

Figure 9.5: Synthesis of unpublished data of honey bee colony losses in Argentina, Brazil, Chile, Colombia, and Uruguay. The loss rates are calculated as the average value per beekeeper (95% Confidence Interval) following vanEngelsdorp et al. (2013). Redraw from Requier et al. (2018b), where more details on the surveys' methods are available.

Color version at the end of the book

reduce participation rates because of the potential overload and discouragement of the beekeepers to answer the same questions for various surveys. Moreover, there is a methodological diversity among the different initiatives that prevents proper spatial and temporal comparisons (van der Zee et al. 2013, vanEngelsdorp et al. 2013), and limits the understanding of the causes of colony losses in Latin America.

This new data reveals a worrying situation of the apiculture in Latin America, associated with some estimates of high loss rates of honey bee colonies in various countries. Moreover, this highlights the need to (i) standardize methods (van der Zee et al. 2013, vanEngelsdorp et al. 2013) and (ii) coordinate among individual survey initiatives (Maggi et al. 2016) to improve the effectiveness of monitoring programs and the understanding of honey bee health in Latin America (Requier et al. in press). The Latin-American Bee Research Association (SOLATINA) was created in 2016 for this purpose, as a large-scale platform to coordinate bee research programs in Latin America (Antúnez et al. 2018). SOLATINA comprises a consortium of researchers from 10 Latin-American countries (Argentina, Bolivia, Brazil, Chile, Colombia, Costa Rica, Ecuador, Mexico, Peru, and Uruguay), representing 90% of its territory, 91% of the beehives, and 92% of the honey exported in 2013 (Requier et al. 2018b).

The "colony losses" working group of the SOLATINA consortium has developed an unified questionnaire of colony losses based on surveys that have proved to be effective in other regions, specifically, those developed by BIP (Kulhanek et al. 2017), COLOSS (Brodschneider et al. 2016), and EPILOBEE (Jacques et al. 2017). This questionnaire was adapted to Latin-American climatic conditions, for example, by taking into account determinants of the season of honey bee low activity other than thermic winter, more representative of tropical and subtropical regions (e.g., dry or rainy season). Moreover, other types of beekeeping activities were included, such as meliponiculture (i.e., the use of stingless bees).

Importance and threats of honey bee populations in Latin America

Latin America produces 28% of the global honey supply with 7.7 million managed honey bee colonies (data for 2013 obtained from FAOSTAT 2018, Requier et al. 2018b). Moreover, Argentina leads the global list of countries in terms of the honey market, with more than 200,000 US dollars intake related to exports of honey in 2013 (subtracting the imports). Five Latin-American countries are also within the 20 largest honey exporting countries in the world (Requier et al. 2018b). This data shows the economic importance of beekeeping and honey production for Latin American countries. Moreover, honey bees are critical for pollination of crops and wild plants (Klein et al. 2007, Rader et al. 2009, Huang et al. 2018), and *Apis mellifera* is regarded as the most important pollinator in farmlands due to its high numbers in a single nest (Rader et al. 2009, Garibaldi et al. 2017). Overall, flower-visiting animals provide a critical ecosystem service, i.e., pollination, with numerous benefits to humans (Potts et al. 2016, Garibaldi et al. 2018, Martin et al. 2019). Bees (in a general term including wild bees and managed bees) visit more than 90% of the globally most important crops, and improve the yield of approximately 75% of them, including most fruits, seeds, and nuts, and several high-value commodity crops, such as coffee, cocoa, and oilseed rape (Klein et al. 2007). This pollination service is estimated to be worth several hundred billion US dollars per year (Lautenbach et al. 2012). An estimated 5–8% of global crop production would be lost without animal pollination (Aizen et al. 2009), likely leading to changes in human diets and a disproportionate expansion of agricultural land in order to fill this shortfall in crop production by volume (Potts et al. 2016). Due to differences in the morphology and functional traits between bee species, a higher diversity of pollinators results in complementary or synergistic effects, improving the quantity and quality of pollination (Blüthgen and Klein 2011, Martin et al. 2019). In fact, recent studies have shown positive effects of wild bee species diversity on the growth and stability of crop yields on top of the service provided by managed pollinators (in particular *Apis mellifera*) (Garibaldi et al. 2013, Rader et al. 2016). However, this critical pollination service provided by bees is currently threatened due to their widespread decline.

Currently, there is a controversy on how to conserve pollinator diversity and pollination service. Managed honey bee colonies comprise numerous bees (i.e., 40,000–60,000 worker bees per hive), which can forage in the range of several square kilometers around the central place of the installed apiary (Winston 1994). Some studies showed negative effects of an increase in honey bee density on richness (Wilms et al. 1996, Gross 2001), reproductive success (Steffan-Dewenter and Tscharntke 2000, Thomson 2004, Hudewenz and Klein 2013), and survival rate of wild bees (Elbgami et al. 2014, Goulson and Sparrow 2008). There is also some evidence for competition between introduced managed honey bees and native local wild pollinators in a context of resource limitation (e.g., Thomson 2004, Goulson and Sparrow 2008, Cane and Tepedino 2017, Wojcik et al. 2018). For instance, a recent study suggests that an apiary of 40 hives could deplete the equivalent in pollen need of 4 million larvae of the wild bee of *Megachile rotundata* over 3 months of foraging activity (Cane and Tepedino 2017). Another recent study converges on this same pattern, showing that the abundance of wild bees in protected areas decreases when approaching an apiary (Henry and Rodet 2018). However, inter-specific competition between managed honey bees and wild bees is not clear-cut (Wojcik et al. 2018), while other studies did not show any effect of the presence of honey bees on native wild species (e.g., Roubik et al. 1986, Steffan-Dewenter and Tscharntke 2000, Roubik and Wolda 2001). On the other hand, pathogen spillover between managed and wild bee pollinators is known to interfere with pollinator health (e.g., Fürst et al. 2014). For instance, introducing managed honey bee pests were found to transfer and reproduce on native *Bombus impatiens* (Hoffmann et al. 2008), and honey bee viruses and diseases generally have a high propensity of spilling-over to other bee-pollinators (Goulson and Hughes 2015).

With respect to pollination and plant reproduction, some studies showed a negative effect of the high density of honey bee flower visits to the abundance of wild bee visits and to plant fruit set (Hargreaves et al. 2009), while other studies show different results, implying complementarity of co-occurrence of managed and wild species (Greenleaf and Kremen 2006, Garibaldi et al. 2013). The competition between managed honey bees and wild bees is therefore complex to apprehend. Nevertheless, recent studies suggest that the presence of massively introduced managed colonies should be regulated in protected areas for the conservation of wild bees (e.g., Thomson 2016, Geslin et al. 2017, Mallinger et al. 2017, Geldmann and González-Varo 2018, Henry and Rodet 2018, Norfolf et al. 2018, Requier et al. 2019). In Latin America, *Apis mellifera* is an exotic species, and detrimental effects to native bee populations can occur. For instance, several studies demonstrate the negative effect of introduced honey bees, in particular, the introduction of the African honey bee, on native solitary bees (Roubik and Villanueva-Guttérez 2009) and native stingless bees (Roubik 1978, 1980, Roubik et al. 1986, Roubik and Wolda 2001).

Conclusion

The western honey bee *Apis mellifera* is exotic in LA, introduced from Europe in the 17th century (Crane 1984). Various European subspecies were first introduced, followed by the African subspecies *A. m. scutellata* in 1957. Since this date, the *A. m. scutellata* populations have spread throughout LA, except in Patagonia (Spivak et al. 1991, Visscher et al. 1997, Moritz et al. 2005, Kono and Kohn 2015, Rangel et al. 2016), with a partial hybridization with the formerly introduced European subspecies. Native *honey-producer* bees also exist in LA: the stingless bees (the tribe of Meliponini). In LA, stingless bees are very diverse in terms of behavior, morphology, and species richness (Velthuis 1997). Some of the stingless bee species are managed for honey production by meliponiculture, a traditional (and cultural) activity in LA. Several studies showed that massively introduced managed species, such as managed colonies of *Apis mellifera* can threaten native species (Thomson 2016, Geslin et al. 2017, Mallinger et al. 2017, Geldmann and González-Varo 2018, Henry and Rodet 2018, Norfolf et al. 2018, Requier et al. 2019). Although usually studied in North Hemisphere for the potential competition between managed honey bees and wild bees, there is similar evidence in LA. The introduced populations of *A. mellifera* threaten the native populations of stingless bees (Roubik 1978, 1980, Roubik et al. 1986, Roubik and Wolda 2001). Too little is known about the health status of stingless bees, however, emerging research programs attempt to survey both the managed populations of stingless bees and honey bee across LA (Antúnez et al. 2018, Requier et al. 2018b, in press). The Latin-American Society for Bee Research, SOLATINA was created for this purpose, as a large-scale platform to coordinate bee research programs in LA (Antúnez et al. 2018, Requier et al. 2018b). Such collaborative organization is of crucial importance to improve knowledge on bee issues in LA, such as bee health, bee diversity, bee conservation, and pollination service. Honey bees and stingless bees play critical socio-ecological roles in LA and more research is required to understand their ecology and to establish adapted and sustainable conservation schemes.

CHAPTER 10

The History of Honey Bees in North America

Seeley, T.D.

Introduction

Until recently, it was widely believed that the history of honey bees in North America began only in the early 1600s, when Europeans started their major colonization of the New World and introduced *Apis mellifera mellifera* (and possibly also *A. m. iberiensis*) to this region. In 2009, however, a team of insect paleontologists reported the stunning discovery of a ca. 14-million-year-old (Middle Miocene) fossil of a worker honey bee preserved in paper shales found in the Stewart Valley Basin, in Nevada, a western state in the United States (Engel et al. 2009). This specimen, shown in Fig. 10.1, is the holotype of a new fossil honey bee species, *Apis nearctica* (Fig. 10.1).

Despite being fragmented, this fossil bee shows numerous morphological features that are finely preserved and show unambiguously that this specimen is a worker honey bee. The drawings in Fig. 10.2 highlight several of the shared, derived morphological characters (synapomorphies) of worker honey bees (tribe *Apis*, genus *Apis*) that are present in this fossil bee. They include compound eyes covered in long, dense setae, mandibles lacking dentition, inner surface of the metabasitarsus with pollen brush (rows of setae), and a sting that is straight and barbed (Fig. 10.2). This discovery greatly changed our understanding of honey bee biogeography, especially with respect to the Western Hemisphere. It also highlights the importance of the fossil record for understanding the biogeographical history of a taxonomic group, such as the genus *Apis*. As the large majority of species are extinct, it is naïve to look only at living species when trying to understand where the members of a taxon have lived.

Cornell University, 332 Hurd Road, *Ithaca*, New York, 14853, USA; tds5@cornell.edu

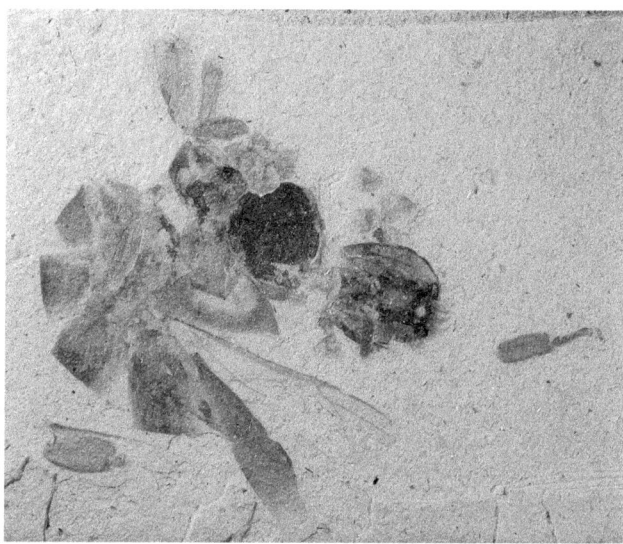

Figure 10.1: The original fossil specimen of a worker of *Apis nearctica*. It is the holotype of the species, i.e., the specimen designated as the type of this species by the authors who named and described the species. It is in the collection of the California Academy of Sciences. From Engel et al. (2009).

Color version at the end of the book

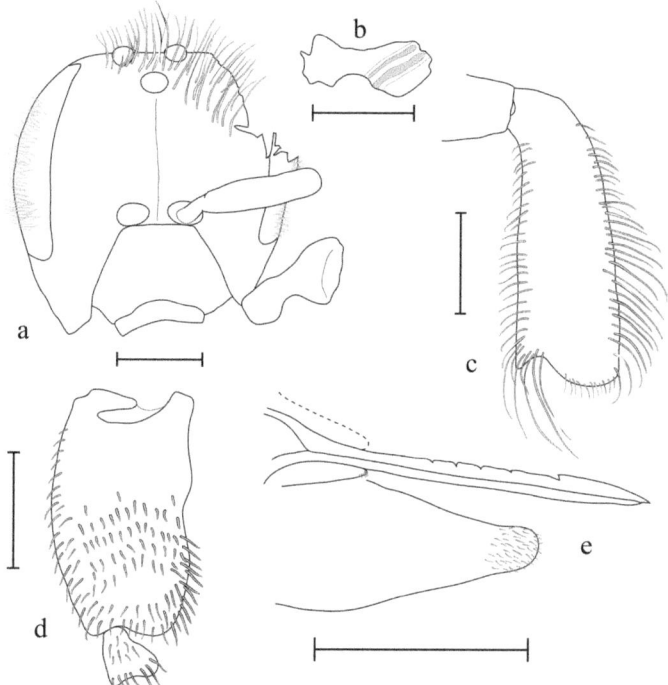

Figure 10.2: Line illustrations of the worker of *Apis nearctica* shown in Fig. 10.1—(A) head and left mandible, (B) right mandible, (C) metabasitarsus, (D) corbicula, (E) sting. All scale bars = 1 mm. From Engel et al. (2009).

It is well established that exchanges of plants and animals occurred between Asia and North America via the Bering Land Bridge during both the Late Cretaceous period (ca. 140 million years ago) and the Miocene period (ca. 24-5 million years ago) (see Hopkins 1967, Kontrimavichus 1985). These periods of biotic exchanges came long before the more famous Pleistocene connection (ca. 2.6 million to 12,000 years ago) that enabled humans to spread from Asia to the Americas. Evidently, *Apis nearctica* (or its ancestor) was sufficiently widespread across Asia during the Miocene that it managed to spread into western North America, along with many other plant and animal species of the time. Later on, when the Bering Land Bridge closed during the late Miocene, *Apis nearctica* became isolated from Asia. We know that there was global cooling during the late Miocene (ca. 7 million years ago, Herbert et al. 2016), and it is widely believed that this global cooling caused the dramatic reductions in the diversity of mammals and plants of western North America that are known from the fossil record (Kürschner et al. 2008). It is likely that these reductions in biodiversity include the extinction of *Apis nearctica*. We need further explorations of late Miocene and Pliocene (ca. 5-2 million years ago) fossils in the western U.S. in order to gain a clearer picture of the diversity and range of *Apis* spp. in North America, and to sharpen our estimates of when these bees disappeared from the fauna (probably the late Miocene or early Pliocene).

What is already clear, however, is that North America was one of the native regions occupied by honey bees (*Apis*), and that honey bees became extinct here until their genus, like that of horses (*Equus*) and gingko trees (*Gingko*), was reintroduced to North America by European colonization of this continent a few hundred years ago.

Earliest introductions of honey bees to North America

The first introduction of *Apis mellifera* into North America, for which there is solid documentation, occurred in the winter of 1622, when several honey bee colonies were sent from England by the Virginia Company of London, via sailing ship, for delivery to the colony of Virginia, in Jamestown (Kingsbury 1933, Kritsky 1991). Evidently, some of these honey bee colonies survived and thrived, because within 60 years several individuals were recorded as having apiaries containing two to 13 hives (Oertel 1976). We also know from historical documents that shortly after the import of hives of bees in Virginia, British colonists in two other locations—Massachusetts in 1640 (Josselyn 1675), and Delaware in 1655 (Johnson 1969)—also imported hives of bees (see Fig. 10.3). Honey bees were also reported in Connecticut in 1644, but it is not known whether the bees in Connecticut had been imported from England or had been brought down from Massachusetts (Oertel 1976). What is certain is that the bees introduced in all four locations were dark European honey bees, *Apis mellifera mellifera*, because the human immigrants to these four sites were primarily from England and Sweden, hence from within the native range of *A. m. mellifera*.

English and Swedish colonists were not the only Europeans to establish settlements in North America. Spanish colonists built a sturdy fort (St. Augustine) and set up short-lived settlements in Florida in 1565, and they established successful settlements in New Mexico (Santa Fe and Taos) in 1609 and 1615. The colonists in New Mexico focused on raising cattle and sheep, relying heavily on the labor

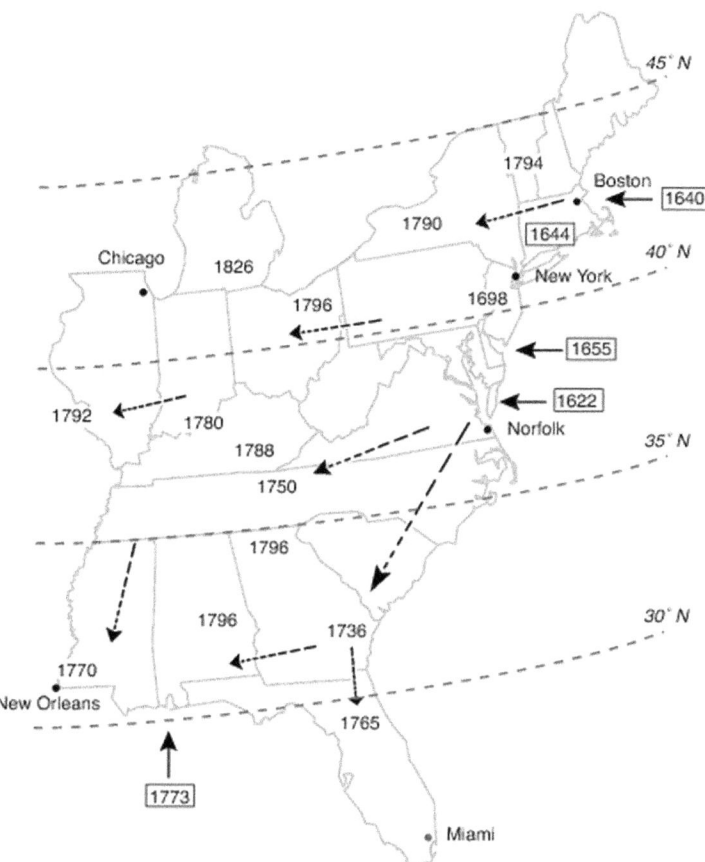

Figure 10.3: The dispersal of the dark European honey bee (*Apis mellifera mellifera*) across eastern North America following introductions (indicated by solid arrows) in Virginia (1622), Massachusetts (1640), Delaware (1655), and perhaps Connecticut (1644). Dashed arrows indicate the honey bees' subsequent spread. From Kritsky (1991).

provided by the Pueblo natives. There are no known records of honey bees being kept at these Spanish settlements, but this bears further investigation by both historians and biologists. Examination of historical Spanish shipping records may reveal that honey bees were indeed shipped to New Spain (which included present-day Mexico) in the 1500s (Brand 1988). Also, renewed genetic analyses of honey bee specimens collected in New Mexico and Arizona before the arrival of the Africanized honey bees—a hybrid between various European subspecies and the African subspecies *Apis mellifera scutellata* (discussed below)—may reveal that their nuclear DNA contains genes derived from *A. m. iberiensis*, the European subspecies of *Apis mellifera* that is native to Spain.

Rapid dispersal of *Apis mellifera mellifera*, 1600s–1800s

What happened to the honey bees after they were introduced to the east coast of North America, in Virginia (1622), Massachusetts (1640), Connecticut (1644), and

Delaware (1655)? Evidently, they cast swarms which established a population of wild colonies living in the woods along the east coast, and then this population of wild colonies gradually expanded across eastern North America. Information about the timing of this dispersal can be gleaned from the writings of the early colonists. In the records of William Penn, the founder of the colony of Pennsylvania, we find no mention of honey bees, even though his writings continue to 1683 (Barton 1792). In the 1698 book by Gabriel Thomas, however, we find that he notes that honey bees were common in the woods in New Jersey and eastern Pennsylvania (Thomas 1698). Beekeepers may have contributed to this expansion of the species' range by moving colonies, but it is likely that most of the dispersal occurred naturally, as colonies living in hollow trees produced swarms. Thomas reports that "the Sweeds often get great store of them [honey bees] where they are free for any Body." No doubt these "free for any Body" bees were living in trees. A few years later, in 1720, Dudley published, in the Philosophical Transactions of the Royal Society of London, a letter titled "An account of a method lately found out in New-England for discovering where the bees hive in the woods, in order to get their honey" (Dudley 1720).

There is strong documentary evidence that swarming was the chief means of honey bee dispersal southward from their "landing place" in Jamestown, Virginia in 1622. This evidence comes from the detailed reports from the 1700s made by the Salzburger Emigrants, who were Protestant refugees expelled from their Austrian homeland by the Catholic Archbishop of Salzburg. They settled in the Georgia Colony (ca. 800 km south of Virginia) in 1734 and made entries in their "Daily Register" that indicate clearly that honey bees were already present when they (the refugees) arrived in their new homeland. A 1739 report states, for example, "Some in the congregation have tried to bring in the bees which are found in the woods in high trees and to place them in barrels resembling beehives or baskets: and several times this has proven successful (Jones and Wilson 1981).

While *Apis mellifera mellifera* was expanding its range north and south along the east coast of North America, it was also spreading westward toward the Mississippi River across a vast region of deciduous forests below the Great Lakes. Fig. 10.3 shows the dates of the earliest written records of honey bees in several locations, along with their likely expansion routes inland. We see that the population that was established in Georgia by 1736 expanded south into Florida and west into Alabama, reaching these regions by 1765 and 1796 (Kritsky 1991). We also see that the population established in Virginia in 1622 expanded west, probably via the Cumberland Pass, and had reached what is now eastern Tennessee by 1750 (Kalm 1772). Furthermore, we see that *A. m. mellifera* expanded its range rather quickly in the north, reaching western New York state by 1790, Ohio by 1796, and Michigan by 1826. In all these regions, swarming, not human import, was almost certainly the main mode of dispersal. Jefferson (1788) included the following observation in his book *Notes on the State of Virginia*: "The bees have generally extended themselves into the country, a little in advance of the white settlers. The Indians, therefore, call them the white man's fly, and consider their approach as indicating the approach of the settlements of the whites." Likewise, Affleck (1841) reported that honey bees were already living in the forests of eastern Ohio when settlers arrived in this region.

Honey bees had crossed the Mississippi River by 1792, as documented by the report of a swarm in St. Louis in that year (Crosby 2015), and they quickly became a commonly found species in the Mississippi River basin. This is documented by entries in the journals of the Lewis and Clark Expedition. For example, on Sunday, 25 March 1804, shortly after the expedition had left St. Louis and was camped along the Kansas River, William Clark wrote: "River rose 14 Inch last night, the men find numbers of Bee Trees, and take great quantities of honey" (quote from Moulton 2002). Similarly, Irving reports in his 1835 book, *A Tour on the Prairies*, which is based on his trip with a surveying party to "Indian Territory" (now the state of Oklahoma), that bee hunting was a pleasurable and profitable pursuit for frontiersman, since the honey plundered from the nests of colonies living in trees was not just a treat, but was also easily bartered and sold (Irving 1835).

Import of seven more subspecies of *Apis mellifera*, 1859–1922

The history of the honey bee in North America becomes more complicated in the second half of the 19th century. In 1859, beekeepers began the import of many thousands of mated queen bees of numerous subspecies (geographical races) from southern Europe, the Middle East, and northern Africa. They did so to obtain queens for breeding and for sale. These imports were made possible by the start of reliable steamship service between Europe and the United States, and they continued for more than 60 years, but then ceased abruptly in 1922. This was the year the U.S. Congress passed the Honey bee Act "To regulate foreign commerce in the importation into the United States of the adult honey bee (*Apis mellifica*). August 31, 1922". Its purpose was "to prevent the introduction and spread of diseases dangerous to the adult honey bee". The main disease of concern at the time was the Isle of Wight disease, an unspecified but highly infectious and lethal disease named for the location of its first reputed outbreak, off the coast of southern England (Bailey and Ball 1991).

Between 1859 and 1922, mated queens of seven more subspecies—in addition to *Apis mellifera mellifera*, and perhaps *A. m. iberiensis*—were shipped to the United States. The seven subspecies represented in these shipments were *A. m. ligustica*: the Italian bee; collected from Italy, starting in 1859; *A. m. lamarckii*: the Egyptian bee, collected from Egypt, starting in 1866; *A. m. carnica*, the Carniolan bee, collected from "Carniola" (a region of Slovenia), starting in 1877); *A. m. cypria*, the Cyprian bee, collected from the island of Cyprus, starting in the early 1880s; *A. m. syriaca*, the Syrian bee, collected from Palestine (the lands that are now mostly in Israel), starting in the early 1880s; *A. m. caucasica*, the Caucasian bee, collected from the Caucasus region (the area between the Black Sea and the Caspian Sea), starting in 1880; and *A. m. intermissa*, the Punic/Tunisian bee, collected from the coastal, mountainous region of North Africa, starting around 1890. The complex history of the introductions of all eight (or nine, if we include *A. m. iberiensis*) subspecies to North America, starting with *A. m. mellifera* is reviewed in detail by Sheppard (1989).

Subspecies composition of honey bees in North America before arrival of the Africanized honey bee

Given the breadth of sources in Europe, the Middle East, and Africa, from where mated queens were shipped to North America between 1859 and 1922, one naturally wonders about the genetic diversity of the bees living on this continent today. Schiff et al. (1994) conducted the first investigation of this subject using modern genetic tools. They performed mitochondrial DNA and allozyme analyses of worker bees collected in the late 1980s and early 1990s from 692 wild (unmanaged) colonies located in 11 southern states. Their collection period preceded the arrival of Africanized honey bees (*A. m. scutellata*) in the United States (see below). The colonies that they sampled occupied trees, caves, bridges, and buildings, plus a few hives in which swarms had been installed, but then had been left unmanaged. Schiff et al. found that 61% of the colonies were matrilineal descendants—as indicated by the haplotypes of their mitochondrial DNA, mtDNA—of *A. m. ligustica* or *A. m. carnica* queens, 37% were matrilineal descendants of *A. m. mellifera* queens, and 2% were matrilineal descendants of *A. m. lamarckii* queens. They found considerable variation in their results, from state to state. For example, in one state, Arizona, the percentage of the 216 colonies sampled that were matrilineal descendants of *A. m. mellifera* was 68%, whereas for the 94 colonies sampled in South Carolina, it was only 27%. This study found no colonies that were matrilineal descendants of queens of the other four subspecies that had been introduced between 1859 and 1922: *A. m. cypria*, *A. m. syriaca*, *A. m. intermissa*, and *A. m. caucasica*. The absence of the first three subspecies is not surprising, because the beekeeping industry in the U.S. never strongly favored them. The absence of evidence for the fourth "missing" subspecies, however, is surprising because some queen breeders have favored this subspecies. Its absence in the Schiff et al. (1994) report may, however, be an artifact of the genetic tools used in this study, which may have been unable to distinguish the matrilineal descendants of *A. m. caucasica* queens from the matrilineal descendants of queens of the two subspecies that Schiff et al. found most often: *A. m. ligustica* and *A. m. carnica*. Recent studies of honey bee mitochondrial DNA have placed *A. m. caucasica* in the same mitochondrial DNA lineage (the C lineage) as *A. m. ligustica* and *A. m. carnica* (Ozdil et al. 2009).

A more recent investigation of the maternal lineages of the wild honey bee colonies in North America has been performed by Magnus et al. (2014) (see also Magnus and Szalanski 2010). This team used DNA sequencing to detect genetic variants of a region in the *COI-COII* gene in the mitochondrial DNA of worker bees. The worker bees that these investigators studied were sampled in 2005–2009, and were collected from 203 wild (unmanaged) colonies and 44 swarms located in both western states (e.g., Utah and Texas) and southern states (e.g., Mississippi and Texas). These investigators found 23 distinct mtDNA haplotypes (distinct variants in the regions of mitochondrial DNA that were studied): 13 haplotypes associated with the C lineage (*A. m. ligustica*, *A. m. carnica*, and *A. m. caucasica*), 5 haplotypes associated with the M lineage (*A. m. mellifera*, *A. m. iberiensis*, and *A. m. intermissa*), and 5 haplotypes associated with the O lineage (*A. m. cypria* and *A. m. anatoliaca*).

It is important to note that the findings just described, from Magnus et al. (2014)—23 mtDNA haplotypes distributed among three lineages, C, M, and O— differ markedly from the findings that the same team of investigators had reported a few years earlier (Magnus et al. 2011) based on bees that were sampled (in 2009) from 140 *managed* colonies owned by 14 queen producers in the western and southern states. In these managed colonies, the researchers found only 7 mtDNA haplotypes, and all were in one lineage—the C lineage (that of *A. m. ligustica*, *A. m. carnica*, and *A. m. caucasica*). Moreover, just 3 of the 7 mtDNA haplotypes that they found accounted for 82% of the 140 colonies sampled. A later study, by Delaney et al. (2009), also looked at the genetics of queen producers' colonies in the western and southern states, and its findings are even starker than those reported by Magnus et al. (2011): just 3 mtDNA haplotypes in the C lineage accounted for more than 97% and 94% of the colonies of the western and southern (respectively) queen producers.

Somehow, more matrilineal lineages (identified by their mtDNA haplotypes) have persisted in the populations of wild (unmanaged) colonies than in the populations of colonies managed by queen producers. This is remarkable, especially when we consider that queen producers in the U.S. "flood" the continent of North America with approximately 1 million queen bees each year, and that each year the 3 million or so managed colonies in the U.S. send forth immense numbers of queens and drones. It is likely that the wild colonies in the U.S. possess higher genetic diversity than the managed colonies, because the additional genetic variants that the wild colonies possess underlie traits that are beneficial (adaptive) to these colonies. It may be, for example, that the population of wild colonies possesses greater genetic diversity than the population of managed colonies because the wild colonies are locally adapted to a wide range of climates, some mild and some harsh. Managed colonies, in contrast, tend to have queens that come from queen producers, and most queen producers in the U.S. live in places with relatively mild climates: the Gulf Coast states (i.e., Florida, Alabama, Mississippi, Louisiana, and Texas), northern California, and Hawaii. As U.S. queen producers breed from queens whose colonies live mainly in mild climates, it is perhaps not surprising that the managed colonies in North America, relative to the wild colonies, have less genetic diversity.

Another possible explanation for the relatively low genetic diversity among the managed colonies in North America, relative to the wild colonies in this continent, is that commercial beekeepers typically replace the queens in their colonies every 1–2 years to ensure colony vigor, and their replacement queens are usually purchased from large-scale queen producers. The queens sold by queen producers are the daughters of a relatively small number of breeder queens (ca. 600 total, for the entire U.S., Delaney et al. 2009), which means there is probably a genetic bottleneck in the population of managed colonies in the U.S. Evidence of this comes from Schiff and Sheppard (1995), who compared the levels of variation in mitochondrial DNA and in six enzyme systems known to be polymorphic in honey bees, between workers collected either from 142 managed colonies (containing breeder queens) or from 692 wild colonies. These researchers found a much lower diversity of mitochondrial DNA haplotypes in the managed colonies than in the wild colonies. Only 4.3% of the former had one of the mtDNA haplotypes associated with *A. m. mellifera/iberiensis*, versus 36.7% of the latter. Also, in their analysis of enzyme polymorphisms, Schiff and Sheppard (1995) found lower levels of polymorphisms in the queen producers' colonies than in the wild colonies. Only one enzyme, malate dehydrogenase, was polymorphic in the queen producers' colonies, whereas this enzyme and three others were polymorphic in the wild colonies.

Africanization of wild colonies in the southern United States

The most recent addition to the amalgamation of honey bees living in North America is the Africanized honey bee. It has entered this continent twice, first in 1987, by way of Florida (the mechanism is unknown), and then a second time in 1990 when swarms flew across the U.S.-Mexico border into Texas (Rubink et al. 1996). Since then, populations of colonies that are hybrids of Africanized and European honey bees have developed in the humid, subtropical parts of the states along the Gulf Coast of the U.S., from Florida to Texas, and in the southernmost parts of New Mexico, Arizona, and California. The Africanized honey bees that live in North America derive from a founder population of a sub-Saharan African subspecies, *A. m. scutellata*, that was introduced to Brazil from South Africa in 1956 for crossbreeding with European subspecies. The aim of this breeding program was to produce a honey bee that was well adapted to tropical living conditions (Kerr 1967). In 1957, several swarms containing *A. m. scutellata* queens escaped from the quarantine apiary in Brazil, and since then this tropically adapted subspecies has established strong populations of wild colonies throughout the American tropics (Winston 1992).

The invasion of the southern U.S. by the Africanized honey bee, and its impact on the resident populations of mainly European derived honey bees living in this region, has been described genetically by a remarkable, 11-year survey of the mitochondrial DNA of a population of wild honey bee colonies living in a wildlife preserve (the Welder Wildlife Refuge) in southern Texas (Pinto et al. 2004, 2005). Each year, from 1991 to 2001, the researchers collected worker bees from colonies occupying tree cavities and bait hives ("swarm traps"). The most informative samples were those collected from new colonies occupying vacant nesting cavities (in hollow trees or bait hives) because these provided the clearest picture each year of the gene flow into the study population. A mitochondrial DNA analysis was made on one worker from each colony to assign it to one of four mitotype categories—African (*A. m. scutellata*), Eastern European (*A. m. ligustica*, *A. m. carnica*, and *A. m. caucasica*), Western European (*A. m. mellifera* and *A. m. iberiensis*), and Egyptian (*A. m. lamarckii*).

Figure 10.4 shows the dynamics in the mitotype composition of the population of new colonies found in vacated cavities and bait hives over the 11 years of the study. It shows dramatic changes over just a few years. Initially, none of the new colonies possessed the *A. m. scutellata* mitotype, but in 1993 one colony was found that had it. The frequency of this African bee mitotype then rose impressively in 1994–1996, as the matriline composition of the population flipped from a predominantly eastern European population to a predominantly African population. What exactly caused this striking pattern of change in the genetics of this population of honey bees? One factor was, of course, the arrival of *A. m. scutellata* in the area in 1993. However, a

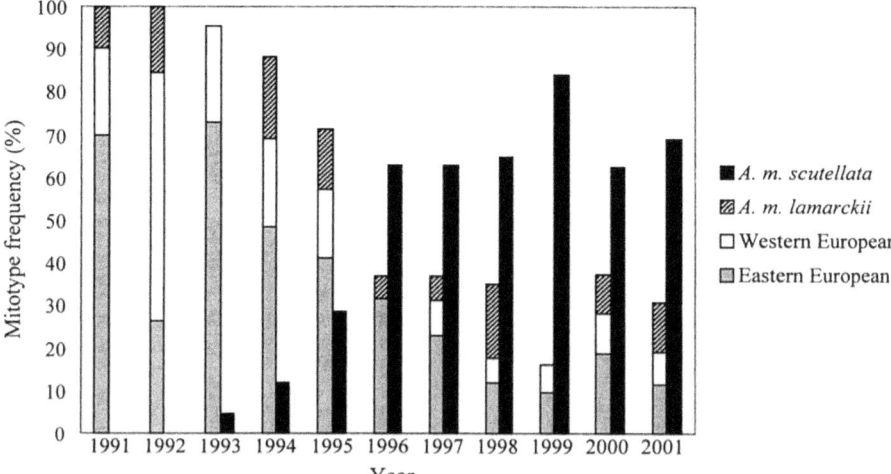

Figure 10.4: Mitotype frequency distributions of newly established colonies (in tree cavities and bait hives) at the Welder Wildlife Refuge, in southern Texas. The stacked columns show the frequencies of the non-*A. m. scutellata* mitotypes and the solid columns show the frequency of the *A. m. scutellata* mitotype. From Pinto et al. (2004).

second factor was the arrival in 1995–1996 of the parasitic mite, *Varroa destructor*. Numerous studies have shown that Africanized honey bee colonies suffer lower mortality from *Varroa destructor* than European honey bee colonies (Guzmán-Novoa et al. 1996). Curiously, the non-*A. m. scutellata* mitotypes were not driven to extinction by the end of this study. A follow-up study was conducted in 2013, in which 28 colonies living in the Welder Wildlife Reserve were again sampled (Rangel et al. 2016). Analysis of the bees' mitochondrial DNA found that the maternal ancestry of 25 of the colonies was African and western European for the other 3 colonies. This study also reports the results of whole-genome analyses of the honey bees living in the Welder Wildlife Refuge. They show that there was strong introgression of African genes into the population between 1991 and 1998, but that since 1998 their proportion has been stable at about 0.90. In contrast, when Mikheyev et al. (2015) performed a whole-genome analysis of the nuclear genes in honey bees living in the Arnot Forest in New York State, which is ca. 1,500 km north of the Welder Wildlife Refuge in Texas, they found that only a tiny fraction, less than 1%, of the ancestry of these bees traces to African honey bees (*A. m. scutellata* and *A. m. jemenitica*). Instead, most of their ancestry (about 80%) traces to honey bees from eastern Europe (*A. m. ligustica*, *A. m. carnica*, and *A. m. caucasica*), and the rest (about 20%) traces to bees from western Europe (*A. m. mellifera*). Thus we see within North America a clear demonstration of how colonies of the African and European subspecies of *Apis mellifera* are markedly different in the climate to which they are best adapted.

Conclusions

The history of honey bees in North America is long and complex. It extends back in time for some 14 million years, to when at least one species, *Apis nearctica*, expanded its range from Asia to North America via a Bering Land Bridge. This history was broken when *Apis nearctica* became extinct, perhaps during a period of global cooling about 7 million years ago. It was reestablished about 400 years ago when European colonists brought *Apis mellifera mellifera* from England to New England in the early 1600s, and perhaps also *Apis mellifera iberiensis* from Spain to New Spain. Over the next 200 years, the population of European honey bees expanded westward from its beachhead on the Atlantic coast, often in advance of the European settlers, and by the early 1800s, it had crossed the Mississippi River. The next major chapter in this biological success story unfolded between 1859 to 1922, a 63-year period during which beekeepers in North America imported honey bees of seven additional subspecies from eastern Europe, western Asia, and northern Africa, with the aim of finding a better bee. These days, it is stocks of just three of these subspecies—*A. m. ligustica, A. m. carnica,* and *A. m. caucasica,* all in the mitochondrial DNA lineage C—that are favored by commercial beekeepers and queen producers in North America. Nevertheless, as of 2009, the population of wild (unmanaged) colonies in North America also includes bees in the mitochondrial DNA lineages M (*A. m. mellifera, A. m. iberiensis,* and *A. m. intermissa*) and O (*A. m. cypria* and *A. m. anatoliaca*), not just C. It remains unknown why the wild colonies in the U.S. possess higher genetic diversity than the managed colonies, but it may be because the wild colonies are better adapted to a broad range of climates than are the managed colonies. The latest chapter in this still-unfolding story started in the late 1980s and early 1990s when colonies of *A. m. scutellata,* the African honey bee, entered the southernmost parts of the U.S. and so expanded its range to include parts of North America. It is noteworthy that North America, which for thousands of years had no honey bees, following the extinction of *Apis nearctica,* is now a special "melting pot" of the genes of *Apis mellifera.*

Conclusion

The honey bee species *A. mellifera* and *A. cerana* are of crucial importance for humanity. One-third of our food is dependent on the pollination of fruits, nuts, and vegetables by bees and other insects. Extensive losses of honey bee colonies in recent years are a major cause for concern. Honey bees face threats from disease, climate change, and management practices. In order to combat these threats, it is important to understand the evolutionary history of honey bees and how they are adapted to different environments across the world. Evolution, as well as phylogenetics of honey bees, remains unclear. Different aspects of honey bee evolution, adaptation, taxonomy, phylogenetics, and methods of their study are discussed in this book.

In the first chapter, the regularities of the colony appearance in the superfamily of bees are analyzed. It is shown that the transformation of highly organized eusocial bee species of the colony into a biological unit is associated with the female-foundress improvement of the instincts of caring for the offspring and the increase in the duration of her life with the reinforcement of various forms of dominance. The idea is that the development of dominant females' offspring under similar conditions, excluding intra-breeding competition as well as the strengthening of interrelation and interdependence among colony members, is associated with the appearance of colony selection. The consolidating role of the unique acoustic signaling in the bee colony reinforcement is shown.

In the second chapter, the phylogenetics based on molecular characteristics of inter- and intraspecies of Asian honey bees are investigated. The present distribution of the honey bee may have resulted from historical geological motions. The first fossil of a recognizable honey bee was linked to the Oligocene Epoch. Phylogenetics of Asian honey bees inferred through nuclear and mitochondrial genes added five new *Apis* species, namely, *A. andreniformis*, *A. koschevnikovi*, *A. laboriosa*, *A. nigrocincta*, and *A. nuluensis*. The greatest distribution and diversity level of *A. cerana* based on classic and geometric morphometrics and molecular analyses were reviewed. The tropical honey bee species from the Borneo (*A. nuluensis*, *A. koschevnikovi*), Sulawesi, and Adjacent islands (*A. nigrocincta*, *A. d. binghami*), Indo-China and Sundaland (*A. andreniformis*) were analyzed.

In the third chapter, a brief description of the origin of European bees and their intraspecific biodiversity were provided. The properties of honey bee subspecies distributed in Russia *A. m. carnica*, *A. m. cypria*, *A. m. iberiensis*, *A. m. ligustica*, *A. m. mellifera*, *A. m. carpathica*, *A. m. acervorum*, *A. m. caucasica*, *A. m. remipes* and Far-Eastern line were presented. The mass import and increasing practice of queen trade and colony movements can endanger regional ecotypes by promoting hybridization with bees with a greater commercial interest as *A. m. carnica* and *A. m. ligustica*. The original geographic distribution pattern of European subspecies is being dissolved in the European Union. The conservation and technical programs attempt to improve the economic value of colonies and preserve locally adapted populations and subspecies in Europe. There is a need to encourage regional breeding efforts to preserve local adaptation and to maintain local strains in isolated conservation apiaries.

In the fourth chapter, the variability of the Siberian honey bees in Russia was investigated using morphometric, mitochondrial DNA (mtDNA) and microsatellite methods. It was found that 65% of bee colonies living in Siberia originate on the maternal line from the *A. m. mellifera* subspecies. The genetic structure and molecular diversity of Siberian honey bee populations were investigated with 11 microsatellite markers. A comparative analysis of the genetic diversity on microsatellite loci in honey bee of two subspecies belonging to different evolutionary lineages *A. m. mellifera* (M lineage), and *A. m. carpathica* (C lineage) was conducted. The mass hybridization of dark European honey bees with south and east European honey bee subspecies are found. The importance of the study molecular biodiversity among bee populations for conservation local gene pool was emphasized.

In the fifth chapter, the roles of the morphology methods to study the phylogeny of *A. mellifera* and *A. cerana* were discussed. The natural distribution of *A. cerana* covers Central Asian, South- and North-eastern Asian countries. The natural distribution of *A. cerana* covers all Europe, Africa, the Middle East, and western Asia. The contemporary taxonomy of *A. cerana* and *A. mellifera* was presented. The classic and geometric morphometry of Asian and European honey bees in taxonomical studies was shown. Diversity of honey bees in genus *Apis*, speciation and distribution of *Apis cerana* in Asia were studied. The consideration of samples for morphometrics of *Apis cerana* was made: the samples of five colonies per locality with 10 bees per colony have shown to be adequate in the analysis of morphological data in studies of honey bee populations.

In the sixth chapter, the current drivers of taxonomic biodiversity loss in Asian and European bees were showed. Loss of taxonomic biodiversity of honey bees can be attributed to colony losses. Elevated losses of *A. mellifera* colonies have been observed since over a decade now in numerous countries. The drivers of colony losses that add to the loss of taxonomic biodiversity of bees are most often pests and pathogens, especially the relatively newly introduced ones, such as the Microsporidia *N. ceranae*, parasitic mite *V. destructor*, parasitic fungi *A. Apis*, small hive beetle *A. tumida*, European foulbrood *M. plutonius*, American foulbrood *P. larvae*, yellow legged hornet *V. velutina*. The loss of taxonomic biodiversity in honey bees could somehow be controlled by breeding local bee strains rather than imported ones.

In the seventh chapter, the methods for measuring the exterior features of honey bee subspecies and for statistical analysis of the results used at FSBSI "Federal Beekeeping Research Center" (Russia) were presented. The systematization and analysis of the software used for classic and geometric morphometry were carried out. An important cause of death of honey bee colonies in Russia by poisoning with pesticides was shown. They are capable of causing not only the quick death of bees, but also the accumulation of contaminated feed reserves, the pollution of wax, which leads to a long course of poisoning. Pesticides can increase the sensitivity of bees to other pathogenic factors and reduce the strength of the colony. Twelve breeds and breed types were included in the state register of selection achievements of the Russian Federation. The biological and economic characteristics of bee colonies of different breeds and types were presented.

In the eighth chapter, the way of breeding better and healthy honey bees to save native biodiversity was discussed. The collective defense against parasites is termed a social immunity. The ability to remove foreign particles and pollen from the body is an important behavior for an individual bee when grooming herself or one bee may groom another bee: autogrooming or allogrooming behavior. Auto grooming has a significant impact as a defense mechanism for genetic resistance to the tracheal mite, *A. woodi*, and *V. destructor*. Disease resistance is correlated with the "hygienic behavior" of worker bees. Hygienic behavior was originally defined as the ability of honey bees to detect and remove brood infected by *P. larvae* from the nest. In Europe, there are attempts to preserve pure honey bee races and to stimulate commercial queen rearing activities. Different selection tools have been incorporated into beekeeping practice to select hygienic behavior lineages in a certain honey bee population.

In the ninth chapter, the properties of Africanized honey bees in Latin America are discussed. The western honey bee *A. mellifera* is exotic in Latin America, introduced from Europe in the 17th century. Various European subspecies were first introduced, followed by the African subspecies *A. m. scutellata* in 1957. Since this date, the *A. m. scutellata* populations have spread throughout Latin America. Native honey-producer bees also exist in Latin America: the stingless bees Meliponini. In Latin America, stingless bees are very diverse in terms of behavior, morphology, and species richness. Some of the stingless bee species are managed for honey production by meliponiculture. Honey bees and stingless bees play critical socio-ecological roles in Latin America, and more research is required to understand their ecology and to establish adapted and sustainable conservation schemes.

In the tenth chapter, the history of honey bees in North America is shown. Over the next 200 years, the population of European honey bees expanded westward from its beachhead on the Atlantic coast, often in advance of the European settlers, and by the early 1800s, it had crossed the Mississippi River. The next major chapter in this biological success story unfolded between 1859 to 1922, a 63-year period during which beekeepers in North America imported honey bees of seven additional subspecies from eastern Europe, western Asia, and northern Africa, with the aim of finding a better bee. These days, it is stocks of just three of these subspecies— *A. m. ligustica*, *A. m. carnica*, and *A. m. caucasica*, all in the mitochondrial DNA lineage C- that are favored by commercial beekeepers and queen producers in North America. Nevertheless, as of 2009, the population of wild (unmanaged) colonies in North America also includes bees in the mitochondrial DNA lineages M (*A. m. mellifera*, *A. m. iberiensis*, and *A. m. intermissa*) and O (*A. m. cypria* and *A. m. anatoliaca*), not just C.

Thus, the book sheds light on many unexplored issues in the biology of bees. Many features of honey bee phylogeny, taxonomy, diseases, and evolution are presented in the book, which can be useful for conservation biodiversity of honey bees. Also, economical development of our countries due to beekeeping allows an increasing income in agriculture without demolition of nature and ecology. Agriculture joint with beekeeping will be able to lead to the successful economic growth of our country and nation. It will be great if information provided in the book will change the current serious situation in beekeeping.

References

Abdulov, T.F., G.S. Shingareva and A.M. Ishemgulov. 2008. Bashkirian honey bees: history. Materials of the International conference "Beekeeping of XXI century. Dark bee (*Apis mellifera mellifera* L.) in Russia". (in Russian). Pishchepromizdat, Moscow: 152–154.

Abrol, D.P. 2013. Asiatic honey bee *Apis cerana*: Biodiversity conservation and agricultural production. Heidelberg: Springer Verlag.

Adams, D.C., F.J. Rohlf and D.E. Slice. 2004. Geometric morphometrics: Ten years of progress following the "revolution". Italian Journal of Zoology. 71: 5–16. doi: 10.1080/11250000409356545.

Adams, J., E.D. Rothman, W.E. Kerr and Z.L. Paulino. 1977. Estimation of sec alleles and queen matings from diploid male frequencies in a population of *Apis mellifera*. Genetics. 86: 583–596.

Affleck, T. 1841. Bee-breeding in the West. Lucas Press, Cincinnati.

Agrochemical Service. 2000. Wood Mackenzie Consultants Limited, Edinburgh, UK.

Aguiar, A.P., A.R. Deans, M.S. Engel, M. Forshage, J.T. Huber, J.T. Jennings et al. 2013. Order hymenoptera. pp. 51–62. *In*: Zhang, Z.-Q. [ed.]. Animal Biodiversity: An Outline of Higher-level Classification and Survey of Taxonomic Richness (*Addenda* 2013). Zootaxa. 3703: 1–82. doi: 10.11646/zootaxa.3703.1.12.

Aizen, M.A. and L.D. Harder. 2009. The global stock of domesticated honey bees is growing slower than agricultural demand for pollination. Current Biology. 19: 1–4.

Aizen, M.A., L.A. Garibaldi, S.A. Cunningham and A.M. Klein. 2009. How much does agriculture depend on pollinators? Lessons from long-term trends in crop production. Annals of Botany. 103: 1579–1588.

Akratanakul, P., K. Saen and N. Pathom. 1990. Beekeeping in Asia. FAO Agricultural Services Bulletin. 68: 4.

Akyol, E., H. Yeninar and O. Kaftanoglu. 2008. Live weight of queen honey bees (*Apis mellifera* L.) predicts reproductive characteristics. Journal of the Kansas Entomological Society 81: 92–100.

Alaux, C., F. Ducloz, D. Crauser and Y. Le Conte. 2010b. Diet effects on honey bee immunocompetence. Biology Letters. 6: 562–565.

Alaux, C., J.L. Brunet, C. Dussaubat, F. Mondet, S. Tchamitchan, M. Cousin et al. 2010a. Interactions between *Nosema* microspores and a neonicotinoid weaken honey bees (*Apis mellifera*). Environmental Microbiology. 12: 774–782. doi: 10.1111/j.1462-2920.2009.02123.x.

Alaux, C., C. Dantec, H. Parrinello and Y. Le Conte. 2011. Nutrigenomics in honey bees: digital gene expression analysis of pollen's nutritive effects on healthy and *varroa*-parasitized bees. BMC Genomics. 12: 496.

Alburaki, M., S. Moulin, S. Legout, A. Alburaki and L. Garnery. 2011. Mitochondrial structure of Eastern honey bee populations from Syria, Lebanon and Iraq. Apidologie. 42: 628–641. doi: 10.1007/s13592-011-0062-4.

Alexander, B.A. 1991. Phylogenetic analysis of the genus *Apis* (Hymenoptera : Apidae). Annals of the Entomological Society of America. 84: 137–149.

Alford, D. 1970. The incipient stages of development of bumble bee colonies. Insectes Sociaux. 171: 1–10.

Alford, D.V. 1971. Egg laying by bumble bee queens at the beginning of colony development. Bee World. 52: 11–18.

Alford, D.V. 1975. Bumble Bees. London: Davis-Poynter.

Aliouane, Y., A.K. EL Hassani, V. Gary, C. Armengaud, M. Lambin and M. Gauthier. 2009. Subchronic exposure of honey bees to sublethal doses of pesticides: effects on behavior. Environmental Toxicology and Chemistry. 28: 113–122. doi: 10.1897/08-110.1.

Allen, M.F., B.V. Ball and B.A. Underwood. 1990. An isolate of *Melissococcus pluton* from *Apis laboriosa*. Journal of Invertebrate Pathology. 55: 439–440.

Allen, M.F. and B.V. Ball. 1996. The incidence and world distribution of honey bee viruses. Bee World. 77: 141–162.

Alpatov, W.W. 1927. Biometrical study on bees of Middle and Southern Russia (in Russian). Revue Zoologique Russe. 7: 31–74.

Alpatov, W.W. 1929. Biometrical studies on variation and races of the honey bee *Apis mellifera* L. The Quarterly Review of Biology. 4: 1–58. doi: 10.1086/394322.

Alpatov, W.W. 1935. Contribution to the study of variation in honey bee. III. The cubital cell on the wings of different forms of the genus *Apis* and its taxonomical and evolutionary significance (in Russian). Revue Zoologique Russe. 14: 664–673.

Alpatov, W.W. 1945. The races of honey bees as the basis of its breeding. MSU Press, Moscow.

Alpatov, W.W. 1948. The races of honey bees and their use in agriculture (in Russian). Sredi Prirody. 4. MOIP Press, Moscow.

Andere, C., M.A. Palacio, E. Rodriguez, M.T. Dominguez, E. Figini and E. Bedascarrasbure. 2000. Evaluation of honey bees defensive behavior in Argentina. I. A field method. American Bee Journal. 140: 975–978.

Anderson, D.L. 1994. Non-reproduction of *Varroa jacobsoni* in *Apis mellifera* colonies in Papua New Guinea and Indonesia. Apidologie. 25: 412–421.

Anderson, D.L. and J.W.H. Trueman. 2000. *Varroa jacobsoni* (Acari: Varroidae) is more than one species. Experimental and Applied Acarology. 24: 165–189.

Anderson, D.L. 2010. Control of Asian honey bees in the Solomon Islands. Bruce, ACT: Australian Centre for International Agricultural Research (ACIAR).

Anderson, D.L., N. Annand, M. Lacey and S. Ete. 2012. Control of Asian honey bees in Solomon Islands. Canberra, ACT: Australian Centre for International Agricultural Research (ACIAR).

Annand, N. 2009. The Solomon Experience with Asian Honey Bees. The Australasian Beekeeper. [online]. http://www.theabk.com/article/solomon-experience-asian-honey-bees.

Antontseva, P.V. 1975. Seasonal changes in external characters of local bees of the Dzhungarskü Alatau (mountains). Proceedings of Alma-Ata and Semipalatinsk Zooveterenary Institutes. 36: 31–32.

Antúnez, K., R. Martin-Hernandez, L. Prieto, A. Meana, P. Zunino, M. Higes et al. 2009. Immune suppression in the honey bee (*Apis mellifera*) following infection by *Nosema ceranae* (Microsporidia). Environmental Microbiology. 11: 2284–2290.

Antúnez, K., F. Requier, P. Aldea, M. Basualdo, B. Branchiccela, R. Calderón et al. 2018. SOLATINA: a Latin-American Bee Research Association to foster the interactions between scientists and coordinate large-scale research programs. Bee World. 95: 124–127.

Arathi, H.S., I. Burns and M. Spivak. 2000. Ethology of hygienic behaviour in the honey bee *Apis mellifera* L. (Hymenoptera: Apidae). Ethology. 106: 365–379.

Arena, M. and F. Sgolastra. 2014. A meta-analysis comparing the sensitivity of bees to pesticides. Ecotoxicology. 23: 324–334.

Arias, M.C. and W.S. Sheppard. 1996. Molecular phylogenetics of honey bee subspecies (*Apis mellifera* L.) inferred from mitochondrial DNA sequence. Molecular Phylogenetics and Evolution. 5: 557–566. doi: 10.1006/mpev.1996.0050.

Arias, M.C., S. Tingek, A. Kelitu and W. Sheppard. 1996. *Apis nulensis* Tingek, Koeniger and Koeniger, 1996 and its genetic relationship with sympatric species inferred from DNA sequences. Apidologie. 27: 415–422.

Arias, M.C. and W.S. Sheppard. 2005. Phylogenetic relationships of honey bees (Hymenoptera: Apinae: Apini) inferred from nuclear and mitochondrial DNA sequence data. Molecular Phylogenetics and Evolution. 37: 25–35. doi: 10.1016/j.ympev.2005.02.017.

Arias, M.C., T.E. Rinderer and W.S. Sheppard. 2006. Further characterization of honey bees from the Iberian Peninsula by allozyme, morphometric and mtDNA haplotype analyses. Journal of Apicultural Research. 45: 188–196.

Aston, D. 2010. Honey bee winter loss survey for England, 2007–2008. Journal of Apicultural Research. 49: 111–112.

Atwal, A.S. 2000. Essentials of beekeeping and pollination. Kalyani Publishers, Ludhiana.

Avetisyan, G.A. 1961. The relation between interior and exterior characteristics of the queen and fertility and productivity of the bee colony. XVIII International Beekeeping Congress, Madrid, Spain: 44–53.

Avetisyan, G.A. 1982. Beekeeping (in Russian). Kolos, Moscow.

Avetisyan, G.A. and A. Cherevko. 2001. Beekeeping (in Russian). Academy Publisher, Moscow.

Aytekin, A.M., M. Terzo, P. Rasmont and N. Çağatay. 2007. Landmark-based geometric morphometric analysis of wing shape in *Sibiricobombus Vogt* (Hymenoptera: Apidae: *Bombus latreille*). Annales Societe Entomologique de France. 43: 95–102. doi: 10.1080/00379271.2007.10697499.

Bacandritsos, N., A. Granato, G. Budge, I. Papanastasiou, E. Roinioti and M. Caldon. 2010. Sudden deaths and colony population decline in Greek honey bee colonies. Journal of Invertebrate Pathology. 105: 335–340.

Badino, G., G. Celebrano and A. Manino. 1982. Genetic variability of *Apis mellifera ligustica* Spin. in a marginal area of its geographical distribution. Experientia. 38: 540–541.

Badino, G., G. Celebrano and A. Manino. 1983. Population structure and *MDH-1* locus variation in *Apis mellifera ligustica*. Journal of Heredity. 74: 443–446.

Badino, G., G. Celebrano and A. Manino. 1984. Population genetics of Italian honey bee *Apis mellifera ligustica* Spin. and its relationships with neighbouring subspecies. Bollettino Del Museo Di Scienze Naturali Di Torino. 2: 571–584.

Bailey, L. 1956. A etiology of European foulbrood: A disease of the larval honey-bee. Nature. 178: 1130.

Bailey, L. 1957. The cause of European foulbrood. Bee World. 38: 85–89.

Bailey, L. 1966. The effect of acid-hydrolyzed sucrose on honey bees. Journal of Apicultural Research. 5: 127–136.

Bailey, L. 1983. *Melissococcus pluton*, the cause of European foulbrood of honey bees (*Apis* spp.). Journal of Applied Bacteriology. 55: 65–69.

Bailey, L. and B.V. Ball. 1991. Honey Bee Pathology. Academic Press, London.

Baitala, T.V., P. Faquinello, V.A.A. Toledo, C.A. Mangolin, E.N. Martins and M.C.C. Ruvolo-Takasusuki. 2010. Potential use of major royal jelly proteins (MRJPs) as molecular markers for royal jelly production in Africanized honey bee colonies. Apidologie. 41: 160–168.

Banaszak, J. 1995. Changes in fauna of wild bees in Europe. Pedagogical University.

Barcode of Life Database. 2018. The Barcode of Life Database, the worldwide repository of DNA barcode data. [online]. www.boldsystems.org.

Bar-Cohen, R., G. Alpern and R. Bar-Anan. 1978. Progeny testing and selecting Italian queen for brood area and honey production. Apidologie. 9: 95–100.

Barour, C. and M. Baylac. 2016. Geometric morphometric discrimination of the three African honey bee subspecies *Apis mellifera intermissa*, *A. m. sahariensis* and *A. m. capensis* (Hymenoptera, Apidae): Fore wing and hind wing landmark configurations. Journal of Hymenoptera Research. 52: 61–70. doi: 10.3897/jhr.52.8787.

Barrett, S.C.H. and J.R. Kohn. 1991. Genetic and evolutionary consequences of small population sizes in plants: Implications for conservation. pp. 3–30. *In*: Falk, D.A. and K.A. Holsinger [eds.]. Genetics and conservation of rare plants. Oxford University Press, New York.

Barry, S., D. Cook, R. Duthie, D. Clifford and D. Anderson. 2010. Future surveillance needs for honey bee biosecurity. Canberra: Rural Industries Research and Development Corporation.

Barton, B.S. 1792. An inquiry into the question, whether the *Apis mellifica*, or true honey-bee, is a native of America. Transactions of the American Philosophical Society. 3: 241–263.

Basumallick, L. and J. Rohrer. 2001. Determination of Hydroxymethylfurfural and Honey and Biomass. Thermo Fisher Scientific, Sunnyvale, CA, USA. (online source) http://www.dionex.com/en-us/webdocs/109807-AN270-IC-HMF-Honey-Biomass-AN70488_E.pdf (May 7, 2013).

Batra, S.W.T. 1970. Behavior of the alkali bee, *Nomia melanderi*, within the nest (Hymenoptera: Halictidae). Annals of the Entomological Society of America. 63: 400–406.

Batra, S.W.T. 1980. Nest of the solitary bee, *Anthophora antiope*, in Punjab, India. Journal of the Kansas Entomological Society. 53: 112–114.

Baylac, M., L. Garnery, D. Tharavy, J. Pedraza-Acosta, A. Rortais and G. Arnold. 2008. ApiClass, an automatic wing morphometric expert system for honey bee identification [online]. http://apiclass.mnhn.fr.

Bazhenova, O.I. 2006. Landscape-climatic types of systems of exogenous relief formation in subarid regions of southern Siberia. Geography and Natural Resources. 4: 57–65.

Beetsma, J. and. K. Zonneveld. 1992. Observations on the initiation and stimulation of oviposition of the *Varroa* mite. Experimental and Applied Acarology. 16: 303–312.

Belloy, L., A. Imdorf, I. Fries, E. Forsgren, H. Berthoud and R. Kuhn. 2007. Spatial distribution of *Melissococcus plutonius* in adult honey bees collected from apiaries and colonies with and without symptoms of European foulbrood. Apidologie. 38: 136–140.

Belzunces, L.P., S. Tchamitchian and J.-L. Brunet. 2012. Neural effects of insecticides in the honey bee. Apidologie. 43: 348–370.

Bernal, J., E. Garrido-Bailón, M.J. Del-Nozal, A.V. González-Porto, R. Martín-Hernández, J.C. Diego et al. 2010. Overview of pesticide residues in stored pollen and their potential effect on bee colony (*Apis mellifera*) losses in Spain. Journal of Economic Entomology. 103: 1964–1971. doi: 10.1603/EC10235.

Berthoud, H., A. Imdorf, M. Haueter, S. Radloff, P. Neumann. 2010. Virus infections and winter losses of honey bee colonies (*Apis mellifera*). Journal of Apicultural Research. 49: 60–65. doi: 10.3896/IBRA.1.49.1.08.

Bienefeld, K. 1996. Factors affecting duration of the post capping period in brood of the honey bee (*Apis mellifera carnica*). Journal of Apicultural Research. 35: 11–17.

Bieńkowska, M., P. Wegrzynowicz, B. Panasiuk, D. Gerula and K. Loc. 2008. Influence of the age of honey bee queens and dose of semen on condition of instrumentally inseminated queens kept in cages with 25 worker bees in bee colonies. Journal of Apicultural Science. 52: 23–34.

Biesmeijer, J.C., S.P.M. Roberts, M. Reemer, R. Ohlemuller, M. Edwards, T. Peeters et al. 2006. Parallel declines in pollinators and insect-pollinated plants in Britain and the Netherlands. Science. 313: 351–354.

Bilash, G.D. and N.I. Krivtsov. 1983. Measuring the exterior of bees. NIIP Press, Rybnoe.

Blagoveshchenskaya, N.N. 1983. Study of wild bees in nature and laboratory. Ulyanovsk Pedagogical Institute, Ulyanovsk.

Bloch, G., T.M. Francoy, I. Wachtel, N. Panitz-Cohen, S. Fuchs and A. Mazar. 2010. Industrial apiculture in the Jordan valley during Biblical times with Anatolian honey bees. Proceedings of the National Academy of Sciences of the United States of America. 107: 11240–11244. doi: 10.1073/pnas.1003265107.

Blüthgen, N. and A.M. Klein. 2011. Functional complementarity and specialisation: The role of biodiversity in plant-pollinator interactions. Basic and Applied Ecology. 12: 282–291.

Boecking, O. and M. Spivak. 1999. Behavioural defenses of honey bees against *Varroa jacobsoni* Oud. Apidologie. 30: 141–158.

Boecking, O. and E. Genersch. 2008. Varroosis-the ongoing crisis in bee keeping. Journal of Consumer Protection and Food Safety. 3: 221–228.

Boitsenyuk, L. 2008. The choice of breeds. (in Russian). Beekeeping. 7: 18–19.

Bonelli, B. 1977. Osservazioni eto-ecologiche sugli Imenotteri Aculeati dell'Ettopia. VII. *Xylocopa* (*Mesotrichia*) *combusta* Smith (Hymenoptera-Anthophridae). Bollettino dell'Istituto di Entomologia della Universita di Bologna. 33: 1–31.

Bonmatin, J.M., I. Moineau, R. Charvet, M.E. Colin, C. Fléché and E.R. Bengsch. 2005. Behaviour of Imidacloprid in fields. Toxicity for honey bees. pp. 438–494. *In*: Lichtfouse, E., J. Schwarzbauer, R. Didier [eds.]. Environmental chemistry. Springer, Berlin, Germany. doi: 10.1007/3-540-26531-7_44.

Bookstein, F.L. 1991. Morphometric Tools for Landmark Data: Geometry and Biology. Cambridge University Press, New York. doi: 10.1017/cbo9780511573064.

Bookstein, F.L. 1996. Combining the tools of geometric morphometrics. pp. 131–152. *In*: Marcus, L., M. Corti, A. Loy and D. Slice [eds.]. Advances in Morphometrics. Plenum Press, NY, L.

Boot, W.J., J.N.M. Calis and J. Beetsma. 1995a. Does time spent on adult bees affect reproductive success of *Varroa* mites? Entomologia Experimentalis et Applicata. 75: 1–7.

Boot, W.J., J. Schoenmaker, J.N.M Calis and J. Beetsma. 1995b. Invasion of *Varroa jacobsoni* into drone brood cells of the honey bee, *Apis mellifera*. Apidologie. 26: 109–118.

Borodachev, A.V. and N.I. Krivtsov. 2000. New breed type of bees "Priokskiy". Bulletin of the RAAS. 4: 70–72.

Borodachev, A.V. and L.N. Savushkina. 2012. The conservation and sustainable use of gene pool of breeds of honey bees (in Russian). Beekeeping. 4: 3–5.

Borodachev, A.V., L.N. Savushkina and V.A. Borodachev. 2015. Maintaining and improving the assessment of the gene pool of honey bees. (in Russian). Beekeeping. 10: 15–17.

Borodachev, A.V., L.N. Savushkina and V.A. Borodachev. 2016. Breeds of bees and breeding farms for their breeding. Zootechny. 8: 4–6.

Borodachev, A.V., L.N. Savushkina and V.A. Borodachev. 2017. Breeding and characteristics of the breed of bees type of bees Priokskiy. Bulletin of Russian Agricultural Science. 1: 62–65.

Bortolotti, L., A.G. Sabatini, F. Mutinelli, M. Astuti, A. Lavazza and R. Piro. 2010. Spring honey bee losses in Italy. Julius-Kühn-Archiv. 148.

Botías, C., R. Martín-Hernández, L. Barrios, A. Meana and M. Higes. 2013. *Nosema* spp. infection and its negative effects on honey bees (*Apis mellifera iberiensis*) at the colony level. Veterinary Research. 44: 1–14.

Bouga, M. and F. Hatjina. 2005. Genetic variability in Greek honey bee (*A. mellifera* L.) populations using geometric morphometrics analysis. Balkan Scientific Conference of Biology Plovdiv, Bulgaria. 598–602.

Bouga, M., G. Kilias, P.C. Harizanis, V. Papasotiropoulos and S. Alahiotis. 2005. Allozyme variability and phylogenetic relationships in honey bee (Hymenoptera: Apidae: *Apis mellifera*) populations from Greece and Cyprus. Biochemical Genetics. 43: 471–483.

Bouga, M., C. Alaux, M. Bienkowska, R. Büchler, N.L. Carreck, E. Cauia et al. 2011. A review of methods for discrimination of honey bee populations as applied to European beekeeping. Journal of Apicultural Research. 50: 51–84. doi: 10.3896/IBRA.1.50.1.06.

Bowen-Walker, P.L., S.J. Martin and A. Gunn. 1999. The transmission of deformed wing virus between honey bees (*Apis mellifera* L.) by the Ectoparasitic Mite (*Varroa jacobsoni* Oud.). Journal of Invertebrate Pathology. 73: 101–106.

Bowen-Walker, P.L. and A. Gunn. 2001. The effect of the ectoparasitic mite, *Varroa destructor* on adult worker Honey bee (*Apis mellifera*) emergence weights, water, protein, carbohydrate and lipid levels. Entomologia Experimentalis et Applicata. 101: 207–217.

Brand, D.D. 1988. The honey bee in New Spain and Mexico. Journal of Cultural Geography. 9: 71–82.

Brandorf, A. and M. Ivoilova. 2014b. Problems of *A. m. mellifera* conservation in Russia. Materials of 20 Int. Apicultural congress SICAMM and BIBBA, Langollen (UK): 31–32.

Brandorf, A. 2016. Flower specialization of the Dark Bee in the North-Eastern region of European Russia. Materialy (Abstract) of 21 international Apicultural congress SICAMM and DDB, Lunteren (Netherlands): 8–9.

Brandorf, A. and M. Ivoilova. 2016. Production of Royal jelly from the honey bees of the Central Russian breed (*Apis mellifera mellifera*). Biomics. 8. 2: 69–72.

Brandorf, A.Z. and M.M. Ivoilova. 2014a. Valuable gene pool of Russia-*Apis mellifera mellifera* L. Problems and prospects of preservation of the gene pool of honey bees in modern conditions: Materials of 1st International scientifically-practical conference devoted to the 145 anniversary from the birthday of M.A. Dernov. Kirov: NIISH Severo-Vostoka: 42–51.

Brandorf, A.Z. and M.M. Ivoilova. 2015. The breeding evaluation honey bee *Apis mellifera mellifera* L. (black bee). NIISH Severo-Vostoka, Kirov. doi: 10.13140/RG.2.2.28180.81288.

Breed, M.D. and G.J. Gamboa. 1977. Control of worker activities by queen behavior in a primitively eusocial bees. Science. 195: 694–696.

Brian, A.D. 1951. Brood development in *Bombus agrorum* (Hymenoptera, Bombidae). Entomologist's Monthly Magazine. 87: 207–212.

Brian, A.D. 1952. Division of labour and foraging in *Bombus agrorum* Fabricius. Journal of Animal Ecology. 21: 223–240.

Brian, M. 1986. Public Insects. Ecology and behavior. Moscow: World.

Brian, M. 1986. Social insects. Ecology and behavior. Mir, Moscow.

Brodschneider, R. and K. Crailsheim. 2010. Nutrition and health in honey bees. Apidologie. 41: 278–294.

Brodschneider, R., R. Moosbeckerhofer and K. Crailsheim. 2010. Surveys as a tool to record winter losses of honey bee colonies: a two year case study in Austria and South Tyrol. Journal of Apicultural Research. 49: 23–30.

Brodschneider, R., A. Gray, R. van der Zee, N. Adjlane, V. Brusbardis, J.D. Charrière et al. 2016. Preliminary analysis of loss rates of honey bee colonies during winter 2015/16 from the COLOSS survey. Journal of Apicultural Research. 55: 375–378.

Brodschneider, R., A. Gray, N. Adjlane, A. Ballis, V. Brusbardis, J.-D. Charrière et al. 2018. Multi-country loss rates of honey bee colonies during winter 2016/2017 from the COLOSS survey. Journal of Apicultural Research. 57: 452–457.

Brothers, D.J. 1975. Phylogeny and classification of the aculeate Hymenoptera, with special reference to Mutillidae. The University of Kansas science bulletin. 50: 483–648.

Bruckner, S., N. Steinhauer, K. Rennich, S. Aurell, D.M. Caron, J.D. Ellis et al. 2018. Honey Bee Colony Losses 2017–2018: Preliminary Results [online]. https://beeinformed.org/results/honey-bee-colony-losses-2017–2018-preliminary-results.

Büchler, R. and W. Drescher. 1990. Variance and heritability of the capped developmental stage in european *Apis mellifera* L. and it's correlation with increased *Varroa jacobsoni* Oud. infestation. Journal of Apicultural Research. 29: 172–176.

Büchler, R., W. Drescher and I. Thornier. 1992. Grooming behaviour of *Apis cerana, Apis mellifera* and *Apis dorsata* and its effect on the parasitic mites *Varroa jacobsoni* and *Tropilaelaps clareae*. Experimental and Applied Acarology. 16: 313–319.

Büchler, R., S. Berg and Y. Le Conte. 2010. Breeding for resistance to *Varroa destructor* in Europe. Apidologie. 41: 393–408.

Büchler, R., C. Costa, F. Hatjina, S. Andonov, M.D. Meixner, Y. Le Conte et al. 2014. The influence of genetic origin and its interaction with environmental effects on the survival of *Apis mellifera* L. colonies in Europe. Journal of Apicultural Research. 53: 205–214.

Budnikova, N.V., D.V. Mitrofanov, L.A. Burmistrova and V.N. Kosarev. 2018a. DDT migration and its metabolites in the system of "soil-honey plants-bees-products of beekeeping" (in Russian). Beekeeping. 9: 4–5.

Budnikova, N.V., D.V. Mitrofanov, L.A. Burmistrova and V.N. Kosarev. 2018b. Hazard of contamination of beekeeping products with HCCH and its isomers. (in Russian). Beekeeping. 10: 8–9.

Burkle, L.A., J.C. Marlin and T.M. Knight. 2013. Plant-pollinator interactions over 120 years: loss of species, co-occurrence and function. Science. 339: 1611–1615.

Butani, D.K. 1950. An *Apis dorsata* colony in New Delhi. Indian Bee Journal. 12: 115.

Butler, C.G. 1954. The World of the Honey bee. Collins, London.

Butler, K.G. 1969. Honey bee family and its evolution. Bee and hive. Moscow: Kolos.

Butler, K.J. 1990. Honey bee world. Moscow: Kolos.

Camazine, M.E. 1986. Differential reproduction of the mite, *Varroa jacobsoni* (Mesostigmata: Varroidae), on africanized and European honey bees (Hymenoptera: Apidae). Annals of the Entomological Society of America. 79: 801–880.

Cane, J.H. and V.J. Tepedino. 2017. Gauging the effect of honey bee pollen collection on native bee communities. Conservation Letters. 10: 205–21.

Cánovas, F., P. De la Rúa, J. Serrano and J. Galián. 2008. Geographical patterns of mitochondrial DNA variation in *Apis mellifera iberiensis* (Hymenoptera: Apidae). Journal of Zoological Systematics and Evolutionary Research. 46: 24–30.

Cao, L., H. Zheng, C. Hu, S. He, H. Kuang and F. Hu. 2012. Phylogeography of *Apis dorsata* (Hymenoptera: Apidae) from China and neighboring Asian areas. Annals of the Entomological Society of America. 105: 298–304.

Carreck, N.L., B.V. Ball and S.J. Martin. 2010a. The epidemiology of cloudy wing virus infections in honey bee colonies in the UK. Journal of Apicultural Research. 49: 66–71.

Carreck, N.L., B.V. Ball and S.J. Martin. 2010b. Honey bee colony collapse and changes in viral prevalence associated with *Varroa destructor*. Journal of Apicultural Research. 49: 93–94.

Carvalho, F. P. 2006. Agriculture, pesticides, food security and food safety. Environmental Science & Policy. 9: 685–692.

Carvell, C., D.B. Roy, S.M. Smart, R.F. Pywell, C.D. Preston and D. Goulson. 2006. Declines in forage availability for bumble bees at a national scale. Biological Conservation. 132: 481–489.

Cauia, E., D. Usurelu, L.M. Magdalena, D. Cimponeriu, P. Apostol, A. Siceanu et al. 2008. Preliminary researches regarding the genetic and morphometric characterization of honey bee (*A. mellifera* L.) from Romania. Scientific Papers Animal Sci. Biotechnologies. 41: 278–286.

Cermák, K. and F. Kaspar. 2000. A method of classifying honey bee races by their body characters. Pszczelnicze Zeszyty Naukowe. 2: 81–85.

Cermák, K. and F. Kaspar. 2019. Výzkumný ústav včelařský v dole. [online]. www.beedol.cz.

Chaimanee, V., J.D. Evans, Y. Chen, C. Jackson and J.S. Pettis. 2016. Sperm viability and gene expression in honey bee queens (*Apis mellifera*) following exposure to the neonicotinoid insecticide imidacloprid and the organophosphate acaricide coumaphos. Journal of Insect Physiology. 89: 1–8. doi: 10.1016/j.jinsphys.2016.03.004.

Chaimanee, V., P. Chantawannakul, Y. Chen, J.D. Evans and J.S. Pettis. 2012. Differential expression of immune genes of adult honey bee (*Apis mellifera*) after inoculated by *Nosema ceranae*. Journal of Insect Physiology. 58: 1090–1095.

Chaud-Netto, J. and J.H. Bueno. 1979. Number of ovarioles in workers of *Apis mellifera adansonii* and *Apis mellifera ligustica*: A Comparative Study. Journal of Apicultural Research. 18: 260–263.

Chauzat, M.P., J.P. Faucon, A.C. Martel, J. Lachaize, N. Cougoule and M. Aubert. 2006. A survey of pesticide residues in pollen loads collected by honey bees in France. Journal of Economic Entomology. 99: 253–262. doi: 10.1093/jee/99.2.253.

Chauzat, M.P., P. Carpentier, A.C. Martel, S. Bougeard, N. Cougoule, P. Porta et al. 2009. Influence of pesticide residues on honey bee (Hymenoptera: Apidae) colony health in France. Journal of Environmental Entomology. 38: 514–523. doi: 10.1603/022.038.0302.

Chauzat, M.P., P. Carpentier, F. Madec, S. Bougeard, N. Cougoule and P. Drajnudel. 2010. Role of infectious agents and parasites in the health of honey bee colonies in France. Journal of Apicultural Research. 49: 31–39.

Chen, C., Z. Liu, Q. Pan, X. Chen, H. Wang, H. Guo et al. 2016. Genomic analyses reveal demographic history and temperate adaptation of the newly discovered honey bee subspecies *Apis mellifera sinisxinyuan* n. ssp. Mol. Biol. Evol. 33: 1337–1348. doi: 10.1093/molbev/msw017.

Chen, C., L. Zhiguang, L. Yuexiong, X. Zheng, W. Shunhai, Z. Xuewen et al. 2017. Managed Honey bee colony losses of the Eastern Honey bee (*Apis cerana*) in China (2011–2014). Apidologie. 48: 692–702.

Chen, L., G. Wang, Y.N. Zhu, H. Xiang and W. Wang. 2016. Advances and perspectives in the application of CRISPR/Cas9 in insects. Zoological Research. 37: 220–228. doi: 10.13918/j.issn.2095-8137.2016.4.220.

Chen, Y., J.D. Evans, I.B. Smith and J.S. Pettis. 2008. *Nosema ceranae* is a long-present and wide-spread microsporidian infection of the European honey bee (*Apis mellifera*) in the United States. Journal of Invertebrate Pathology. 97: 186–188.

Chen, Y.P. and R. Siede. 2007. Honey bee viruses. Advances in Virus Research. 70: 33–80.

Cherevko, J.A. 2006. Natural selection and pure-bred breeding (in Russian). Beekeeping. 10: 10–12.

Choi, Y.S., M.Y. Lee, I.P. Hong, N.S. Kim, H.K. Kim, K.G. Lee et al. 2010. Occurrence of sacbrood virus in Korean apiaries from *Apis cerana* (Hymenoptera: Apidae). Journal of Apiculture (Korea). 25: 187–191.

Clarke, K.E., T.E. Rinderer, P. Franck, J.G. Quezada-Euan and B.P. Oldroyd. 2002. The Africanization of honey bees (*Apis mellifera* L.) of the Yucatan: A study of a massive hybridisation event across time. Evolution. 56: 1462–1474.

Claudianos, C., H. Ranson, R.M. Johnson, S. Biswas, M.A. Schuler, M.R. Berenbaum et al. 2006. A deficit of detoxification enzymes: pesticide sensitivity and environmental response in the honey bee. Insect Molecular Biology. 15: 615–636. doi: 10.1111/j.1365-2583.2006.00672.x.

Cockerell, T.D.A. 1908. Descriptions and records of bees. Annals and Magazine of Natural History. 8: 323–334.

Collins, A.M. and M.S. Blum. 1982. Bioassay of compounds derived from the honey bee sting. Journal of Chemical Ecology. 8: 463–470.

Collins, A.M. 1985. Africanized honey bees in Venezuela: defensive behaviour. Proceedings of the Third International Conference on Apiculture in Tropical Climates. International Bee Research Association, Nairobi, Kenya: 117–122.

Comparini, A. and A. Biasiolo. 1991. Genetic discrimination of Italian bee, *Apis mellifera ligustica versus* Carniolan bee, *Apis mellifera carnica* by allozyme variability analysis. Biochemical Systematics and Ecology. 19. 3: 189–194.

Corlett, R.T. 2001. Pollination in a degraded tropical landscape: a Hong Kong case study. Journal of Tropical Ecology. 17: 155–161.

Corlett, R.T. 2011. Honey bees in natural ecosystems. pp. 215–225. *In*: Hepburn, H.R. and S.E. Radloff [eds.]. Honey Bees of Asia. Springer, Berlin.

Cornman, R.S., D.R. Tarpy, Y.P. Chen, L. Jeffreys, D. Lopez, J.S. Pettis et al. 2012. Pathogen Webs in Collapsing Honey Bee Colonies. PLOS ONE. 7: 15.

Cornuet, J.-M. 1982. The MDH polymorphism in some west Mediterranean honey bee populations. In Proceedings of the 9th Congress of the International Union for the Study of Social Insects. Boulder. Colorado. USA: 5–6.

Cornuet, J.-M., A. Daoudi and C. Chevalet. 1986. Genetic pollution and number of matings in a black honey bee (*Apis mellifera mellifera*) population. Theoretical and Applied Genetics. 73: 223–227.

Cornuet, J.-M. and J. Fresnaye. 1989. Etude biométrique de colonies d'abeilles d'Espagne et du Portugal. Apidologie. 20: 93–101.

Cornuet, J.-M. and L. Garnery. 1991. Mitochondrial DNA variability in honey bees and its phylogeographic implications. Apidologie. 22: 627–642.

Cox-Foster, D.L., S. Conlan, E.C. Holmes, G. Palacios, J.D. Evans, N.A. Moran et al. 2007. A metagenomic survey of microbes in honey bee colony collapse disorder. Science. 318: 283–287.

Crailsheim, K., R. Brodschneider and P. Neumann. 2009. The COLOSS puzzle: filling in the gaps. pp. 46–47. *In*: Proceedings of the 4th COLOSS Conference. Zagreb, Croatia.

Crane, E. 1984. Honey bees. pp. 403–415. Mason, I.L. [ed.]. Evolution of Domesticated Animals. Longman Group, London.

Crane, E. 1990. Bees and beekeeping: science, practice and World Resources. pp. 109–111. Cornell University Press, Ithaca, New York.

Crane, E. 1995. History of beekeeping in Asia. pp. 3–18. *In*: Kevan, P.G. [Ed.]. 1995. The Asiatic Hive Bee: Apiculture, Biology, and Role in Sustainable Development in Tropical and Subtropical Asia. Enviroquest Ltd., Cambridge, Ontario, 315 pp.

Crane, E. 1999. The World History of Beekeeping and Honey Hunting. Routledge, New York.

Cremer, S. and M. Sixt. 2009. Analogies in the evolution of individual and social immunity. Philosophical Transactions of the Royal Society B: Biological Sciences. 364: 129–142. doi: 10.1098/rstb.2008.0166.

Crone, E.E. and N.M. Williams. 2016. Bumble bee colony dynamics: quantifying the importance of land use and floral resources for colony growth and queen production. Ecology Letters. 19: 460–468.

Crosby, A.W. 2015. Ecological Imperialism, the Biological Expansions of Europe, 900–1900. Cambridge University Press, New York.

Crozier, R.H. and Y.C. Crozier. 1993. The mitochondrial genome of the honey bee *Apis mellifera*: complete sequence and genome organization. Genetics. 133: 97–117.

Currie, R.W., S.F. Pernal and E. Guzmán-Novoa. 2010. Honey bee colony losses in Canada. Journal of Apicultural Research. 49: 104–106.

Currie, R.W.S.C. and Jay. 1988. The influence of a colony's queen state, time of the year and drifting behaviour, on the acceptance and longevity of adult drone honey bees (*Apis mellifera* L.). Journal of Apicultural Research. 27: 219–226.

Cuthbertson, A.G.S. and M.A. Brown. 2009. Issues affecting British honey bee biodiversity and the need for conservation of this important ecological component. International Journal of Environmental Science and Technology. 6: 695–699.

Cuthbertson, A.G.S., M.E. Wakefield, M.E. Powell, G. Harris, H. Anderson, G.E. Budge et al. 2013. The small hive beetle *Aethina tumida*: a review of its biology and control measures. Current Zoology. 59: 644–653.

Dahle, B. 2010. Role of *Varroa destructor* for honey bee colony losses in Norway. Journal of Apicultural Research. 49: 124–125.

Dainat, B., D. vanEngelsdorp and P. Neumann. 2012a. Colony collapse disorder in Europe. Environmental Microbiology Reports. 4: 123–125.

Dainat, B., J.D. Evans, Y.P. Chen, L. Gauthier and P. Neumann. 2012b. Predictive markers of honey bee colony collapse. PLOS ONE. 7: e32151.

Dainat, B. and P. Neumann. 2013. Clinical signs of deformed wing virus infection are predictive markers for honey bee colony losses. Journal of Invertebrate Pathology. 112: 278–280.

Daly, H.V. and S.S. Balling. 1978. Identification of Africanized honey bees in the western hemisphere by discriminant analysis. Journal of the Kansas Entomological Society. 51: 857–869.

Daly, H.V., K. Hoelmer, P. Norman and T. Allen. 1982. Computer assisted measurement and identification of honey bees. Annals of the Entomological Society of America. 75: 591–594. doi: 10.1093/aesa/75.6.591.

Daly, H.V. 1991. Systematics and identification of Africanized honey bees. pp. 13–44. *In*: Spivak, M., D.J.C. Fletcher and M.D. Breed [eds.]. The "African" Honey Bee. Westview, Boulder.

Daly, H.V. 1992. A statistical and empirical evaluation of some morphometric variables of honey bee classification. pp. 127–255. *In*: Sorenson, J.T. and R. Foottit [eds.]. Ordination in the Study of Morphology, Evolution. and Systematics of Insects: Applications and Quantitative Genetic Rationales. Elsevier, Amsterdam.

Dams, L.R. 1978. Bees and honey-hunting scenes in the Mesolithic Rock Art of Eastern Spain. Bee World. 59: 45-53.

Damus, M.S. and G.W. Otis. 1997. A morphometric analysis of *Apis cerana* F. and *Apis nigrocincta* Smith populations from Southeast Asia. Apidologie. 28: 309–323.

Danforth, B.N., S. Sipes, J. Fang and S.G. Brady. 2006. The history of early bee diversification based on five genes plus morphology. Proceedings of the National Academy of Sciences of the United States of America. 103: 15118–15123. doi: 10.1073/pnas.0604033103.

Danforth, B.N., S.C. Cardinal, C. Praz, E. Almeida and D. Michez. 2013. Impact of molecular data on our understanding of bee phylogeny and evolution. Ann. Rev. Entomol. 58: 57–78.

Danka, R.G., R.L.I. Hellmich and T.E. Rinderer. 1992. Nest usurpation, supersedure and colony failure contribute to Africanization of commercially managed European honey bees in Venezuela. Journal of Apicultural Research. 31: 119–123.

Danka, R.G. and J.D. Villa. 2000. Inheritance of resistance to *Acarapis woodi* (Acari: Tarsonemidae) in first-generation crosses of honey bees (Hymenoptera: Apidae). Journal of Economic Entomology. 93: 1602–1605.

Danka, R.G., L.I. de Guzman, T.E. Rinderer, H.A. Sylvester, C.M. Wagener, A.L. Bourgeois et al. 2012. Functionality of *Varroa*-resistant honey bees (Hymenoptera: Apidae) when used in migratory beekeeping for crop pollination. Journal of Economic Entomology. 105: 313–321.

Dar, S.A. and S.B. Ahmad. 2013. Occurrence of Nosemosis (*Nosema Apis*) affecting Honey bee (*Apis mellifera* L.) Colonies in Kashmir. Applied Biological Research. 5: 53–56.

Darchen, R. 1969. Sur la biologie de *Trigona* (Apotrigona) *nebulata komiensis* Cock. Biologia Gabonica. 5: 151–187.

Darchen, R. 1973. Essal d'interpretation du determinisme castes chez les trigones et. les melipones. Comptes Rendus de l'Académie des Sciences. 276: 607–609.

Darchen, R. and B. Delage-Darchen. 1975. Contribution a l'étude d'une abeille du Mexique *Melipona beecheii* B. (Hymenoptère: Apidae). Apidologie. 6: 295–339.

de Guzman, L.I., M. Forbes, C.R. Cervancia, T.E. Rinderer and S. Somera. 1992. *Apis andreniformis* Smith in Palawan, Philippines. Journal of Apicultural Research. 31: 111.

de Guzman, L.I., T.E. Rinderer and V.A. Lancaster. 1995. A short test evaluating larval attractiveness of honey bees to *Varroa jacobsoni*. Journal of Apicultural Research. 34: 89–92. doi: 10.1080/00218839.1995.11100892.

De La Riva, J., F. Le Pont, V. Ali, A. Matias, S. Mollinedo and J.P. Dujardin. 2001. Wing geometry as a tool for studying the *Lutzomyia longipalpis* (Diptera: Psychodidae) complex. Memórias do Instituto Oswaldo Cruz. 96: 1089–1094. doi: 10.1590/S0074-02762001000800011.

De la Rúa, P., J. Galian, J. Serrano and R.F.A. Moritz. 2001a. Molecular characterization and population structure of the honey bees from the Balearic islands (Spain). Apidologie. 32: 417–427.

De la Rúa, P., J. Galián, J. Serrano and R.F.A. Moritz. 2001a. Genetic structure and distinctness of *Apis mellifera* L. populations from the Canary Islands. Molecular Ecology. 10: 1733–1742.

De la Rúa, P., J. Galián, J. Serrano and R.F.A. Moritz. 2001b. Molecular characterization and population structure of the honey bees from the Balearic Islands (Spain). Apidologie. 32: 417–427.

De la Rúa, P., J. Serrano and J. Galian. 2002a. Biodiversity of *Apis mellifera* populations from Tenerife (Canary Islands) and introgressive hybridization with East European races. Biodiversity and Conservation. 11: 59–67.

De la Rúa, P., J. Galian, J. Serrano and R.F.A. Moritz. 2002b. Microsatellite analysis of non-migratory colonies of *Apis mellifera iberica* from south-eastern Spain. Journal of Zoological Systematics and Evolutionary Research. 40: 164–168.

De la Rúa, P., J. Galian, J. Serrano and R.F.A. Moritz. 2003. Genetic structure of Balearic honey bee populations based on microsatellite polymorphism. Genetics Selection Evolution. 35: 339–350.

De la Rúa, P., Y. Jimenez, J. Galián and J. Serrano. 2004. Evaluation of the biodiversity of honey bee (*Apis mellifera*) populations from eastern Spain. Journal of Apicultural Research. 43: 162–166.

De la Rúa, P., R. Hernandez-Garcia, Y. Jimenez, J. Galian and J. Serrano. 2005. Biodiversity of *Apis mellifera iberica* (Hymenoptera: Apidae) from north-eastern Spain assessed by mitochondrial analysis. Insect Systematics and Evolution. 36: 21–28.

De la Rúa, P., J. Galián, B.V. Pedersen and J. Serrano. 2006. Molecular characterization and population structure of *Apis mellifera* from Madeira and the Azores. Apidologie. 37: 699–708.

De la Rúa, P., R. Jaffe, R. Dall'Olio, I. Munoz and J. Serrano. 2009. Biodiversity, conservation and current threats to European honey bees. Apidologie. 40: 263–284. doi: 10.1051/apido/2009027.

De Miranda, J.R. and E. Genersch. 2010. Deformed wing virus. Journal of Invertebrate Pathology. 103: 48–61.

De Ruijter, A. 1987. Reproduction of *Varroa jacobsoni* during successive brood cycles of the honey bee. Apidologie. 18: 321–326.

Decourtye, A., E. Mader and N. Desneux. 2010. Landscape enhancement of floral resources for honey bees in agro-ecosystems. Apidologie. 41: 264–277.

Decourtye, A., J. Devillers, P. Aupinel, F. Brun, C. Bagnis, J. Fourrier et al. 2011. Honey bee tracking with microchips: a new methodology to measure the effects of pesticides. Ecotoxicology. 20: 429–437.

DeGrandi-Hoffman, G., G. Wardell, F. Ahumada-Segura, T.E. Rinderer, R. Danka and J. Pettis. 2008. Comparisons of pollen substitute diets for honey bees: consumption rates by colonies and effects on brood and adult populations. Journal of Apicultural Research. 47: 265–270. doi: 10.1080/00218839.2008.11101473.

DeGrandi-Hoffman, G., Y.P. Chen, E. Huang and M.H. Huang. 2010. The effect of diet on protein concentration, hypopharyngeal gland development and virus load in worker honey bees (*Apis mellifera* L.). Journal of Insect Physiology. 56: 1184–1191. doi: 10.1016/j.jinsphys.2010.03.017.

Delaney, D.A., M.D. Meixner, N.M. Schiff and W.S. Sheppard. 2009. Genetic characterization of commercial honey bee (Hymenoptera: Apidae) populations in the United States by using mitochondrial and microsatellite markers. Annals of the Entomological Society of America. 102: 666–673.

Delgado, M. and S. del Amo. 1984. Dianámica de poblaciones en una zona tropical húmeda. Biotica. 9: 351–365.

Deredec, A. and F. Courchamp. 2003. Extinction thresholds in host-parasite dynamics. Ann. Zool. Fennici. 40: 115–130.

Desneux, N., A. Decourtye and J.M. Delpuech. 2007. The sublethal effects of pesticides on beneficial arthropods. Annual Review of Entomology. 52: 81–106.

Di Pasquale, G., M. Salignon, Y. Le Conte, L.P. Belzunces, A. Decourtye, A. Kretzschmar et al. 2013. Influence of pollen nutrition on honey bee health: do pollen quality and diversity matter? PLOS ONE. 8: e72016.

Di Prisco, G., V. Cavaliere, D. Annoscia, P. Varricchio, E. Caprio, F. Nazzi et al. 2013. Neonicotinoid clothianidin adversely affects insect immunity and promotes replication of a viral pathogen in honey bees. Proc. Natl. Acad. Sci. USA. 110: 18466–18471.

Dianov, P.A. 1977. Chronographic variability of parameters of honey bees from the mountainous zone of Kazakhstan. Izvestiya Akademii Nauk Kazakhstana seriya Biologicheskaya. 1: 38–40.

Diao, Q., L. Sun, H. Zheng, Z. Zeng, S. Wang, S. Xu et al. 2018. Genomic and transcriptomic analysis of the Asian honey bee *Apis cerana* provides novel insights into honey bee biology. Scientific Reports. 8: 1–14. doi: 10.1038/s41598-017-17338-6.

Dietemann, V., J. Pflugfelder, D. Anderson, J.D. Charrière, N. Chejanovski and B. Dainat. 2012. *Varroa destructor*: Research avenues towards sustainable control. Journal of Apicultural Research. 51: 125–132.

Dietz, A. 1986. Evolution. pp. 3–21. *In*: Rinderer, T.E. [ed.]. Bee Genetics and Breeding. Academic Press, London.

Dietz, A. 1992. Honey bees of the world. pp. 23–71. *In*: Graham, J.M. [ed.]. The Hive and the Honey Bee. Dadant and Sons, Inc. Hamilton, Illinois, USA.

Diniz-Filho, J.A.F., H.R. Hepburn, S. Radloff and S. Fuchs. 2000. Spatial analysis of morphometrical variation in African honey bees (*Apis mellifera* L.) on a continental scale. Apidologie. 31: 191–204. doi: 10.1051/apido:2000116.

Djukic, M., E. Brzuszkiewicz, A. Fünfhaus, J. Voss, K. Gollnow, L. Poppinga et al. 2014. How to kill the honey bee larva: Genomic potential and virulence mechanisms of *Paenibacillus larvae*. PLOS ONE. doi: 10.1371/journal.pone.0090914.

Dogantzis, K.A. and A. Zayed. 2019. Recent advances in population and quantitative genomics of honey bees. Current Opinion in Insect Science. 31: 93–98.

Dolati, L., N.J. Rafie and H. Khalesro. 2013. Landmark-based morphometric study in the fore and hind wings of an Iranian race of European honey bee (*Apis mellifera meda*). Journal of Apicultural Research. 57: 187–197. doi: 10.2478/jas-2013-0028.

Dolgov, L.A. 1982. Osobennosti biologii shmelej, zaselyayushch iskusstvennye gnezda. Nasekomye-opyliteli selskoxozyajstvennych kultur. (in Russian). Novosibirsk.

Donze, G., M. Herrmann, B. Bachofen and P.M. Guerin. 1996. Effect of mating frequency and brood cell infestation rate on the reproductive success of the honey bee parasite *Varroa jacobsoni*. Ecological Entomology. 21: 17–26.

Dorland, W.A. 1990. Dorland's illustrated medical dictionary. W.B. Saunders Company, London.

Drescher, N., H.M. Wallace, M. Katouli, C.F. Massaro and S.D. Leonhardt. 2014. Diversity matters: how bees benefit from different resin sources. Oecologia. 176: 943–953.

Duchateau, M.J., H.W. Velthuis and J.J. Boomsma. 2004. Sex ratio variation in the bumble bee *Bombus terrestris*. Behavioral Ecology. 15: 71–82.

Dudley, P. 1720. An account of a method lately found out in New-England, for discovering where the bees hive in the woods, in order to get their honey. Philosophical Transactions of The Royal Society of London Series B. 31: 148–150.

Dunn, K.J. 1992. Exotic Asian bee detected in Torres Strait. Bee Briefs. 9: 18–19.

Duo, D., B. Roger and J. Penh. 1975. Comparison using radioisotope exchange of food between bees (*Apis mellifera ligustica* S.) of three castes. Proceedings of 25-th International Conference in Beekeeping. Grenoble: Apimondia.

DuPraw, E.J. 1964. Non-Linnean taxonomy. Nature. 202: 849–852. doi: 10.1038/202849a0.

DuPraw, E.J. 1965a. Non-Linnean taxonomy and the systematics of honey bees. Systematic Zoology. 14: 1–24. doi: 10.2307/2411899.

DuPraw, E.J. 1965b. The recognition and handling of honey bee specimens in non-Linnean taxonomy. Journal of Apicultural Research. 4: 71–84.

Dutton, R. and J. Simpson. 1977. Producing honey with *Apis florea* in Oman. Bee world. 58: 71–76.

Dutton, R.W., F. Ruttner, A. Berkeley and M.J.D. Manley. 1981. Observations on the morphology, relationships and ecology of *Apis mellifera* of Oman. Journal of Apicultural Research. 20: 201–214. doi: 10.1080/00218839.1981.11100498.

Eguaras, M., J. Marcangeli and N.A. Fernandez. 1994. Influence of "parasitic intensity" on *Varroa jacobsoni* Oud. reproduction. Journal of Apicultural Research. 33: 155–159.

Eickwort, G.C. and K.R. Eickwort. 1973. Aspects of the biology of Costa Rican halictine bees, V. *Augochlorella edentata* (Hymenoptera: Halictidae). Journal of the Kansas Entomological Society. 46: 3–16.

Eimanifar, A., R.T. Kimball, E.L. Braun, S. Fuchs, B. Grünewald and J.D. Ellis. 2017. The complete mitochondrial genome and phylogenetic placement of *Apis nigrocincta* Smith (Insecta: Hymenoptera: Apidae), an Asian, cavity-nesting honey bee. Mitochondrial DNA Part B Resources. 2: 249–250.

Eischen, F.A., R.H. Graham and R. COX. 2005. Regional distribution of *Paenibacillus larvae* subspecies larvae, the causative organism of American foulbrood, in honey bee colonies of the western United States. Journal of Economic Entomology. 98: 1087–1093.

Elbgami, T., W.E. Kunin, W.O.H. Hughes and J.C. Biesmeijer. 2014. The effect of proximity to a honey bee apiary on bumble bee colony fitness, development and performance. Apidologie. 45: 504–513.

Ellis, J.D. 2004. The ecology and control of small hive beetle (*Aethina tumida* Murray). PhD dissertation, Rhodes University, Grahamstown, South Africa.

Ellis, J.D., J.D. Evans and J. Pettis. 2010a. Reviewing colony losses and Colony Collapse Disorder in the United States. Journal of Apicultural Research. 49: 134–136.

Ellis, J.D., J.D. Evans and J. Pettis. 2010b. Colony losses, managed colony population decline and Colony Collapse Disorder in the United States. Journal of Apicultural Research. 49: 134–136.

Ellis, J.D., P. Neumann, R. Hepburn and P.J. Elzen. 2002. Longevity and reproductive success of *Aethina tumida* (Coleoptera: Nitidulidae) fed different natural diets. Journal of Economic Entomology. 95: 902–907.

Emmanouel, N.G., L.A. Santas, D.G. Tambouratzis, D.A. Griffiths and C.E. Bowman. 1984. *Varroa* disease and its control in Greece. Acarologia. 4: 1099–1106.

Engel, M.S. and T.R. Schultz 1997. Phylogeny and behavior in honey bees (Hymenoptera: Apidae). Ann. Ent. Soc. Am. 90: 43–53.

Engel, M.S. 1998. Fossil honey bees and evolution in the genus *Apis* (Hymenoptera, Apidae). Apidologie. 29: 265–281.

Engel, M.S. 1999. The taxonomy of recent and fossil honey bees (Hymenoptera Apidae *Apis*). Journal of Hymenoptera Research. 8: 165–196. doi: 10.1007/978-1-4614-4960-7_18.

Engel, M.S. 2006. A giant honey bee from the middle Miocene of Japan (Hymenoptera: Apidae). American Museum Novitates. 3504: 1–12.

Engel, M.S., I.A. Hinojosa-Diaz and A.P. Rasnitsin. 2009. A honey bee from the Miocene of Nevada and the biogeography of *Apis* (Hymenoptera: Apidae, Apini). Proceedings of the California Academy of Sciences. 60: 23–38.

Engel, M.S., U. Kotthoff and T. Wappler. 2011. *Apis armbrusteri* Zeuner, 1931 (Insecta, Hymenoptera): proposed conservation by designation of a neotype. Bulletin of Zoological Nomenclature. 68: 117–121. doi: 10.21805/bzn.v68i2.a12.

Engel, M.S. 2012. The honey bees of Indonesia (Hymenoptera: Apidae). Treubia. 39: 41–49.

Engel, M.S., B. Wang, A.S. Alqarni, L.B. Jia, T. Su, Z.K. Zhou and T. Wappler. 2018. A primitive honey bee from the Middle Miocene deposits of southeastern Yunnan, China (Hymenoptera, Apidae). Zookeys. 775: 117–129. doi: 10.3897/zookeys.775.24909.

Eskov, E.K. 1969. The sound apparatus of bees. Biophysics. 14: 158–166.

Eskov, E.K. 1972. Correction of the bees of the acoustic signal range under the influence of sound interference (in Russian). Reports of Academy of Sciences of USSR. 202: 211–213.

Eskov, E.K. 1974. Generation and perception of electric fields by bees (in Russian). Zoologicheskii Zhurnal. 53: 800–802.

Eskov, E.K. 1975. Honey bee phonoreceptors (in Russian). Biofizika. 20: 646–651.

Eskov, E.K. 1979. Acoustic alarm of social insects (in Russian). Nauka, Moscow.

Eskov, E.K. and L.V. Dolgov. 1986.Temperature regulation in the nest and its role in the life of a bumblebee family. Russian Journal of Zoology. 65: 1500–1507.

Eskov, E.K. 1992. Ethology of the honey bee (in Russian). Kolos, Moscow.

Eskov, E.K. 1995. Honey bee Ecology (in Russian). Russkoe slovo, Ryazan.

Eskov, E.K. and M.D. Eskova. 2001. Diameter and symmetry of bee cells (in Russian). Pchelovodstvo. 1: 6–7.

Eskov, E.K. and V.A. Toboev. 2011. Changes in the structure of sounds generated by bee colonies in the process of sociotomy (in Russian). Zoologicheskii zhurnal. 90: 59–61.

Eskov, E.K. and M.D. Eskova. 2012. Factors affecting the size of bee honeycomb cells (in Russian). Pchelovodstvo. 8: 19–21.

Eskov, E.K. 2013. Generation, perception and use of acoustic and electric fields in honey bee communication. Biophysics. 58: 827–836.

Eskov, E.K. 2018. Phylogenetic conditionality of acoustic signaling diversity in honey bee related to the development of sociality (in Russian). Bulletin Reviews. 8: 309–318.

Estoup, A., L. Garnery, M. Solignac and J. Cornuet. 1995. Microsatellite variation in honey bee (*Apis mellifera* L.) populations: Hierarchical genetic structure and tests of infinite allele and stepwise mutation models. Genetics. 140: 679–695.

Evans, J.D. and M. Spivak. 2010. Socialized medicine: individual and communal disease barriers in honey bees. Journal of Invertebrate Pathology. 103: 62–72. doi: 10.1016/j.jip.2009.06.019.

Evans, J.D. and R.S. Schwarz. 2011. Bees brought to their knees: microbes affecting honey bee health. Trends in Microbiology. 19: 614–620.

FABIS (Fast Africanized Bee Identification System). 2019. USDA-ARS laboratory. [online]. http://www.ars.usda.gov/Research/docs.htm?docid=11053.

Fabre, J.A. 1963. Life of insects. Moscow: Uchpedgiz.

Falconer, D.S. 1989. Introduction to Quantitative Genetics. Longman, New York.

Falconer, D.S. and T.F.C. Mackay. 1996. Introduction to quantitative genetics. Longmans Green, Harlow, Essex, UK. Introduction to quantitative genetics. Longmans Green, Harlow, Essex, UK.

FAOSTAT (Food and Agriculture Organization of the United Nations). 2018. [online] http://www.fao.org/faostat.

Farooqui, T. 2013. A potential link among biogenic amines-based pesticides, learning and memory and colony collapse disorder: A unique hypothesis. Neurochemistry International. 62: 122–136. doi: 10.1016/j.neuint.2012.09.020.

Faucon, J.P., L. Mathieu, M. Ribiere, A.C. Martel, P. Drajnudel, S. Zeggane et al. 2002. Honey bee winter mortality in France in 1999 and 2000. Bee World. 83: 14–23.

Feigenbaum, C. and D. Naug. 2010. The influence of social hunger on food distribution and its implications for disease transmission in a honey bee colony. Insectes Sociaux. 57: 217–222.

Feon, V., A. Schermann-Legionnet, Y. Delettre, S. Aviron, R. Billeter, R. Bugter et al. 2010. Intensification of agriculture, landscape composition and wild bee communities: A large scale study in four European countries; Agriculture Ecosystems and Environment. 137: 143–150.

Ferguson, A.W. and J.B. Free. 1980. Queen pheromone transfer within honeybee colonies. Physiological Entomology. 5: 359–366.

Ferreira, K.M., O. Line e Salvia, M.C. Arias and M.A. Del Lima. 2009. Cytochrome b variation in *Apis mellifera* samples and its association with *COI-COII* patterns. Genetica. 135: 149–155.

Fischer, P. and C.M. Grozinger. 2008. Pheromonal regulation of starvation resistance in honey bee workers (*Apis mellifera*). Naturwissenschaften. 95: 723–729.

Fitriya, J., R. Raffiudin, T. Atmowidi and R. Hepburn. 2012. *Apis koschevnikovi*: Distribution in South Kalimantan and cytochrome b mitochondrial DNA variations. Journal of Insect Biodiversity. 8: 23–30.

Fletcher, D.J.C. 1977a. A preliminary analysis of rapid colony development in *Apis mellifera adansonii* L. Eighth International Congress of the IUSSI, Pudoc, Wageningen: 144–145.

Fletcher, D.J.C. 1977b. Evaluation of introductions of European honey-bees into southern and eastern Africa. Eighth International Congress of the IUSSI, Pudoc, Wageningen: 146–147.

Fletcher, D.J.C. 1978. The African bee, *Apis mellifera adansonii*, in Africa. Annual Review of Entomology. 23: 151–171.

Floris, I., A. Satta, L. Ruiu and F. Buffa. 2007. Searching for the origin of Sardinian honey bees. Morphometric comparison between samples from Sardinia and northern Tunisia. Redia, XC: 105–108.

Flynn, D.F.B., M. Gogol-Prokurat, T. Nogeire, N. Molinari, B.T. Richers, B.B. Lin et al. 2009. Loss of functional diversity under land use intensification across multiple taxa. Ecology Letters. 1: 22–33.

Fontana, P., C. Costa, G. Di Prisco, E. Ruzzier, D. Annoscia, A. Battisti et al. 2018. Appeal for biodiversity protection of native honey bee subspecies of *Apis mellifera* in Italy (San Michele all'Adige declaration).

Forsgren, E. 2010. European foulbrood in honey bees. Journal of Invertebrate Pathology. 103: 5–9.

Forsgren, E. and I. Fries. 2013. Temporal study of *Nosema* spp. in a cold climate. Environmental Microbiology Reports. 5: 78–82.

Foti, N., M. Lungu, C. Pelimon, I. Barac, M. Copaitici and E. Marza. 1965. Researches on morphological characteristics and biological features of the bee population in Romania. Proceedings of XXth Jubiliar International Congress of Beekeeping Apimondia. Apimondia, Bucharest, Romania: 171–176.

Francis, R.M., S.L. Nielsen and P. Kryger. 2013. *Varroa*-Virus Interaction in Collapsing Honey Bee Colonies. PLOS ONE. 8: 9.

Franck, P., L. Garnery, M. Solignac and J.-M. Cornuet. 1998. The origin of west European subspecies of honey bees (*Apis mellifera*): new insights from microsatellite and mitochondrial data. Evolution. 52: 1119–1134. doi: 10.2307/2411242.

Franck, P., L. Garnery, G. Celebrano, M. Solignac and J.-M. Cornuet. 2000a. Hybrid origins of honey bees from Italy (*Apis mellifera ligustica*) and Sicily (*A. m. sicula*). Molecular Ecology. 9: 907–921.

Franck, P., L. Garnery, A. Loiseau, M. Solignac and J.-M. Cornuet. 2000b. Molecular confirmation of a fourth lineage in honey bees from Middle-East. Apidologie. 31: 167–180. doi: 10.1051/20000114.

Franck, P., L. Garnery, A. Loiseau, B.P. Oldroyd, H.R. Hepburn, M. Solignac et al. 2001. Genetic diversity of the honey bee in Africa: microsatellite and mitochondrial data. Heredity. 86: 420–430. doi: 10.1046/j.1365-2540.2001.00842.x.

Francoy, T.M., P.R.R. Prado, L.S. Gonçalves, L. Da Fontoura Costa and D. De Jong. 2006. Morphometric differences in a single wing cell can discriminate *Apis mellifera* racial types. Apidologie. 37: 91–97. doi: 10.1051/apido:2005062.

Francoy, T.M., D. Wittmann, M. Drauschke, S. Müler, V. Steinhage, M.A.F. Bezerra-Laure et al. 2008. Identification of Africanized honey bees through wing morphometrics: two fast and efficient procedures. Apidologie. 39: 488–494. doi: 10.1051/apido:2008028.

Francoy, T.M., D. Wittmann, V. Steinhage, M. Drauschke, S. Müler, D.R. Cunha et al. 2009. Morphometric and genetic changes in a population of *Apis mellifera* after 34 years of Africanization. Genet. Mol. Res. 8: 709–717. doi: 10.4238/vol8-2kerr019.

Free, J.B. and C.G. Butler. 1959. Bumble bees. Collins, London.

Free, J.B. 1970. Effect of flower shapes and nectar guides on the behavior of foraging honeybees. Behaviour. 37: 269–285.

Freitas, B.M., V.L. Imperatriz-Fonseca, L.M. Medina, A.D.M.P. Kleinert, L. Galetto, G. Nates-Parra et al. 2009. Diversity, threats and conservation of native bees in the Neotropics. Apidologie. 40: 332–346.

Fries, I., S. Camazine and J. Sneyd. 1994. Population dynamics of *Varroa jacobsoni*: a model and a review. Bee World. 75: 5–28.

Fries, I. and P. Rosenkranz. 1996. Number of reproductive cycle of *Varroa jacobsoni* in honey bees (*Apis mellifera*) colonies. Experimental and Applied Acarology. 20: 103–112. doi: 10.1007/BF00051156.

Fuchs, S. 1992. Choice in *Varroa jacobsoni* Oud. between honey bee drone or worker brood cells for reproduction. Behavioral Ecology and Sociobiology. 3: 429–435.

Fürst, M.A., D.P. McMahon, J.L. Osborne, R.J. Paxton and M.J.F. Brown. 2014. Disease associations between honey bees and bumble bees as a threat to wild pollinators. Nature. 506: 364–366.

Gaidar, V.A. and V.P. Pilipenko. 1989. Carpathian bees. Uzhgorod, Karpaty.

Gajger, I.T., O. Vugrek, D. Grilec and Z. Petrinec. 2010. Prevalence and distribution of *Nosema ceranae* in Croatian Honey bee colonies. Veterinarni medicina. 55: 457–462.

Garibaldi, L.A., F. Requier, O. Rollin and G.K.S. Andersson. 2017. Towards an integrated species and habitat management of crop pollination. Curr. Opin. Insect. Sci. 21: 1–10.

Garibaldi, L.A., G.K.S. Andersson, F. Requier, T.P.M. Fijen, J. Hipólito, D. Kleijn et al. 2018. Complementarity and synergisms among ecosystem services supporting crop yield. Glob. Food. Sec. 17: 38–47.

Garibaldi, L.A., I. Steffan-Dewenter, R. Winfree, M.A. Aizen, R. Bommarco, S.A. Cunningham et al. 2013. Wild pollinators enhance fruit set of crops regardless of honey bee abundance. Science. 339: 1608–1611.

Garnery, L., J.-M. Cornuet and M. Solignac. 1992. Evolutionary history of the honey bee *Apis mellifera* inferred from mitochondrial DNA analysis. Molecular Ecology. 1: 145–154.

Garnery, L., M. Solignac, G. Celebrano and J.M. Cornuet. 1993. A simple test using restricted PCR-amplified mitochondrial DNA to study the genetic structure of *Apis mellifera* L. Experientia. 49: 1016–1021. doi: 10.1007/BF02125651.

Garnery, L., E.H. Mosshine, B.P. Oldroyd and J.M. Cornuet. 1995. Mitochondrial DNA variation in Moroccan and Spanish honey bee populations. Molecular Ecology. 4: 465–471.

Garnery, L., P. Franck, E. Baudry, D. Vautrin, J.-M. Cornuet and M. Solignac. 1998a. Genetic diversity of the west European honey bee (*Apis mellifera mellifera* and *A. m. iberica*). I. Mitochondrial DNA. Genetics. Selection Evolution. 30: 31–47. doi: 10.1186/1297-9686-30-S1-S31.

Garnery, L., P. Franck, E. Baudry, D. Vautrin, J.M. Cornuet and M. Solignac. 1998b. Genetic diversity of the West European honey bee (*Apis mellifera mellifera* and *A. m. iberica*). II. Microsatellite loci. Genetics Selection Evolution. 30: S49–S74.

Garofalo, C.A. 1978. Bionomics of *Bombus* (*Fervidobombus*) *morio* (Swederus). 2. Body size length of life of workers. Journal of Apicultural Research. 17: 130–136.

Gauld, I.D., B. Bolton, T. Huddleston, M.G. Fitton, M.R. Shaw, J.S. Noyes et al. 1988. The Hymenoptera. British Museum (Natural History). Oxford University Press, New York.

Gayeva, D.V. 2015. Pollination as an ecosystem service in agricultural use. Vestnik Baltiyskogo Federalnogo Universiteta. 1: 19–34.

GBIF, The Global Biodiversity Information Facility. 2018a. doi: 10.15468/dl.ahbcwc.

GBIF, The Global Biodiversity Information Facility. 2018b. doi: 10.15468/dl.lf1ioq.

Geldmann, J. and J.P. González-Varo. 2018. Conserving honey bees does not help wildlife. Science. 359: 392–393.

Genersch, E. 2010. American Foulbrood in Honey bees and its causative agent, *Paenibacillus larvae*. Journal of Invertebrate Pathology. 103: 10–19.

Genersch, E., E. Forsgren, J. Pentikäinen, A. Ashiralieva, S. Rauch and J. Kilwinski. 2006. Reclassification of *Paenibacillus larvae* subsp. *pulvifaciens* and *Paenibacillus larvae* subsp. *larvae* as *Paenibacillus larvae* without subspecies differentiation. International Journal of Systematic and Evolutionary Microbiology. 56: 501–511.

Genersch, E., W. von der Ohe, H. Kaatz, A. Schroeder and C. Otten. 2010. The German bee monitoring project: a long term study to understand periodically high winter losses of honey bee colonies. Apidologie. 41: 332–352.

Gerber, H.S. and S.C. Klostermeyer. 1972. Factors affecting the sex ratio and nesting behavior of the Sinaloa leafcutters bee. Washington: Agric. Exper. Station Techn. Bull. 73: 11.

Geslin, B., B. Gauzens, M. Baude, I. Dajoz, C. Fontaine, M. Henry et al. 2017. Massively Introduced Managed Species and Their Consequences for Plant-Pollinator Interactions. Adv. Ecol. Res. 57: 147–199.

Giannini, T.C., W.F. Costa, G.D. Cordeiro, V.L. Imperatriz-Fonseca, A.M. Saraiva, J. Biesmeijer et al. 2017. Projected climate change threatens pollinators and crop production in Brazil. PLOS ONE. 12: e0182274.

Giersch, T., T. Berg, F. Galea and M. Hornitzky. 2009. *Nosema ceranae* infects honey bees (*Apis mellifera*) and contaminates honey in Australia. Apidologie. 40: 117–123.

Gilley, D.C., D.R. Tarpy and B.B. Land. 2003. The effect of queen quality on the interactions of workers and dueling queen honey bees (*Apis mellifera* L.). Behavioral Ecology and Sociobiology. 55: 190–196.

Gilliam, M., S. Taber and J.B. Rose. 1978. Chalkbrood disease of honey bees, *Apis mellifera* L.: a progress report. Apidologie. 9: 75–89.

Gilliam, M., S.I.I.I. Taber and G.V. Richardson. 1983. Hygienic behavior of honey bees in relation to chalkbrood disease. Apidologie 14: 29–39.

Gilliam, M., S.I.I.I. Taber, B.J. Loren and D.B. Prest. 1988. Factors affecting development of chalkbrood disease in colonies of honey bees, *Apis mellifera*, fed pollen contaminated with *Ascosphaera apis*. Journal of Invertebrate Pathology. 52: 314–325.

Girolami, V., L. Mazzon, A. Squartini, N. Mori, M. Marzaro, A. Di Bernardo et al. 2009. Translocation of neonicotinoid insecticides from coated seeds to seedling guttation drops: a novel way of intoxication for bees. Journal of Economic Entomology. 102: 1808–1815. doi: 10.1603/029.102.0511.

Girolami, V., M. Marzaro, L. Vivan, L. Mazzon, M. Greatti, C. Giorio et al. 2012. Fatal powdering of bees in flight with particulates of neonicotinoids seed coating and humidity implication. Journal of Applied Entomology. 136: 17–26. doi: 10.1111/j.1439-0418.2011.01648.x.

Gisder, S., V. Schüler, L.L. Horchler, D. Groth and E. Genersch. 2017. Long-term temporal trends of *Nosema* spp. infection prevalence in north-east Germany: continuous spread of *Nosema ceranae*, an emerging pathogen of honey bees (*Apis mellifera*), but no general replacement of *Nosema Apis*. Frontiers in Cellular and Infection Microbiology. 7: 301.

Glinski, Z. 1982. Studies on pathogenicity of *Ascosphaera apis* for larvae of the honey bee *Apis mellifera* L. Part II. Relationships between biochemical types and virulence of *A. apis*. Annales Universitatis Marie Curie-Sklodowska. 37: 69–77.

Goblirsch, M., Z.Y. Huang and M. Spivak. 2013. Physiological and behavioral changes in honey bees (*Apis mellifera*) induced by *Nosema ceranae* infection. PLOS ONE. 8: e58165.

Goetze, G.K.L. 1964. Die Honigbiene in Natürlicher und Künstlicher Zuchtauslese. Parey, Hamburg.

Gomeh, H., J.N. Rafie and M. Modaber. 2016. Comparison of standard and geometric morphometric methods for discrimination of honey bees populations (*Apis mellifera* L.) in Iran. Journal of Entomology and Zoology Studies. 4: 47–53.

Gonçalves, L.S., A.C. Stort and D. De Jong. 1991. Beekeeping. in Brazil. pp. 359–372. *In*: Spivak, M., D.J.C. Fletcher and M.D. Breed [eds.]. The "African" Honey Bee. Westview Press, San Francisco, California.

Gonçalves, L.S. and K.P. Gramacho. 2003. Aplicación del Comportamiento Higiénico en el mejoramiento genético de dlas abejas. pp. 79–81. *In*: Memoria del 10 Congreso Internacional de Actualización Apícola. Tlaxcala-México: Instituto Tecnológico Agropecuário. 29.

Gorbachev, K.A. 1916. Caucasian grey mountain bee and its place among other bees. Tiflis.

Goulson, D. and K.R. Sparrow. 2008. Evidence for competition between honey bees and bumble bees; effects on bumble bee worker size. Journal Insect Conserv. 13: 177–181.

Goulson, D., G.C. Lye and B. Darvill. 2008. Decline and conservation of bumble bees. Annual Review of Entomology. 53: 191–208.

Goulson, D. and W.O.H. Hughes. 2015. Mitigating the anthropogenic spread of bee parasites to protect wild pollinators. Biological Conservation. 191: 10–19.

Goulson, D., E. Nicholls, C. Botías and E.L. Rotheray. 2015. Bee declines driven by combined stress from parasites, pesticides and lack of flowers. Science. 347: 1255957.

Gradish, A.E., J. van der Steen, C.D. Scott-Dupree, A.R. Cabrera, G.C. Cutler, D. Goulson et al. 2019. Comparison of Pesticide Exposure in Honey Bees (Hymenoptera: Apidae) and Bumble Bees (Hymenoptera: Apidae): Implications for Risk Assessments. Environmental Entomology. 48: 12–21.

Gramacho, K.P. 1999. Fatores que interferem no comportamento higiênico das abelhas *Apis mellifera*. Ph.D. thesis, Faculdade de Filosofia, Ciências e Letras de Ribeirão Preto, USP, Ribeirão Preto, SP, Brazil.

Grankin, N.N. 1981. The variability of the characteristics of the Central Russian Queen bees (in Russian). Beekeeping. 11: 11.

Grankin, N.N. 1998. Selection and reproduction of Central Russian bees (in Russian). Beekeeping. 6: 11–13.

Grankin, N.N. 2006. Sanitizing activity of *A. m. mellifera* L. (in Russian). Beekeeping. 3: 25–26.

Grankin, N.N. 2008. "Orlovskiy" breed type of *A. m. mellifera* L. (in Russian). Beekeeping. 4: 8–9.

Gray, A., M. Peterson and A. Teale. 2010. An update on recent colony losses in Scotland from a sample survey covering 2006–2008. Journal of Apicultural Research. 49: 129–131.

Gregorc, A., V. Lokar and M.I. Smodiš Škerl. 2008. Testing of the isolation of the Rog-Ponikve mating station for Carniolan (*Apis mellifera carnica*) honey bee qeens. Journal of Apicultural Research. 47: 138–142.

Greenleaf, S.S. and C. Kremen. 2006. Wild bees enhance honey bees" pollination of hybrid sunflower. Proceedings of the National Academy of Sciences of the United States of America. 103: 13890–13895.

Gromisz, M. 1962. Season variations of wing measurements and cubital index of honey bee (*Apis mellifera* L.). Pszczelnicze Zeszyty Naukowe. 6: 113–120.

Gross, C.L. 2001. The effect of introduced honey bees on native bee visitation and fruit-set in Dillwynia juniperina (Fabaceae) in a fragmented ecosystem. Biological Conservation. 102: 89–95.

Grozdanić, S. 1971. Biologische Untersuchungen an den Bienen (Apoidea, Hymenoptera). Deutsche Entomologische Zeitschrift. 18: 217–226. doi: 10.1002/mmnd.19710180115.

Gubin, A.F. 1947. The honey bees and the pollination of the red clover. The Ministry of Agriculture of the RSFSR. Selkhozgiz, Moscow.

Gubin, V.A. 1983. Features of behavior of the Carpathian bees (in Russian). Beekeeping. 2: 10–12.

Gupta, R.K. 2014. Taxonomy and distribution of different honey bee species. pp. 63–103. *In*: Beekeeping. for Poverty Alleviation and Livelihood Security. Springer, Dordrecht.

Guzmán-Novoa, E. and R. Page. 1994. Genetic dominance and worker interactions affect honey bee colony defense. Behavioral Ecology. 5: 91–97.

Guzmán-Novoa, E., A. Sanchez, R.E. Page, Jr. and T. Garcia. 1996. Susceptibility of European and Africanized honey bees (*Apis mellifera* L.) and their hybrids to *Varroa jacobsoni* Oud. Apidologie. 27: 93–103.

Guzmán-Novoa, E., L. Eccles, Y. Calvete, J. McGowan, P.G. Kelly and A. Correa-Benitez. 2010. *Varroa destructor* is the main culprit for the death and reduced populations of overwintered honey bee (*Apis mellifera*) colonies in *Ontario*, Canada. Apidologie. 41: 443–450.

Gvozdetsky, N.A. and N.I. Mikhailov. 1963. Physical geography of the USSR. Geografgiz, Moscow, USSR.

Haddad, N., A. Bataeneh, I. Albaba D. Obeid and S. Abdulrahman. 2009a. Status of colony losses in the Middle East. Proceedings of the 41st Apimondia Congress Montpellier, France: 36.

Haddad, N., S. Fuchs, H.R. Hepburn and S.E. Radloff. 2009b. *Apis florea* in Jordan: Source of the founder population. Apidologie. 40: 508–512. doi: 10.1051/apido/2009011.

Hadisoesilo, S., G.W. Otis and M. Meixner. 1995. Two distinct populations of cavity-nesting honey bees (Hymenoptera, Apidae) in South Sulawesi, Indonesia. Journal of the Kansas Entomological Society. 68: 399–407.

Hadisoesilo, S. and G.W. Otis. 1996. Drone flight times confirm the species status of *Apis nigrocincta* Smith, 1861 to be a species distinct from *Apis cerana* F, 1793, in Sulawesi, Indonesia. Apidologie. 27: 361–369.

Hadisoesilo, S. and G.W. Otis. 1998. Differences in drone cappings of *Apis cerana* and *Apis nigrocincta*. Journal of Apicultural Research. 37: 11–15.

Hadisoesilo, S., M. Meixner and F. Ruttner. 1999. Geographic variation within *Apis koschevnikovi* Buttel-Reepen, 1906, in Borneo. Treubia. 31: 305–311.

Hadisoesilo, S. 2001. Diversity of Indonesia original honey bee. Biodiversitas. 2: 123–128.

Hadisoesilo, S., R. Raffiudin, W. Susanti, T. Atmowidi, C. Hepburn, S.E. Radloff et al. 2008. Morphometric analysis and biogeography of *Apis koschevnikovi* Enderlein (1906). Apidologie. 39: 495–503.

Hall, H.G. 1990. Parental analysis of introgressive hybridizations between African and European honey bees using nuclear RFLPs. Genetics. 125: 611–622.

Hall, H.G. and D.R. Smith. 1991. Distinguishing African and European honey bee matrilines using amplified mitochondrial DNA. Proceedings of the National Academy of Sciences of the United States of the United States of America. 88: 4548–4552.

Hall, R. 1998. The plate tectonics of Cenozoic SE Asia and the distribution of land and sea. pp. 99–131. *In*: Hall, R. and J.D. Holloway [eds.]. Biogeography and Geological Evolution of SE Asia. Backhuys Publishers, Leiden, Netherlands.

Han, F., A. Wallberg and M.T. Webster. 2012. From where did the Western honey bee (*Apis mellifera*) originate? Ecology and Evolution. 2: 1949–1957. doi: 10.1002/ece3.312.

Hall, R. 2013. The paleogeography of Sundaland and Wallacea since the Late Jurassic. Journal of Limnology. 72: 1–17.

Harbo, J.R. and J.W. Harris. 1999. Heritability in honey bees (Hymenoptera: Apidae) of characteristics associated with resistance to *Varroa jacobsoni* (Mesostigmata: Varroidae). Journal of Economic Entomology. 92: 261–265.

Harbo, J.R. and J.W. Harris. 2005. Suppressed mite reproduction explained by the behavior of adult bees. Journal of Apicultural Research. 44: 21–23.

Hargreaves, A.L., L.D. Harder and S.D. Johnson. 2009. Consumptive emasculation: the ecological and evolutionary consequences of pollen theft. Biological reviews of the Cambridge Philosophical Society. 84: 259–276.

Harris, J.W. 2007. Bees with *Varroa* sensitive hygiene preferentially remove mite infested pupae aged five days post capping. Journal of Apicultural Research. 46: 134–139. doi: 10.3896/ibra.1.46.3.02.

Harris, J.W. 2008. Effect of brood type on *Varroa*-sensitive hygiene by worker honey bees (Hymenoptera, Apidae). Annals of the Entomological Society of America. 101: 1137–1144.

Hassett, J., K.A. Browne, G.P. McCormack, E. Moore, Native Irish Honey Bee Society, G. Soland et al. 2018. A significant pure population of the dark European honey bee (*Apis mellifera mellifera*) remains in Ireland. Journal of Apicultural Research. 57: 337–350. doi: 10.1080/00218839.2018.1433949.

Hatjina, F., C. Costa, R. Büchler, A. Uzunov, M. Drazic, J. Filipi et al. 2014. Population dynamics of European honey bee genotypes under different environmental conditions. Journal of Apicultural Research. 53: 233–247.

Hatjina, F., M. Bouga, A. Karatasou, A. Kontothanasi, L. Charistos, C. Emmanouil et al. 2010. Pilot survey of honey bee colony losses in Greece (2007 and 2008). Journal of Apicultural Research. 49: 116–118.

Haydak, M.G. 1969. Life Activity of hHoney Bees. Bee and hive. Moscow: Kolos.

Haydak, M.H. 1970. Honey bee nutrition. Annual Review of Entomology. 15: 143–156.

Heaney, L.R. 1986. Biogeography of mammals in SE Asia: estimates of rates of colonization, extinction and speciation. Biological Journal of the Linnean Society. 28: 127–165.

Heard, M.S., J. Baas, J.L. Dorne, E. Lahive, A.G. Robinson, A. Rortais et al. 2017. Comparative toxicity of pesticides and environmental contaminants in bees: are honey bees a useful proxy for wild bee species? Sci. Tot. Environ. 578: 357–365.

Hebert, P.D.N., A. Cywinska, S.L. Ball and J.R. deWaard. 2003. Biological identifications through DNA barcodes. Proceedings of the Royal Society B: Biological Sciences. 270: 313–321.

Henry, M., M. Beguin, F. Requier, O. Rollin, J.F. Odoux, P. Aupinel et al. 2012a. A common pesticide decreases foraging success and survival in honey bees. Science. 336: 348–350. doi: 10.3389/fphys.2013.00037.

Henry, M., M. Beguin, F. Requier, O. Rollin, J.F. Odoux, P. Aupinel et al. 2012b. Response to Comment on "A Common Pesticide Decreases Foraging Success and Survival in Honey Bees". Science. 337: 1453.

Henry, M., C. Bertrand, V. Le Féon, F. Requier, J.F. Odoux, P. Aupinel et al. 2014. Pesticide risk assessment in free-ranging bees is weather and landscape dependent. Nature Communications. 5: 4359.

Henry, H., M.A. Becher, J.L. Osborne, P.J. Kennedy, P. Aupinel, V. Bretagnolle et al. 2017. Predictive systems models can help elucidate bee declines driven by multiple combined stressors. Apidologie. 48: 328–339.

Henry, M. and G. Rodet. 2018. Controlling the impact of the managed honey bee on wild bees in protected areas. Scientific Reports. 8: 9308.

Hepburn, H.R. and S.E. Radloff. 1996. Morphometric and pheromonal analyses of *Apis mellifera* L. along a transect from the Sahara to the Pyrenees. Apidologie. 27: 35–45.

Hepburn, H.R. and S.E. Radloff. 1997. Biogeographical correlates of population variance in the honey bees (*Apis mellifera* L.) of Africa. Apidologie. 285: 243–258.

Hepburn, R.H. and S.E. Radloff. 1998. Honey bees of Africa. Springer Verlag, Berlin.

Hepburn, H.R., D.R. Smith, S.E. Radloff and G.W. Otis. 2001. Infraspecific categories of *Apis cerana*: morphometric, allozymal and mtDNA diversity. Apidologie. 32: 3–23.

Hepburn, H.R. and S.E. Radloff. 2011a. Biogeography of the dwarf honey bees, *Apis andreniformis* and *Apis florea*. Apidologie. 42: 293–300. .

Hepburn, H.R. and S.E. Radloff. 2011b. Honey bees of Asia. Springer-Verlag, Berlin, Heidelberg. New York, USA.

Herbert, T.D., K.T. Lawrence, A. Tzanova, L.C. Peterson, R. Caballero-Gill and C.S. Kelly. 2016. Late Miocene global cooling and the rise of modern ecosystems. Nature. Geoscience. doi: 10.1038/NGEO2813.

Heywood, V.H. 1995. Global biodiversity assessment. Cambridge University Press.

Higes, M., R. Martin and A. Meana. 2006. *Nosema ceranae*, a new microsporidian parasite in Honey bees in Europe. Journal of Invertebrate Pathology. 92: 93–95.

Higes, M., R. Martín-Hernández, C. Botías, E. Garrido Bailon, A.V. Gonzalez-Porto, L. Barrios et al. 2008. How natural infection by *Nosema ceranae* causes honey bee colony collapse. Environmental Microbiology. 10: 2659–2669.

Higes, M., R. Martín-Hernández, E. Garrido-Bailón, A.V. González-Porto, P. García-Palencia, A. Meana et al. 2009. Honey bee colony collapse due to *Nosema ceranae* in professional apiaries. Environmental Microbiology. 1: 110–113.

Hines, H.M. and S.D. Hendrix. 2005. Bumble bee (Hymenoptera: Apidae) diversity and abundance in tallgrass prairie patches: Effects of local and landscape floral resources. Environmental Entomology. 34: 1477–1484.

Hobbs, G.A. 1965. Ecology of species of *Bombus Latr.* (Hymenoptera: Apidae) in southern Alberta. II. Subgenus Bombias Robt. The Canadian Entomologist. 97: 120–128.

Hoffmann, M., T.M. Brooks, G.A.B. da Fonseca, C. Gascon, A.F.A. Hawkins, R.E. James et al. 2008. Conservation planning and the IUCN Red List. Endanger. Species Research. 6: 113–125.

Hooper, D.U., E.C. Adair, B.J. Cardinale, J.E. Byrnes, B.A. Hungate, K.L. Matulich et al. 2012. A global synthesis reveals biodiversity loss as a major driver of ecosystem change. Nature. 486: 105.

Hopkins, D.M. 1967. The Bering Land Bridge. Stanford Univ. Press, Stanford.

Houle, D., J. Mezey, P. Galpern and A. Carter. 2003. Automated measurement of *Drosophila* wings. BMC Evol. Biol. 3: 25. doi: 10.1186/1471-2148-3-25.

Houston, T.E. 1970. Discovery of on apparent male soldier caste in a nest of a Halictinae bee (Hymenoptera: Halictidae), with notes on the nest. Australian Journal of Zoology. 18: 345–351.

Houston, T.E. 1977. Nesting biology of three allodapine bees in the subgenus *Exoneurella* Michener (Hymenoptera: Anthophoridae). Transactions of The Royal Society of South Australia. 101: 99–113.

http://www.flywings.org.uk/MorphoJ_page.htm

http://www3.canisius.edu/~sheets

http://drawwing.org Huang, K.L.J., J.M. Kingston, M. Albrecht, D.A. Holway and J.R. Kohn. 2018. The worldwide importance of honey bees as pollinators in natural habitats. Proceedings of the Royal Society of London. 285: 20172140.

Huang, W.F., J.H. Jiang, Y.W. Chen and C.H. Wang. 2007. A *Nosema ceranae* isolate from the Honey bee *Apis mellifera*. Apidologie. 38: 30–37.

Huang, Z.Y. and G.W. Otis. 1991. Inspection and feeding of larvae by worker honey bees (Hymenoptera: Apidae): effect of starvation and food quality. Journal of Insect Behavior. 4: 305–317.

Hudewenz, A. and A.M. Klein. 2013. Competition between honey bees and wild bees and the role of nesting resources in a nature reserve. Journal of Insect Conservation. 17: 1275–1283.

Huryn, V.M.B. 1997. Ecological impacts of introduced honey bees. The Quarterly Review of Biology. 72: 275–297.

Ibrahim, A. and M. Spivak. 2006. The relationship between hygienic behavior and suppression of mite reproduction as honey bee (*Apis mellifera*) mechanisms of resistance to *Varroa destructor*. Apidologie. 37: 31–40.

Ibrahim, A., G.S. Reuter and M. Spivak. 2007. Field trial of honey bee colonies bred for mechanisms of resistance against *Varroa destructor*. Apidologie 38: 67–76. doi: 10.1051/apido:2006065.

Ihering, R. 1903. Biologie der stachellosen Honigbienen Brasiliens. Zoologische Jahrbücher. Abteilung für Systematik. 19: 179–287.

Ilyasov, A.R., M.L. Lee, M. Proshchalykin, A. Lelej, S.H. Lim, D.I. Kim et al. 2019. Phylogenetic relationships of Russian Far-East *Apis cerana* with other north Asian populations. Journal of Apicultural Science (unpublished).

Ilyasov, R.A., A.V. Poskryakov, A.V. Petukhov and A.G. Nikolenko. 2007. Local honey bee (*Apis mellifera mellifera* L.) populations in the Urals. Russian Journal of Genetics. 43: 709–711.

Ilyasov, R.A. and Z.V. Shareeva. 2014. Effect of fluvalinate and amitraz on the bee colony (in Russian). Beekeeping. 6: 24–25.

Ilyasov, R.A. 2016. Genetic population structure and phylogenetic position of the dark forest bee *Apis mellifera mellifera* L. of the Urals and Volga region. Ph.D. Thesis, Institute of Biochemistry and Genetics. of the Ufa Scientific Center of Russian Academy of Sciences, Ufa, Russia.

Ilyasov, R.A., A.V. Poskryakov, A.V. Petukhov and A.G. Nikolenko. 2016. Molecular genetic analysis of five extant reserves of black honey bee *Apis melifera melifera* in the Urals and the Volga region. Russian Journal of Genetics. 52: 828–839.

Ilyasov, R.A., J. Park, J. Takahashi and H.W. Kwon. 2018. Phylogenetic uniqueness of honey bee *Apis cerana* from the Korean peninsula inferred from the mitochondrial, nuclear and morphological data. Journal of Apicultural Science. 62: 189–214.

Imperatriz-Fonseca, V.L. 1976. Studies on *Paratrigona subnuda* (Moure) Hymenoptera, Apidae, Meliponinae. I: Members of the colony. Revista Brasileira de Entomologia. 20: 101–112.

Invernizzi, C., C. Abud, I.H. Tomasco, J. Harriet, G. Ramallo, J. Campa et al. 2009. Presence of *Nosema ceranae* in Honey bees (*Apis mellifera*) in Uruguay. Journal of Invertebrate Pathology. 101: 150–153.

Irving, W. 1835. A Tour on the Prairies. Carey, Lea and Blanchard, Philadelphia.

Ishemgulov, A.M. 2006. Bashkirian breed. Reserves of increase of efficiency of beekeeping and apitherapy. Ufa.

Ishemgulov, A.M. 2008. Bashkir breed of bees (in Russian). Beekeeping. 7: 8–9.

Ivanova, E.N., T.A. Staykova and M. Bouga. 2007. Allozyme variability in honey bee populations from some mountainous regions in the southwest of Bulgaria. Journal of Apicultural Research. 46: 3–7.

Ivanova, E.N. and P.P. Petrov. 2010. Regional differences in honey bee winter losses in Bulgaria during the period 2006–9. Journal of Apicultural Research. 49: 102–103.

Iwasa, T., N. Motoyama, J.T. Ambrose and R.M.R. Michael Roe. 2004. Mechanism for the differential toxicity of neonicotinoid insecticides in the honey bee *Apis mellifera*. Crop Protection. 23: 371–378. doi: 10.1016/j.cropro.2003.08.018.

Jachimowicz, T. and G. El Sherbiny. 1975. Zur Problematik der Verwendung von Invertzucker fur die Bienenfutterung. Apidologie. 6: 121–143.

Jacques, A., M. Laurent, M. Ribière-Chabert, M. Saussac, S. Bougeard, G.E. Budge et al. 2017. A pan-European epidemiological study reveals honey bee colony survival depends on beekeeper education and disease control. PLOS ONE. 12: e0172591. doi: 10.1371/journal.pone.0172591.

Janke, A., S. Klopfstein, L. Vilhelmsen, J.M. Heraty, M. Sharkey and F. Ronquist. 2013. The Hymenopteran Tree of life: evidence from protein-coding genes and objectively aligned ribosomal data. Proceedings of the National Academy of Sciences of the United States of America. 8: e69344. doi: 10.1371/journal.pone.0069344.

Janzen, D.H. 1971. The ecological significance of an arboreal nest of *Bombus pullatus* in Costa Rica. Journal of the Kansas Entomological Society. 44: 210–216.

Jay, S.C. 1969. The problem of drifting in commercial apiaries. American Bee Journal. 109: 178–179.

Jaycox, E.R. 1961. The effects of various foods and temperatures on sexual maturity of the drone honey bee (*Apis mellifera*). Annals of the Entomological Society of America. 54: 519–523.

Jefferson, T. 1788. Notes on the State of Virginia. Pritchard and Hall, Philadelphia.

Jensen, A.B. and B.V. Pedersen. 2005. Honey bee conservation: a case story from Læsø Island, Denmark. pp. 142–164. *In*: Lodesani, M. and C. Costa [eds.]. Beekeeping and Conserving Biodiversity of Honey Bee. Sustainable Bee Breeding. Northern Bee Books, Mytholmroyd, Hebden Bridge, UK.

Jensen, A.B., K.A. Palmer, J.J. Boomsma and B.V. Pedersen. 2005. Varying degrees of *Apis mellifera ligustica* introgression in protected populations of the black Honey bee, *Apis mellifera mellifera*, in northwest Europe. Molecular Ecology. 14: 93–106.

Johnson, A. 1969. The Swedish Settlements on the Delaware, 1638–1664. Baltimore Genealogy Public Co., Baltimore.

Johnson, R.M., H.S. Pollock and M.R. Berenbaum. 2009. Synergistic Interactions Between In-Hive Miticides in *Apis mellifera*. Journal of Economic Entomology 102: 474–479. doi: 10.1603/029.102.0202.

Jones, G.F. and R. Wilson. 1981. Detailed report on the Salzburger emigrants who settled in America. University of Georgia Press, Athens.

Jones, H. and K. Bienefeld. 2016. The Asian Honey Bee (*Apis cerana*) is significantly in Decline. Bee World. 93: 90–97.

Josselyn, J. 1675. An account of two voyages to New-England, made during 1638, 1663. G. Widdowes, London.

Kaiser, F. 1976. Ricsenhonigbiene im FlugkaFig. besucht. Allgemeine Deutsche Imker-Zeitung. 10: 93–97.

Kalm, P. 1772. Travels into North America. Lowndes, London.

Kaluza, B.F., H.M. Wallace, T.A. Heard, V. Minden, A.M. Klein and S.D. Leonhardt. 2018. Social bees are fitter in more biodiverse environments. Scientific Reports. 8: 12353.

Kammerer, F.X. 1989. Aktueller Stand der Erkenntnisse über die Fütterung von Bienen mit Zucker. Imkerfreund. 1: 12–14.

Kandemir, İ., A. Özkan and S. Fuchs. 2011. Reevaluation of honey bee (*Apis mellifera*) microtaxonomy: a geometric morphometric approach. Apidologie. 42: 618–627. doi: 10.1007/s13592-011-0063-3.

Kandemir, İ., M. Kence and A. Kence. 2005. Morphometric and electrophoretic variation in different honey bee (*Apis mellifera* L.) populations. Turkish Journal of Veterinary and Animal Sciences. 29: 885–890.

Kandemir, İ., M. Kence, W.S. Sheppard and A. Kence. 2006. Mitochondrial DNA variation in honey bee (*Apis mellifera* L.) populations from Turkey. Journal of Apicultural Research. 45: 33–38.

Kandemir, İ., M.G. Moradi, B. Özden and A. Özkan. 2009. Wing geometry as a tool for studying the population structure of dwarf honey bees (*Apis florea* Fabricius 1876) in Iran. Journal of Apicultural Research. 48: 238–246. doi: 10.3896/ibra.1.48.4.03.

Kasprzak, S. and G. Topolska. 2007. *Nosema ceranae* (Eukaryota: Fungi: Microsporea)—nowy pasożyt pszczoły miodnej *Apis mellifera*. Wiad Parazytology. 53: 281–284.

Katayama, E. 1973. Observations on the brood development in *Bombus ignitus* (Hymenoptera, Apidae). II. Brood development and feeding habits. Kontyu. 41: 203–216.

Kauhausen-Keller, D., F. Ruttner and R. Keller. 1997. Morphometric studies on the microtaxonomy of the species *Apis mellifera* L. Apidologie. 28: 295–307. doi: 10.1051/apido:19970506.

Keller, I., P. Fluri and A. Imdorf. 2005. Pollen nutrition and colony development in honey bees-part II. Bee World. 86: 27–34.

Kerr, J.T. 2017. A cocktail of poisons. Science. 356: 1331–1332.

Kerr, W.E. 1974. Sex determination in bees. III. Caste determination and genetic control in Melipona. Insectes Sociaux. 21: 357–367.

Kevan, P.G. and B.F. Viana. 2003. The global decline of pollination services. Biodiversity. 4: 3–8.

Kingsbury, S.M. 1933. The Records of the Virginia Company of London. Library of Congress, Washington, DC.

Kirrane, M.J., L.I. de Guzman, B.A. Holloway, A.M. Frake, T.E. Rinderer and P.M. Whelan. 2015. Phenotypic and genetic analyses of the *varroa* sensitive hygienic trait in Russian honey bee (Hymenoptera: Apidae) colonies. PLOS ONE. 10: e0116672. doi: 10.1371/journal.pone.0116672.

Klee, J., A.M. Besana, E. Genersch, S. Gisder, A. Nanetti, T. Dinh Quyet et al. 2007. Widespread dispersal of the microsporidian *Nosema ceranae*, an emergent pathogen of the western honey bee, *Apis mellifera*. Journal of Invertebrate Pathology. 96: 1–10.

Klein, A.M., B.E. Vaissière, J.H. Cane, I. Steffan-Dewenter, S.A. Cunningham, C. Kremen et al. 2007. Importance of pollinators in changing landscapes for world crops. Proceedings of the Royal Society of London. 274: 303–313.

Klingenberg, C.P., A.V. Badyaev, S.M. Sowry and N.J. Beckwith. 2001. Inferring developmental modularity from morphological integration: Analysis of individual variation and asymmetry in bumble bee wings. Am. Nat. 157: 11–23. doi: 10.1086/317002.

Klingenberg, C.P. 2011. MorphoJ: an integrated software package for geometric morphometrics. Molecular Ecology Resources. 11: 353–357. doi: 10.1111/j.1755-0998.2010.02924.x.

Klochko, R.T. and A.V. Blinov. 2015. Neonicotinoids-a danger to bees (in Russian). Beekeeping. 8: 46–47.

Klochko, R.T. and A.V. Blinov. 2016. Ten reasons for the death of bees in 2015 (in Russian). Beekeeping. 1: 52–55.

Klochko, R.T., S.N. Lugandkiy and A.V. Blinov. 2018. One more time on bee death (in Russian). Beekeeping. 5: 44–46.

Knerer, G. and M. Schwarz. 1976. Halictine social evolution: the Australian enigma. Science. 194: 445–448.

Knerer, G. and M. Schwarz. 1978. Beobachtungen an austrakischen Furchenbienen (Hymenoptera; Halictinae). Zoologischer Anzeiger. 200: 321–333.

Knerer, G. 1980. Biologie and Sozialverhaften von Bienenarten der Gattung Halictus Latreille (Hymenoptera; Halictidae). Zoologische Jahrbücher. Abteilung für Systematik, Ökologie und Geographie der Tiere. 107: 511–536.

Kodes, L.G. and N.I. Popova. 2010. Pedigree and economically useful features of honey bees of the Far East. Monograph. Ussuriysk PSHA.

Kocher, S.D. and R.J. Paxton. 2014. Comparative methods offer powerful insights into social evolution in bees. Apidologie. 45: 289–305.

Koeniger, G., N. Koeniger, S. Tingek and A. Kelitu. 2000. Mating flights and sperm transfer in the dwarf honey bee *Apis andreniformis* (Smith, 1858). Apidologie. 31: 301–311.

Koeniger, N. 1976a. Interspecific competition between *Apis florea* and *Apis mellifera* in the tropics. Bee World. 57: 110–112.

Koeniger, N. 1976b. Neue Aspekte der Phylogenie innerhalb der Gattung *Apis*. Apidologie. 7: 357–366. doi: 10.1051/apido:19760406.

Koeniger, N. 1976c. Die Arten der Honigbiene. Allgemeine Deutsche Imkerzeitung. 10: 89–92.

Koeniger, N., G. Koeniger, S. Tingek, M. Mardan and E. Rinderer. 1988. Reproductive isolation by different time of drone flight between *Apis cerana* Fabricius, 1793 and *Apis vechti* (Maa, 1953). Apidologie. 19: 103–106.

Koeniger, N., G. Koeniger, M. Gries, S. Tingek and A. Kelitu. 1996. Reproductive isolation of *Apis nuluensis* Tingek, Koeniger and Koeniger, 1996 by species-specific mating time. Apidologie. 27: 353–359.

Koeniger, N., G. Koeniger and H. Pechhacker. 2005. The nearer the better? Drones (*Apis mellifera*) prefer nearer drone congregation areas. Insectes Sociaux. 52: 31–35.

Koeniger, N., G. Koeniger and S. Tingek. 2010. Honey bees of Borneo: exploring the centre of *Apis* diversity. Natural History Publication (Borneo), Kota Kinabalu.

Koetz, A. 2013a. The Asian honey bee (*Apis cerana*) and its strains-with special focus on *Apis cerana* Java genotype Literature review. Department of Agriculture, Fisheries and Forestry, Queensland.

Koetz, A.H. 2013b. Ecology, behaviour and control of *Apis cerana* with a focus on relevance to the Australian incursion. Insects. 4: 558–592. doi: 10.3390/insects4040558.

Kokorev, N.M. and B.Y. Chernov. 2005. The Selection work on the apiary. TID Continent-Press, Moscow.

Komlatskiy, V.I. 2018. The development of bees in anthropogenic transformation of the ecological environment. Results of research work for 2017: 253–254.

Kono, Y. and J.R. Kohn. 2015. Range and frequency of Africanized honey bees in California (USA). PLOS ONE. 10: 15.

Kontrimavichus, V.L. 1985. Beringia in the Cenozoic Era. Balkema, Rotterdam.

Konusova, O.L., N.V. Ostroverkhova, A.N. Kucher, D.V. Kurbatskij and T.N. Kireeva. 2016. Morphometric variability of honey bees *Apis mellifera* L., differing in variants of the *COI-COII* mtDNA locus. Vestnik Tomskogo gosudarstvennogo universiteta. Biologiya (TSU J. Biol.). 1: 62–81.

Kosarev, M.N., A.Ya. Sharipov, F.G. Yumaguzhin and L.N. Savushkina. 2001. Selection of breed type "Burzyan tree bee" (in Russian). Beekeeping. 6: 10–13.

Kosarev, M.N., A.Y. Sharipov, F.G. Yumaguzhin and L.N. Savushkina. 2011. The breeding of breed type "Burzyanskaya Bortevaya" (in Russian). Beekeeping. 6: 14–15.

Kosarev, M.N., A. Ya. Sharipov, F.G. Yumaguzhin and L.N. Savushkina. 2012. Biological features of the Burzyan population of the Middle Russian breed of bees. Russian Beekeeping on the path of accession to the WTO. Materials of the International scientific-practical conference. Yaroslavl: 107–110.

Kostoev, M.M. 2009. Breeding work with bees in the Republic of Ingushetia. (in Russian). Beekeeping. 6: 14–15.

Kotthoff, U., T. Wappler, M.S. Engel and J. Ali. 2013. Greater past disparity and diversity hints at ancient migrations of European honey bee lineages into Africa and Asia. Journal of Biogeography. 40: 1832–1838. doi: 10.1111/jbi.12151.

Kozhevnikov, G.A. 1900a. Materials to the natural history of the honey bee (*Apis mellifera* L.). I. (in Russian). Société Des Amateurs De Sciences Naturelles. XCIX. Moscow.

Kozhevnikov, G.A. 1900b. Races of Caucasian bees. Vestnik Russkogo Obshchestva Pchelovodstva, Saint-Petersburg.

Kozhevnikov, G.A. 1905. Materials to the natural history of the honey bee (*Apis mellifera* L.). II. (in Russian). Société Des Amateurs De Sciences Naturelles. XCIX. Moscow.

Kraus, B. and R.E. Page. 1995. The impact of *Varroa jacobsoni* Oud. on feral bees (*Apis mellifera*) of California. Environ. Entomol. 24: 1473–1480.

Kraus, F.B. 2005. Requirements for local pop lation conservation and breeding. pp. 87–107. *In*: Lodesani, M. and C. Costa [eds.]. Beekeeping and Conserving Biodiversity of Honey Bees. Northern Bee Books, Hebden Bridge, UK.

Kremen, C., N.M. Williams, M.A. Aizen, B. Gemmill Herren, G. LeBuhn, R. Minckley et al. 2007. Pollination and other ecosystem services produced by mobile organisms: a conceptual framework for the effects of land-use change. Ecology Letters. 10: 299–314.

Kritsky, G. 1991. Lessons from history, the spread of the honey bee in North America. The American Bee Journal. 131: 367–370.

Krivtsov, N., G.D. Bilash and A.V. Borodachev. 1999. Breeding and improvement of breeding and productive qualities of bees: guidelines. Inform Agrotekh, Moscow.

Krivtsov, N. and N.N. Grankin. 2004. Middle Russian bees and their breeding. GNU research Institute of beekeeping.

Krivtsov, N.A., S.S. Sokolskiy, L.N. Savushkina and E.M. Lyubimov. 2008. Selection of breed type "Krasnopolyanskiy" of Grey Mountain Caucasian bee. Bulletin of the RAAS. 5: 69–71.

Krivtsov, N.I. and V.I. Lebedev. 1995. And the Maintenance of bee colonies with the basics of breeding. Kolos, Moscow.

Krivtsov, N.I. and S.S. Sokolskiy. 2002. Gray mountain Caucasian bees. Omega Print, Rostov-on-Don.

Krivtsov, N.I. and. S.S. Sokolskiy and E.M. Lyubimov. 2009a. Gray mountain Caucasian bees. Poligraf-Yug, Sochi.

Krivtsov, N.I., S.S. Sokolskiy, L.N. Savushkina and E.M. Lyubimov. 2009b. Economic and biological signs of bee colonies of breed type "Krasnopolyanskiy" of Grey Mountain Caucasian bees. Materials of the coordination meeting and the 9th scientific-practical conference Intermed. GNU research Institute of beekeeping: 47–51.

Kshirsagar, K.K. 1983. Morphometric studies on the Indian hive bee, *Apis cerana indica* F. 1. Morphometric characters useful in identification of intraspecific taxa. pp. 254–261. *In*: Proceedings of the 2nd International Conference on Apiculture in Tropical Climates, New Delhi.

Kuldna, P., K. Peterson, H. Poltimaee and J. Luig. 2009. An application of DPSIR framework to identify issues of pollinator loss. Ecological Economics. 69: 32–42.

Kulhanek, K., N. Steinhauer, K. Rennich, D.M. Caron, R.R. Sagili, J.S. Pettis et al. 2017. A national survey of managed honey bee 2015-2016 annual colony losses in the USA. Journal of Apicultural Research. 56: 328–340.

Kunieda, T., T. Fujiyuki, R. Kucharski, S. Foret, S.A. Ament, A.L. Toth et al. 2006. Carbohydrate metabolism genes and pathways in insects : insights from the honey bee genome. Insect Mol. Biol. 15: 563-576.

Kürschner, W.M., Z. Kvaček and D.L. Dilcher. 2008. The impact of Miocene atmospheric carbon dioxide fluctuations on climate and the evolution of terrestrial ecosystems. P. Natl. Acad. Sci. USA. 105: 449–453.

Lahjie, A.M. and B. Seibert. 1990. Honey gathering by people in the interior of East Kalimantan. Bee World. 71: 153–157.

Latreille, P.A. 1825. Familles naturelles du règne animal, exposé succinctement et dans un ordre analytique avec l'indication de leurs genres. Jean-Baptiste Baillière. Paris. 570.

Lautenbach, S., R. Seppelt, J. Liebscher and C.F. Dormann. 2012. Spatial and Temporal Trends of Global Pollination Benefit. PLOS ONE. 7: e35954.

Le Conte, Y., M. Ellis and W. Ritter. 2010. *Varroa* mites and honey bee health: can *Varroa* explain part of the colony losses? Apidologie. 41: 353–363.

Lebedinskiy, I.A., A.Y. Lavrskiy and A.V. Petukhov. 2016. Volumetric parameters of rectal pads as morphophysiological indicators of adaptable potential of honey bee's (*Apis mellifera*) colonies. Biomics. 8: 147–153.

LeBlanc, B.W., O.K. Davis, S. Boue, A. DeLucca and T. Deeby. 2009. Antioxidant activity of sonoran desert bee pollen. Food Chemistry. 115: 1299–1305.

Lecocq, A., A.B. Jensen, P. Kryger and J.C. Nieh. 2016. Parasite infection accelerates age polyethism in young honey bees. Scientific Reports, 6, 22042. doi: 10.1038/srep22042.

Lee, H.S. and S. Nagy. 1990. Relative reactivities of sugars in the formation of 5-hydroxymethylfurfural in sugar-catalyst model systems. Journal of Food Processing and Preservation. 14: 171–178.

Lekishvili, M.A. and A.A. Khidesheli. 1967. Characteristics of the main populations of the grey Georgian bee. XXI International Congress on beekeeping. Bucharest. Apimondia. 451–453.

Leonhardt, S.D., N. Blüthgen and T. Schmitt. 2009. Smelling like resin: terpenoids account for species-specific cuticular profiles in Southeast-Asian stingless bees. Insectes Sociaux. 56: 157–170.

Leonhardt, S.D., N. Blüthgen and T. Schmitt. 2011a. Chemical profiles of body surfaces and nests from six Bornean stingless bee species. Journal of Chemical Ecology. 37: 98–104.

Leonhardt, S.D., T. Schmitt and N. Blüthgen. 2011b. Tree resin composition, collection behavior and selective filters shape chemical profiles of tropical bees. PLOS ONE. 6: e23445.

Leonhardt, S.D., H.M. Wallace, N. Blüthgen and F. Wenzel. 2015. Potential role of environmentally derived cuticular compounds in stingless bees. Chemoecology. 25: 159–167.

Li, J., H. Qin, J. Wu, B.M. Sadd, X. Wang, J.D. Evans et al. 2012. The prevalence of parasites and pathogens in Asian honey bees *Apis cerana* in China. PLOS ONE. 7: e47955. doi: 10.1371/journal. pone.0047955.

Li, W. 1998. Introducing *Apis mellifera* threatens *Apis cerana*. Journal of Bee. 23: 4–6.

Lin, N. and C.D. Michener. 1972. Evolution of sociality in insects. The Quarterly Review of Biology. 47: 131–159.

Lindauer, M. 1971. The functional significance of the honey bee waggle dance. The American Naturalist. 105: 89–96.

Lindauer, M. and W.E. Kerr. 1960. Communication between the workers of stingless bees. Bee World. 41: 29–71.

Lipp, J. 1994. Der Honig. Ulmer, Stuttgart.

Lo, N., R.S. Gloag, D.L. Anderson and B.P. Oldroyd. 2010. A molecular phylogeny of the genus *Apis* suggests that the giant honey bee of the Philippines, *A. breviligula* Maa and the plains honey bee of southern India, *A. indica fabricius*, are valid species. Systematic Entomology. 35: 226–233. doi: 10.1111/j.1365-3113.2009.00504.x.

Lobell, D.B., W. Schlenker and J. Costa-Roberts. 2011. Climate trends and global crop production since 1980. Science. 333 (6042): 616–620.

Locke, B., Y. Le Conte, D. Crauser and I. Fries. 2012. Host adaptations reduce the reproductive success of *Varroa destructor* in two distinct European honey bee populations. Ecology and Evolution. 2: 1144–1150.

Lord, W.G. and S.K. Nagi. 1987. *Apis florea* discovered in Africa. Bee World. 68: 39–40. doi: 10.1080/0005772X.1987.11098907.

Lowe, T.M. and S.R. Eddy. 1997. tRNAscan-SE : a program for improved detection of transfer RNA genes in genomic sequence. Nucleic Acid Research. 25: 955–964.

Lundgren, J.G. 2017. Predicting both obvious and obscure effects of pesticides on bees. pp. 39–59. *In*: Beekeeping—From Science to Practice. Springer, Cham.

Lyubimov, E.M. and L.N. Savushkina. 2010. Biological signs of bee Queens and drones of breed type of grey Mountain Caucasian breed "Krasnopolyanskiy". The main directions of development of beekeeping at the present stage. Materials of scientific and practical conference. VPO RGATU, Ryazan: 54–57.

Maa, T.C. 1944. On the classification and phylogeny of the Chinese honey bees. Entomology Shaowuana. 1: 4–5.

Maa, T.C. 1953. An enquiry into the systematics of the tribus Apidini or honey bees (Hymenoptera). Treubia. 21: 525–640.

Maciel, A.C.M. 1976. Notes sur labiologie *d'Heriades truncorum* L. (Hymenoptera Megahilidae). Apidologie. 7: 169–187.

Maeta, Y. 1979. Tohoku nogyo schikenio kencyu hokoku. Bulletin of the Tohoku National Agricultural Experiment Station (Morioka). 61: 59–68.

Maggi, M., K. Antúnez, C. Invernizzi, P. Aldea, M. Vargas, P. Negri et al. 2016. Honey bee health in South America. Apidologie. 47: 835–854.

Magnus, R.M. and A.L. Szalanski. 2010. Genetic evidence for honey bees (*Apis mellifera* L.) of Middle Eastern lineage in the United States. Sociobiology 55: 285–296.

Magnus, R.M., A.D. Tripodi and A.L. Szalanski. 2011. Mitochondrial DNA diversity of honey bees, *Apis mellifera* L. (Hymenoptera: Apidae) from queen breeders in the United States. Journal of Apicultural Science. 55: 5–14.

Magnus, R.M., A.D. Tripodi and A.L. Szalanski. 2014. Mitochondrial DNA diversity of honey bees (*Apis mellifera*) from unmanaged colonies and swarms in the United States. Biochem. Genet. 52: 245–257.

Malkova, S.A. and N.P. Vasilenko. 2008. Breed type "Maikopskiy" of Carpathian bees (in Russian). Beekeeping. 3: 8–10.

Mallinger, R.E., H.R. Gaines-Day and C. Gratton. 2017. Do managed bees have negative effects on wild bees? PLOS ONE. 12: e0189268.

Malyshev, S.I. 1928. The life story of bumble bee colony. Pchelovodnoe delo. 6: 292–296.

Manino, A. and F. Marletto. 1984. Il sistema enzimatico MDH in popolazioni di *Apis mellifera* L. della Valle d'Aosta. L'apicoltore Moderno. 75: 89–94.

Marletto, F., A. Manino and P. Pedrini. 1984. Intergradazione tra sotto-species di *Apis mellifera* in Liguria. L'apicoltore Moderno. 75: 159–163.

Martimianakis, S., E. Klossa-Kilia, M. Bouga and G. Kilias. 2011. Phylogenetic relationships of Greek *Apis mellifera* subspecies based on sequencing of mtDNA segments (*COI* and *ND5*). Journal of Apicultural Research. 50: 42–50.

Martin, E.A., B. Feit, F. Requier, H. Friberg and M. Jonsson. 2019. Assessing the resilience of biodiversity-driven functions in agroecosystems under environmental change. Adv. Ecol. Res. 60: 59–123.

Martin, S. 1995. Ontogenesis of the mite *Varroa jacobsoni* Oud. in drone brood of the honey bee *Apis mellifera* L. under natural conditions, Experimental and Applied Acarology. 19: 199–210.

Martin, S. 1998. A population model for the parasitic mite *Varroa jacobsoni* in honey bee (*Apis mellifera*) colonies. Ecological Modelling. 109: 267–281.

Martin, S.J. and C. Cook. 1996. Effect of host brood type on the number of offspring laid by the honey bee parasite *Varroa jacobsoni*. Experimental and Applied Acarology 20: 387–390. doi: 10.1007/BF00130551.

Martin, S.J. 2001. The role of *Varroa* and viral pathogens in the collapse of honey bee colonies: a modelling approach. The Journal of Applied Ecology. 38: 1082–1093.

Martin, S.J., B.V. Ball and N.L. Carreck. 2010. Prevalence and persistence of deformed wing virus (DWV) in untreated or acaricide-treated *Varroa destructor* infested honey bee (*Apis mellifera*) colonies. Journal of Apicultural Research. 49: 72–79.

Marzaro, M., L. Vivan, A. Targa, L. Mazzon, N. Mori, M. Greatti et al. 2011. Lethal aerial powdering of honey bees with neonicotinoids from fragments of maize seed coat. Bulletin of Insectology. 64: 119–126.

Masterman, R., R. Ross, K. Mesce and M. Spivak. 2001. Olfactory and behavioral response thresholds to odors of diseased brood differ between hygienic and non hygienic honey bees (*Apis mellifera* L.). Journal of Comparative Physiology. 187: 441–452.

Matheson, A. 1993. World bee health report. Bee World. 74: 176–212.

Mathias, D.M.S., C. Borgemeister and H. von Wehrden. 2017. Thinking beyond Western commercial honey bee hives: towards improved conservation of honey bee diversity. Biodiversity and Conservation. 14: 3499–3504.

Mattila, H.R. and G.W. Otis. 2006. The effects of pollen availability during larval development on the behaviour and physiology of spring-reared honey bee workers. Apidologie. 37: 533–546.

Mattila, H.R. and B.H. Smith. 2008. Learning and memory in workers reared by nutritionally stressed honey bee (*Apis mellifera* L.) colonies. Physiology and Behavior. 95: 609–616.

Mattu, V.K. and L.R. Verma. 1984. Morphometric studies on the Indian honey bee, *Apis cerana indica* F. effect of seasonal variations. Apidologie. 15: 63–74.

Maurizio, A. 1958. Pollen feeding and life processes of a honey bee. New in beekeeping. Moscow: State Agricultural Publishing House.

Mayack, C. and D. Naug. 2009. Energetic stress in the honey bee *Apis mellifera* from *Nosema ceranae* infection. Journal of Invertebrate Pathology. 100: 185–188.

Mayer, D.F. and C.A. Johansen. 1976. Biological observations on *Anthophora urbana* Cresson (Hymenoptera: Anthophoridae). The Pan-Pacific Entomologist. 52: 120–125.

Mayhew, P.J. 2007. Why are there so many insect species? Perspectives from fossils and phylogenies. Biological Reviews. 82: 425–454. doi: 10.1111/j.1469-185X.2007.00018.x.

Meana, A. 2009. Histopathology of *Nosema* infected bees. Workshop: *Nosema* disease: Lack of knowledge and work standardization. COLOSS, Guadalajara.

Medina, L.M. and S.J. Martin. 1999. A comparative study of *Varroa jacobsoni* reproduction in worker cells of honey bees (*Apis mellifera*) in England and Africanized bees in Yucatan, Mexico. Experimental and Applied Acarology. 23: 659–667.

Medrzycki, P., R. Montanari, L. Bortolotti, A.G. Sabatini, S. Maini and C. Porrini. 2003. Effects of imidacloprid administered in sublethal doses on honey bee behaviour. Laboratory tests. Bulletin of Insectology. 56: 59–62.

Medrzycki, P., F. Sgolastra, L. Bortolotti, G. Bogo, S. Tosi, E. Padovani et al. 2010. Influence of brood rearing temperature on honey bee development and susceptibility to poisoning by pesticides. Journal of Apicultural Research. 49: 52–59. doi: 10.3896/IBRA.1.49.1.07.

Meixner, M.D, W.S. Sheppard and J. Poklukar. 1993. Asymmetrical distribution of a mitochondrial DNA polymorphism between two introgressing honey bee races. Apidologie. 24: 147–153.

Meixner, M.D, R. Francis, A. Gajda, P. Kryger, S. Andonov, A. Uzunov et al. 2014. Occurrence of parasites and pathogens in honey bee colonies used in a European genotype-environment interactions experiment. Journal of Apicultural Research. 53: 215–229.

Meixner, M.D, R. Büchler, C. Costa, S. Andonov, M. Bieńkowska, M. Bouga et al. 2015. Looking for the Best Bee- An Experiment about Interactions Between Origin and Environment of Honey Bee Strains in Europe. The American Bee Journal. 155: 663–666.

Meixner, M.D., C. Costa, P. Kryger, F. Hatjina, M. Bouga, E. Ivanova et al. 2010. Conserving diversity and vitality for honey bee breeding. Journal of Apicultural Research. 49: 85–92.

Meixner, M.D., M.A. Leta, N. Koeniger and S. Fuchs. 2011. The honey bees of Ethiopia represent a new subspecies of *Apis mellifera*: *Apis mellifera simensis* n. ssp. Apidologie. 42: 425–437. doi: 10.1007/s13592-011-0007-y.

Meixner, M.D., M.A. Pinto, M. Bouga, P. Kryger, E. Ivanova and S. Fuchs. 2013. Standard methods for characterising subspecies and ecotypes of *Apis mellifera*. Journal of Apicultural Research. 52: 1–28. doi: 10.3896/ibra.1.52.4.05.

Mikhailov, A.S. 1927a. Seasonal variability of bees (in Russian). Opytnaya Paseka. 6: 180–183.

Mikhailov, A.S. 1927b. Über die Saison-Variabilität der Honigbiene. Archiv fir Bienenkunde. 8: 16–24.

Michener, C.D. 1944. Comparative external morphology and a classification of the bees (Hymenoptera). Bulletin of the American Museum of Natural History. 82: 151–326.

Michener, C.D. 1965. A classifications of the bees of the Australian and South Pacific regions. Bulletin of the American Museum of Natural History. 130: 362.

Michener, C.D. 1974. The social behaviour of the bees. Harvard Univ. Press, Cambridge, Massachusetts.

Michener, C D. and D.J. Brothers. 1974. Were workers of eusocial Hymenoptera initially altruistic or oppressed? Proceedings of the National Academy of Sciences. 71: 671–674.

Michener, C.D. 1975. A taxonomic study of African allodapine bees (Hymenoptera: Anthophoridae, Ceratinini). Bulletin of the American Museum of Natural History. 155: 67–240.

Michener, C.D. and M. Amir. 1977. The seasonal cycle and habitat of a tropical bumble-bee. Pacific Insects. 17: 234–240.

Michener, C.D. 1979. Biogeography of the bees. Annals of the Missouri Botanical Garden. 66: 277–347.

Michener, C.D. 2000. The Bees of the World. Johns Hopkins University Press.

Michener, C.D. 2007. The bees of the world. John Hopkins University Press, Baltimore.

Michez, D., M. Vander Planck and M.S. Engel. 2012. Fossil bees and their plant associates. pp. 103–164. *In*: Patiny, S. (ed.). Evolution of Plant-pollinator Relationships. Cambridge University Press: Cambridge.

Miguel, I., M. Iriondo, L. Garnery, W.S. Sheppard and A. Estonba. 2007. Gene flow within the M evolutionary lineage of *Apis mellifera*: role of the Pyrenees, isolation by distance and post-glacial re-colonization routes in the Western Europe. Apidologie. 38: 141–155.

Miguel, I., M. Baylac, M. Iriondo, C. Manzano, L. Garnery and A. Estonba. 2011. Both geometric morphometric and microsatellite data consistently support the differentiation of the *Apis mellifera* M evolutionary branch. Apidologie. 42: 150–161. doi: 10.1051/apido/2010048.

Mikheyev, A.S., M.M.Y. Ti, J. Arora and T.D. Seeley. 2015. Museum samples reveal rapid evolution by wild honey bees exposed to a novel parasite. Nature Communications. 6: 7991. doi: 10.1038/ncomms8991.

Mizis, A.P. 1976. External morphological characters of the Lithuanian honey bee and their correlation variability. pp. 153–156. *In*: Proceedings of International Symposium on Bee Genetics, Selection and Reproduction, Moscow.

Mogga, J. and F. Ruttner. 1988. *Apis florea* in Africa: Source of the founder population. Bee World. 69: 100–103. doi: 10.1080/0005772X.1988.11098960.

Mojica, M.J.J. 2011. Why beekeeping is for keeps. Bureau of Agricultural Research Digest. 13: 2.

Monteiro, L.R., J.A.F. Diniz-Filho, S.F. Dos Reis and E.D. Araújo. 2002. Geometric estimates of heritability in biological shape. Evolution. 56: 563–572. doi: 10.1111/j.0014-3820.2002.tb01367.x.

Morales, C.L. 2009. Pollination requirement of raspberry in SW Argentina. Preliminary results. The International Journal of Plant Reproductive Biology. 1: 195–198.

Moretto, G., L.S. Goncalves, D. De Jong and M.Z. Bichuette. 1991. The effects of climate and bee race on *Varroa jacobsoni* Oud. infestations in Brazil. Apidologie. 22: 197–203.

Moretto, G., L.S. Goncalves and D. De Jong. 1993. Heritability of Africanized and European honey bee defensive behaviour against the mite Varroa jacobsoni. Revue Brasilia Genetica 16: 71–77.

Moritz, R.F.A. 1985. The effects of multiple mating on the worker-queen conflict in *Apis mellifera* L. Behavioral Ecology and Sociobiology. 16: 375–377.

Moritz, R.F.A. and E.E. Southwick. 1992. Bees as Superorganisms. Springer-Verlag, Berlin.

Moritz, R.F.A. and M. Jordan. 1992. Selection of resistance against *Varroa jacobsoni* across caste and sex in the honey bee (*Apis mellifera*). Experimental and Applied Acarology. 16: 345–353.

Moritz, R.F.A., S. Hartel and P. Neumann. 2005. Global invasions of the western honey bee (*Apis mellifera*) and the consequences for biodiversity. Ecoscience. 12: 289–301.

Moritz, R.F.A., F.B. Kraus, P. Kryger and R.M. Crewe. 2007. The size of wild Honey bee populations (*Apis mellifera*) and its implications for the conservation of Honey bees. Journal of Insect Conservation. 11: 391–397.

Morse, R.A. and F.M. Laigo. 1969. *Apis dorsata* in the Philippine, Monograph of the Philippine Association of Entomologists, Inc., University of the Philippine, The College, Laguna. 1: 96.

Morse, R.A. 1978. Honey Bee Pests, Predators and Diseases. Cornell University Press, *Ithaca*, NY, USA.

Mossagegh, M.S. 1993. New geographical distribution line of *Apis* florea in Iran. pp. 64–66. *In*: Conner, L.J., T. Rinderer, H.A. Sylvester and S. Wongsiri [eds.]. Asian Apiculture. Wicwas Press, Cheshire.

Moulton, G.E. 2002. The Definitive Journals of Lewis and Clark. Univ. of Nebraska Press, Lincoln, Nebraska.

Moure, J.S. and J.M.F. Camargo. 1995. Melipona (Michinelia) capixaba, uma nova especie de Meliponinae do Sudoeste do Brasil. Revista Brasileira de Zoologia. 11: 289–296.

Mullin, C.A., M. Frazier, J.L. Frazier, S. Ashcraft, R. Simonds, D. vanEngelsdorp et al. 2010. High Levels of Miticides and Agrochemicals in North American Apiaries: Implications for Honey Bee Health. PLOS ONE. 5: e9754.

Muñoz, I., R. Dall'olio, M. Lodesani and P. De la Rua. 2009. Population genetic structure of coastal Croatian honey bees (*Apis mellifera carnica*). Apidologie. 40: 617–626.

Muñoz, I., M.J. Madrid-Jiménez and P. De la Rúa. 2011. Temporal genetic analysis of an introgressed island honey bee population (Tenerife, Canary Islands, Spain). Journal of Apicultural Research. 51: 144–146.

Muñoz, I., J. Stevanović, Z. Stanimirović and P. De la Rúa. 2012. Genetic variation of *Apis mellifera* from Serbia inferred from mitochondrial analysis. Journal of Apicultural Science. 56: 59–69.

Muñoz, I., M.A. Pinto and P. De la Rua. 2013. Temporal changes in mitochondrial diversity highlights contrasting population events in Macaronesian honey bees. Apidologie. 44: 295–305.

Muñoz, I., R. Dall'olio, M. Lodesani and P. De la Rúa. 2014. Estimating introgression in *Apis mellifera siciliana* populations: are the conservation islands really effective? Insect Conserv. Divers. 7: 563–571.

Muñoz, I., D. Henriques, J.S. Johnston, J. Chávez-Galarza, P. Kryger and M.A. Pinto. 2015. Reduced SNP panels for genetic identification and introgression analysis in the dark honey bee (*Apis mellifera mellifera*). PLOS ONE. 10: e0124365.

Murray, T.E., M. Kuhlmann and S.G. Potts. 2009. Conservation ecology of bees: populations, species and communities. Apidologie. 40: 211–236.

Muszynska, J. 1979. Troche o trutniach. Pszczelarstwo. 30: 5–6.

Mutinelli, F., C. Costa, M. Lodesani, A. Baggio, P. Medrzycki, G. Formato et al. 2010. Honey bee colony losses in Italy. Journal of Apicultural Research. 49: 119–120.

Nagir, M.T., T. Atmowidi and S. Kahono. 2016. The distribution and nest-site preference of *Apis dorsata binghami* at Maros Forest, South Sulawesi, Indonesia. Journal of Insect Biodiversity. 4: 1–14.

Nath, S., P. Roy, R. Leo and M. John. 1994. Honey hunters and beekeepers of Tamil Nadu. pp. 86. A survey document. Keystone, Pondicherry.

Nath, S. and K. Sharma. 2007. Honey trails in the Blue Mountains: Ecology, people and livelihood in the Nilgiri Biosphere Reserve, India. Keystone Foundation, Kotagiri.

National Research Council and National Academies Press. 2007. Status of pollinators in North America. The National Academies Press.

Naug, D. 2009. Nutritional stress due to habitat loss may explain recent honey bee colony collapses. Biological Conservation. 142: 2369–2372.

Naug, D. and A. Gibbs. 2009. Behavioral changes mediated by hunger in honey bees infected with *Nosema ceranae*. Apidologie. 40: 595–599.

Nawrocka, A., İ. Kandemir, S. Fuchs and A. Tofilski. 2018. Computer software for identification of honey bee subspecies and evolutionary lineages. Apidologie. 49: 172–184. doi: 10.1007/s13592-017-0538-y.

Nazarova, N.P. 2009. On the effect of pesticides on the activity of honey bees. 2009. Vestnik Moskovskogo Gosudarstvennogo Oblastnogo Universiteta. 4: 134–137.

Nazzi, F. and N. Milani. 1996. The presence of inhibitors of the reproduction of *Varroa jacobsoni* Oud (Gamasida: Varroidae) in infested cells. Experimental and Applied Acarology. 20: 617–623.

Nedić, N., R.M. Francis, L. Stanisavljević, I. Pihler, N. Kezić, C. Bendixen and P. Kryger. 2014. Detecting population admixture in the honey bees of Serbia. Journal of Apicultural Research. 53: 303–313.

Nei, M. 1975. Molecular population genetics and evolution. Holland Press, Amsterdam.

Neumann, P., R.F.A. Moritz and D. Mautz. 2000. Colony evaluation is not affected by drifting of drone and worker honey bees (*Apis mellifera* L.) at a performance testing apiary. Apidologie. 31: 67–79. doi: 10.1051/apido:2000107.

Neumann, P. and N. Carreck. 2010. Honey bee colony losses: a global perspective. Journal of Apicultural Research. 49: 1–6.

Nguyen, B.K., M. Ribiere, D. vanEngelsdorp, C. Snoeck, C. Saegerman, A.L. Kalkstein et al. 2011. Effects of honey bee virus prevalence, *Varroa destructor* load and queen condition on honey bee colony survival over the winter in Belgium. Journal of Apicultural Research. 50: 195–202.

Nieto, A., S.P.M. Roberts, J. Kemp, P. Rasmont, M. Kuhlmann, M. García Criado et al. 2014. European Red List of Bees. Publication Office of the European Union, Belgium, Luxembourg.

Nikolenko, A.G., R.G. Farkhutdinov, R.A. Ilyasov and A.V. Poskryakov. 2010. The area of the Burzyan population of the dark forest bee *Apis mellifera* L. (Hymenoptera: Apidae). Proceedings of the Russian Entomological Society. 81: 202–208.

Nikonorov, Yu.M., G.V. Ben'kovskaya, A.V. Poskryakov, A.G. Nikolenko and V.A. Vakhitov. 1998. The use of the PCR technique for control of pure-breeding of honey bee (*Apis mellifera mellifera* L.) colonies from the Southern Ural. Russian Journal of Genetics. 34: 1344–1347.

Nixon, H.L. and C.R. Ribbands. 1952. Food transmission within the honeybee community. Proceedings of the Royal Society B: Biological Sciences. 140: 43.

Norden, B., S.W.T. Batra, H.M. Fales, A. Hefetz, G.J. Shaw. 1980. *Anthophora* bees: unusual glycerides from maternal Dufour's glands serve as larval food and cell lining. Science. 207: 1095–1097.

Norfolf, O., F. Gilbert and M.P. Eichhorn. 2018. Alien honey bees increase pollination risks for range-restricted plants. Diversity and Distributions. 24: 705–713.

Oertel, E. 1976. Bicentennial bees. The American Bee Journal. 116: 70–290.

OIE—World Organization for Animal Health. 2004. Manual of diagnostic tests and vaccines for terrestrial animals. 5th Edition. OIE, Paris, France. [online]. http://www.oie.int/standard-setting/terrestrial-manual.

OIE—World Organization for Animal Health. 2016. Manual of diagnostic tests and vaccines for terrestrial animals. 7th Edition. OIE, Paris, France. [online]. http://www.oie.int/standard-setting/terrestrial-manual.

Okada, I. 1986. Biological characteristics of the Japanese Honey bee, *Apis cerana japonica*. pp. 119–122. *In*: Proceedings of the 30th International Apiculture Congress, Apimondia, Nagoya.

Okuyama, H., S. Tingek and J. Takahashi. 2017. The complete mitochondrial genome of the cavity-nesting honey bee, *Apis cerana* (Insecta: Hymenoptera: Apidae) from Borneo. Mitochondrial DNA Part B Resources. 2: 475–476.

Oldroyd, B.P. and C. Moran. 1983. Heritability of worker characters in the honey bee (*Apis mellifera*). Australian Journal of Biological Sciences. 36: 323–332.

Oldroyd, B.P., T.E. Rinderer and S.M. Buco. 1991. Heritability of morphological characters used to distinguish European and Africanized honey bees. Theoretical and Applied Genetics. 82: 499–504.

Oldroyd, B.P. 1999. Coevolution while you wait. *Varroa jacobsoni*, a new parasite of western honey bees. Trends in Ecology and Evolution. 14: 312–315.

Oldroyd, B.P. and S. Wongsiri. 2006. Asian Honey bees: Biology, Conservation and Human Interactions. Harvard University Press, Cambridge.

Oldroyd, B.P. 2007. What's killing American honey Bees? PLOS Biology. 5: 1195–1199.

Oleksa, A., I. Chybicki, A. Tofilski and J. Burczyk. 2011. Nuclear and mitochondrial patterns of introgression into native dark bees (*Apis mellifera mellifera*) in Poland. Journal of Apicultural Research. 50: 116–129.

Oleksa, A. and A. Tofilski. 2015. Wing geometric morphometrics and microsatellite analysis provide similar discrimination of honey bee subspecies. Apidologie. 46: 49–60. doi: 10.1007/s13592-014-0300-7.

Onions, G.W. 1912. South African fertile-worker bees. Agricultural Journal of South Africa. 3: 720–728.

Osborne, J.L., I.H. Williams and S.A. Corbet. 1991. Bees, pollination and habitat change in the European Community. Bee World. 72: 99–116.

Ostroverkhova, N.V., O.L. Konusova, A.N. Kucher, Yu.L. Pogorelov, E.A. Belikh and A.A. Vorotov. 2013. Population genetic structure of honey bee (*Apis mellifera* L.) in the village of Leboter in Chainskiy district of Tomsk region. Vestnik Tomskogo gosudarstvennogo universiteta. Biologiya (TSU J. Biol.). 1: 161–172.

Ostroverkhova, N.V., O.L. Konusova, A.N. Kucher, T.N. Kireeva, A.A. Vorotov and E.A. Belikh. 2015. Genetic diversity of the locus *COI-COII* of mitochondrial DNA in honey bee populations (*Apis mellifera* L.) from the Tomsk region. Russian Journal of Genetics. 51: 80–90. doi: 10.1134/S102279541501010X.

Ostroverkhova, N.V., O.L. Konusova, A.N. Kucher and I.V. Sharakhov. 2016. A comprehensive characterization of the honey bees in Siberia (Russia). pp. 1–37. *In*: Chambó, E.D. [ed.]. Beekeeping and Bee Conservation-Advances in Research. InTech, Grotia. doi: 10.5772/62395.

Ostroverkhova, N.V., A.N. Kucher, O.L. Konusova, T.N. Kireeva and I.V. Sharakhov. 2017. Genetic diversity of honey bees in different geographical regions of Siberia. International Journal of Environmental Studies. 74: 771–781.

Ostroverkhova, N.V., A.N. Kucher, O.L. Konusova, E.S. Gushchina, V.V. Yartsev and Y.L. Pogorelov. 2018a. Dark-colored forest bee *Apis mellifera* in Siberia, Russia: current state and conservation of populations. pp. 157–180. *In*: Şen, B. and O. Grillo [eds.]. Selected Studies in Biodiversity. InTech Open, London, UK.

Ostroverkhova, N.V., A.N. Kucher, O.L. Konusova, I.V. Sharakhov. 2018b. The *MRJP3* microsatellite marker: determination of honey bee subspecies or/and royal jelly productivity of bee colony. Far Eastern Entomologist. 353: 24–28.

Ostroverkhova, N.V., A.N. Kucher, N.P. Babushkina, O.L. Konusova and I.V. Sharakhov. 2018c. Variability and structure of the repetitive region of the major royal jelly protein gene *MRJP3* in honey bee *Apis mellifera* of different evolutionary branches. Journal of Molecular Biology Research. 8: 122–131.

Ostroverkhova, N.V., A.N. Kucher, E.P. Golubeva, S.A. Rosseykina and O.L. Konusova. 2019a. Study of *Nosema* spp. in the Tomsk region, Siberia: co-infection is widespread in honey bee colonies. Far Eastern Entomologist. 378: 12–22.

Ostroverkhova, N.V., A.N. Kucher, O.L. Konusova, T.N. Kireeva, S.A. Rosseykina, V.V. Yartsev et al. 2019b. Genetic diversity of honey bee Apis mellifera in Siberia. *In*: Ilyasov, R.A. and H.W. Kwon. [eds.]. Phylogenetics of Bees. CRC Press, Taylor and Francis Group, Boca Raton, FL., USA (in press).

Otis, G.W. 1991. A review of the diversity of species within *Apis*. pp. 29–50. *In*: Smith, D.R. [ed.]. Diversity in the Genus *Apis*. Westview Press, Oxford, UK.

Otis, G.W. 1996. Distributions of recently recognized species of honey bees (Hymenoptera, Apidae-*Apis*) in Asia. Journal of the Kansas Entomological Society. 69: 311–333.

Otis, G.W., O.R. Taylor and M.L. Winston. 2002. Colony size affects reproductive attributes of African honey bees (*Apis mellifera* L.). Proceedings of the 2nd International Conference on Africanized Honey Bees and Bee Mites. Root, Medina, Ohio: 25–32.

Owen, R.E., F.H. Rodd and R.C. Plowright. 1980. Sex ratios in bumble bee colonics: Complications due to orphaning? Behavioral Ecology and Sociobiology. 7: 287–291.

Özdil, F.A., A.Y. Mehmet, A. Yildiz and H.G. Hall. 2009. Molecular characterization of Turkish honey bee populations (*Apis mellifera*) inferred from mitochondrial DNA RFLP and sequence results. Apidologie. 40: 570–576.

Özkan Koca, A., M.M. Gharleko, B. Özden and İ. Kandemir. 2009. Multivariate morphometric study on Apis florea distributed in Iran. Turkish Journal of Zoology. 33: 93–102.

Özkan, A. and İ. Kandemir. 2010. Discrimination of western honey bee populations in Turkey using geometric morphometric methods. Bee World. 87: 24–26. doi: 10.1080/0005772x.2010.11417341.

Özkan Koca, A. 2012. Analysis of *Apis mellifera* (Hymenoptera: Apidae) Subspecies Distributed in the Middle East by Using Geometric Morphometric Methods. Ph.D. Thesis, Ankara University, Ankara, Turkey.

Özkan Koca, A. and İ. Kandemir. 2013. Comparison of two morphometric methods for discriminating honey bee (*Apis mellifera* L.) populations in Turkey. Turkish Journal of Zoology. 37: 205–210. doi: 10.3906/zoo-1104-10.

Pager, H. 1973. Rock Paintings in Southern Africa Showing Bees and Honey Hunting. Bee World. 54: 61–68.

Paini, D.R. 2004. Impact of the introduced honey bee, *Apis mellifera* (Hymenoptera: Apidae) on native bees: A review. Austral Ecology. 29: 399–407.

Palacio, M.A., E. Figini, S. Ruffinengo, E. Rodriguez, M. Del Hoyo and E.L. Bedascarrasbure. 2000. Changes in a population of Apis mellifera L. selected for hygienic behaviour and its relation to brood disease tolerance. Apidologie. 31: 471–478.

Palmer, M.R., D.R. Smith and O. Kaftanoglu. 2000. Turkish honey bees: genetic variation and evidence for a fourth lineage of *Apis mellifera* mtDNA. Journal of Heredity. 91: 42–46.

Papachristoforou, A., A. Rortais, M. Bouga, G. Arnold and L. Garnery. 2013. Genetic characterization of the cyprian honey bee (*Apis mellifera cypria*) based on microsatellites and mitochondrial DNA polymorphisms. Journal of Apicultural Science. 57: 127Garnery134.

Papachristoforou, A., J. Sueur, A. Rortais, S. Angelopoulos, A. Thrasyvoulou and G. Arnold. 2008. High frequency sounds produced by Cyprian honey bees *Apis mellifera cypria* when confronting their predator, the Oriental hornet *Vespa orientalis*. Apidologie. 39: 468–474.

Parayeva, L.K. 1970. Honey plants of Western Siberia. Russia, Novosibirsk.

Pareja, L., M. Colazzo, A. Pérez-Parada, S. Niell, L. Carrasco-Letelier, N. Besil et al. 2011. Detection of Pesticides in Active and Depopulated Beehives in Uruguay. International Journal of Environmental Research and Public Health. 8: 3844–3858. doi: 10.3390/ijerph8103844.

Park, W. 1923. Flight studies of the honey bee. The American Bee Journal. 63: 71.

Park, D., J.W. Jung, B.S. Choi, M. Jayakodi, J. Lee, J. Lim et al. 2015. Uncovering the novel characteristics of Asian honey bee, *Apis cerana*, by whole genome sequencing. BMC Genomics. 16: 1–16. doi: 10.1186/1471-2164-16-1.

Partap, U. and L.R. Verma. 1992. Floral biology and foraging behaviour of *Apis cerana* on lettuce crop and its impact on seed production. Progressive Horticulture. 24: 42–47.

Partap, U. and L.R. Verma. 1994. Pollination of radish by *Apis cerana*. Journal of Apicultural Research. 33: 237–241.

Partap, U. 1999. Conservation of endangered Himalayan honey bee, *Apis cerana* for crop pollination. Asian bee Journal. 1: 44–49.

Partap, U. 2011. The pollination role of honey bees. pp. 227–255. *In*: Hepburn, H.R. and S.E. Radloff [eds.]. Honey Bees of Asia. Springer, Berlin.

Pauw, A. and J.A. Hawkins. 2011. Reconstruction of historical pollination rates reveals linked declines of pollinators and plants. Oikos. 120: 344–349.

Pavlinov, L.Ya. 2001. Geometric morphometry-a new analytical approach to the comparison of computer images (in Russian). pp. 65–90. In the book: Information and telecommunication resources in zoology and botany. S.-Petersburg.

Pavlinov, L.Ya. and N.G. Mikeshina. 2002. Principles and Methods of Geometric Morphometrics (in Russian). Journal Obshchei Biologii. 63: 473–493.

Paxton, R., J. Klee, S. Korpela and I. Fries. 2007. *Nosema ceranae* has infected *Apis mellifera* in Europe since at least 1998 and may be more virulent than *Nosema Apis*. Apidologie. 38: 558–565.

Peakall, R. and P.E. Smouse. 2012. GenAlEx 6.5: genetic analysis in Excel. Population genetic software for teaching and research—an update. Bioinformatics. 28: 2537–2539.

Pechhacker, H. and W. Leichtfried. 1991. Leistungsprüfung bei der Honigbiene. Bienenvater. 112: 182–184.

Peng, Y.S., M.E. Nasr and S.J. Locke. 1989. Geographical races of *Apis cerana* Fabricius in China and their distribution. Review of recent Chinese publications and a preliminary statistical analysis. Apidologie. 20: 9–20.

Péntek-Zakar, A., A. Oleksa, T. Borowik and S. Kusza. 2015. Population structure of honey bees in the Carpathian Basin (Hungary) confirms introgression from surrounding subspecies. Ecology and Evolution. 5: 5456–5467.

Perepelova, L.I. 1928. Drones, queen's egg laying and swarming. Opytnaya paseka. 5-6: 214–217.

Pernal, S.F. and R.W. Currie. 2000. Pollen quality of fresh and 1-year-old single pollen diets for worker honey bees (*Apis mellifera* L.). Apidologie. 31: 387–409.

Pesenko, Yu.A. 1982. Alfalfa leaf-cutter Megachile rotunda la and its breeding for pollination of alfalfa. Moscow: Science.

Petrov, E.M. 1983. Bashkir honey bee. (in Russian). Russia, Ufa.

Pettis, J.S., A.M. Collins, R. Wilbanks and M.F. Feldlaufer. 2004. Effects of coumaphos on queen rearing in the honey bee, *Apis mellifera*. Apidologie. 35: 605–610. doi: 10.1051/apido:2004056.

Pettis, J.S., D. van Engelsdorp, J. Johnson and G. Dively. 2012. Pesticide exposure in honey bees results in increased levels of the gut pathogen *Nosema*. Naturwissenschaften. 99: 153–158. doi: 10.1007/s00114-011-0881-1.

Pinto, M.A., W.L. Rubink, R.N. Coulson, J.C. Patton and J.S. Johnston. 2004. Temporal pattern of Africanization in a feral honey bee population from Texas inferred from mitochondrial DNA. Evolution. 58: 1047–1055.

Pinto, M.A., W.L. Rubink, J.C. Patton, R.N. Coulson and J.S. Johnston. 2005. Africanization in the United States: Replacement of feral European honey bees (*Apis mellifera* L.) by an African hybrid swarm. Genetics. 170: 1653–1665.

Pinto, M.A., D. Henriques, J. Cha´vez-Galarza, P. Kryger, L. Garnery, R. van der Zee et al. 2014. Genetic integrity of the dark European honey bee (*Apis mellifera mellifera*) from protected populations: A genome-wide assessment using SNPs and mtDNA sequence data. Journal of Apicultural Research. 53: 269–278. doi: 10.3896/IBRA.1.53.2.08.

Pirk, C.W.W, C. Boodhoo, H. Human and S.W. Nicolson. 2009. The importance of protein type and protein to carbohydrate ratio for survival and ovarian activation of caged honey bees (*Apis mellifera scutellata*). Apidologie. 41: 62–67. doi: 10.1051/apido/2009055.

Pirk, C.W.W., C.L. Sole and R.M. Crewe. 2011. Pheromones. pp. 207–214. *In*: Hepburn, H. and S. Radloff [eds.]. Honey Bees of Asia. Springer, Berlin, Heidelberg.

Pirk, C.W.W., R.M. Crewe and R.F.A. Moritz. 2017. Risks and benefits of the biological interface between managed and wild bee pollinators. Functional Ecology. 31: 47–55.

Plokhinskiy, N.A. 1969. Guide to biometrics for livestock specialists (in Russian). Kolos Press, Moscow.

Plokhinskiy, N.A. 1970. Biometrics. MSU Press, Moscow.

Potts, S.G., B. Vulliamy, S. Roberts, C. O'Toole, A. Dafni and G. Ne'eman. 2005. Role of nesting resources in organizing diverse bee communities in a Mediterranean landscape. Ecological Entomology. 30: 78–85.

Potts, S.G., S.P. Roberts, R. Dean, G. Marris, M. Brown, R. Jones et al. 2009. Declines of managed honey bees and beekeepers in Europe. Journal of Apicultural Research. 49: 15–22.

Potts, S.G., S.P.M. Roberts, R. Dean, G. Marris, M.A. Brown, R. Jones et al. 2010a. Declines of managed honey bees and beekeepers in Europe. Journal of Apicultural Research. 49: 15–22.

Potts, S.G., J.C. Biesmeijer, C. Kremen, P. Neumann, O. Schweiger and W.E. Kunin. 2010b. Global pollinator declines: trends, impacts and drivers. Trends in Ecology and Evolution. 25: 345–353.

Potts, S.G., V. Imperatriz-Fonseca, H.T. Ngo, M.A. Aizen, J.C. Biesmeijer, T.D. Breeze et al. 2016. Safeguarding pollinators and their values to human well-being. Nature. 540: 220–229.

Povrean, A. 1971. Sur le determinisme des castes chez les bourdons (Hymenoptera, Apoidea< Bombus Latr.). Annales de Zoologie et Ecologie Animale. 3: 501–507.

Prado, A., M. Pioz, C. Vidau, F. Requier, J.L. Brunet, M. Jury et al. 2019. Exposure to pollen-bound pesticide mixtures induces longer-lived but less efficient honey bees. Sci. Total Environ. 650: 1250–1260.

Pretorius, E. and C.H. Scholtz. 2001. Geometric morphometrics and the analysis of higher taxa: a case study based on the metendosternite of the Scarabaeoidea (Coleoptera). Biological Journal of the Linnean Society. 74: 35–50. doi: 10.1111/j.1095-8312.2001.tb01375x.

Pretorius, E. 2005. Using geometrics to investigate wing dimorphism in males and females Hymenoptera-a case study based on the genus *Tachysphex* Kohl (Hymenoptera: Sphecidae: Larrinae). Australian Journal of Entomology. 44: 113–121. doi: 10.1111/j.1440-6055.2005.00464.x.

Pritchard, J.K., M. Stephens and P. Donnelly. 2000. Inference of population structure using multilocus genotype data. Genetics. 155: 945–959.

Pritchard, D. 2014. Which is best bee for the North and how would I identify it? Abstracts Conference SICAMM, BIBBA, Llangollen.

Pyaskovskiy, V.M., T.V. Verbelchuk, S.P. Verbelchuk and M.M. Krivoy. 2018. Modern challenges with chemical toxicosis of bees. Pasechnik 172: 14–17.

Quezada-Euán, J.G., E.E. Pérez-Castro and W.J. May-Itzá. 2003. Hybridization between European and African-derived honey bee populations (*Apis mellifera*) at different altitudes in Peru. Apidologie. 34: 217–225. doi: 10.1051/apido:2003010.

R Core Team. 2018. R: A Language and Environment for Statistical Computing. Version 3.3.3. R Foundation for Statistical Computing, Vienna, Austria.

Radchenko, V.G. 1978. New type of nest found in the bee *Metallinella atrocaerulea* (Hymenoptera, Megachilidae) (in Russian). Entomological Review. 53: 515–519.

Radchenko, V.G. 1979. Nesting bee *Nomioides minutissimus* (Rossi) (Hymenoptera, Halictidae) (in Russian). Entomological Review. 58: 762–765.

Radchenko, V.G. and Yu.A. Pesenko. 1994. Biology of bees (Hymenoptera, Apoidea) (in Russian). St.-Petersburg.

Rader, R., B.G. Howlett, S.A. Cunningham, D.A. Westcott, L.E. Newstrom-Lloyd, M.K. Walker et al. 2009. Alternative pollinator taxa are equally efficient but not as effective as the honey bee in a mass flowering crop. Journal of Applied Ecology. 46: 1080–1087.

Rader, R., I. Bartomeus, L.A. Garibaldi, M.P.D. Garratt, B.G. Howlett, R. Winfree et al. 2016. Non-bee insects are important contributors to global crop pollination. Proc. Natl. Acad. Sci. USA. 113: 146–151.

Radloff, S., H.R. Hepburn, C. Hepburn and P. De la Rua. 2001. Morphometric affinities and population structure of honey bees of the Balearic Islands (Spain). Journal of Apicultural Research. 40: 97–104.

Radloff, S.E. and H.R. Hepburn. 1998. The matter of sampling distance and confidence levels in the subspecies classification of honey bees, *Apis mellifera* L. Apidologie. 29: 491–501. doi: 10.1051/apido:19980602.

Radloff, S.E., H.R. Hepburn and J.B. Lindsey. 2003. Quantitative analysis of intracolonial and intercolonial morphometric variance in honey bees, *Apis mellifera* and *Apis cerana*. Apidologie. 34: 339–351. doi: 10.1051/apido:2003034.

Radloff, S.E., H.R. Hepburn and S. Fuchs. 2005a. The morphometric affinities of *Apis cerana* of the Hindu Kush and Himalayan regions of western Asia. Apidologie. 36: 25–30.

Radloff, S.E., H.R. Hepburn, C. Hepburn, S. Fuchs, G.W. Otis, M.M. Sein et al. 2005b. Multivariate morphometric analysis of *Apis cerana* of southern mainland Asia. Apidologie. 36: 127–139.

Radloff, S.E., C. Hepburn, H.R. Hepburn, S. Hadisoesilo, S. Fuchs, K. Tan et al. 2010. Population structure and classification of *Apis cerana*. Apidologie. 41: 589–601. doi: 10.1051/apido/2010008.

Radloff, S.E., H.R. Hepburn and M.S. Engel. 2011. The Asian species of *Apis*. *In*: Hepburn, H.R. and S.E. Radloff [eds.]. Honey Bees of Asia. Springer-Verlag, Berlin, Heidelberg. doi: 10.1007/978-3-642-22-4.

Raffiudin, R. and R.H. Crozier. 2007. Phylogenetic analysis of honey bee behavioral evolution. Molecular Phylogenetics and Evolution. 43: 543–552. doi: 10.1016/j.ympev.2006.10.013.

Raffiudin, R., S. Salmah and J. Jasmi. 2019. Distribution, nesting trees and genetic diversity of *Apis andreniformis* in West Sumatra. *In*: Abrol, D.P. [ed.]. The Future Role of Dwarf Honey Bee in Natural and Agricultural Ecosystems: Taylor and Francis, London, UK (in press).

Rafie, J.N., R. Mohamadi and H. Teimory. 2014. Comparison of two morphometrics methods for discriminating of honey bee (*Apis mellifera meda* Sk.) populations in Iran. International Journal of Zoology and Research. 4: 61–70.

Ragim-Zade, M.C. 1975. ILG in the food-producing activity of honey bees (in Russian). Pchelovodstvo. 2: 14–17.

Raine, N.E. and R.J. Gill. 2015. Ecology: Tasteless pesticides affect bees in the field. Nature. 521: 38.

Rangel, J., M. Giresi, M.A. Pinto, K.A. Baum, W.L. Rubink, R.N. Coulson et al. 2016. Africanization of a feral honey bee (*Apis mellifera*) population in South Texas: does a decade make a difference? Ecology and Evolution. 6: 2158–2169.

Rašić, S., M. Mladenović and L. Stanisavljević. 2015. Use of geometric morphometrics to differentiate selected lines of Carniolan honey bees (*Apis mellifera carnica*) in Serbia and Montenegro. Arch. Biol. Sci. Belgrade. 67: 929–934. doi: 10.2298/abs140224054r.

Rasnitsin, A.P. 1980a. Order Vespidae. Hymenoptera. pp. 122–126. *In*: Rasnitsin, A.P. [ed.]. History of the Development of the Class of Insects (in Russian). Nauka, Moscow.

Rasnitsin, A.P. 1980b. The origin and evolution of the hymenoptera (in Russian). Nauka, Moscow.

Ratnasingham, S. and P.D.N. Hebert. 2007. Bold: The Barcode of Life Data System (http://www.barcodinglife.org). Molecular Ecology Notes. 7: 355–364.

Ratnieks, F.L.W. 1991. Africanized bees: Natural selection for colonizing ability. pp. 119–136. *In*: Ratnieks, F.L.W. [ed.]. The "African" Honey Bee. Westview Press, San Francisco, California.

Rattanawannee, A., A. Chanchao and S. Wongsiri 2007. Morphometric and genetic variation of small dwarf honey bees *Apis andreniformis* Smith, 1858 in Thailand. Insect Science. 14: 451–460.

Rattanawannee, A., C. Chanchao and S. Wongsiri. 2010. Gender and species identification of four native honey bees (Apidae: *Apis*) in Thailand based on wing morphometric analysis. Annu. Ent. Soc. Amer. 103: 965–970.

Ravoet, J., L. De Smet, I. Meeus, G. Smagghe, T. Wenseleers and D.C. de Graaf. 2014. Widespread occurrence of honey bee pathogens in solitary bees. Journal of Invertebrate Pathology. 122: 55–58.

Rennich, K., J. Pettis, D. VanEngelsdorp, R. Bozarth, H. Eversole, K. Roccasecca et al. 2012. 2011–2012 National honey bee pests and diseases survey report. [Online]. http://www.aphis.usda.gov/plant_ health/plant_pest_info/honey_bees/downloads/2011_National_Survey_Report.pdf.

Requier, F. 2013. Dynamique spatio-temporelle des ressources florales et écologie de l'abeille domestique en paysage agricole intensif. Universite de Poitiers.

Requier, F., J.F. Odoux, T. Tamic, N. Moreau, M. Henry, A. Decourtye et al. 2015a. Honey-bee diet in intensive farmland habitats reveals an unexpectedly high flower richness and a critical role of weeds. Ecol. Appl. 25: 881–890.

Requier, F., J.F. Odoux, T. Tamic, N. Moreau, M. Henry, A. Decourtye et al. 2015b. Floral resources used by honey bees in agricultural landscapes. Bull. Ecol. Soc. Am. 96: 487–491.

Requier, F., G. Garcia, G.K.S. Andersson, F. Oddi and L.A. Garibaldi. 2016. Honey bee colony losses: what's happening in South America? American Bee Journal. 156: 1247–1250.

Requier, F., J.F. Odoux, M. Henry and V. Bretagnolle. 2017a. The carry-over effects of pollen shortage decrease the survival of honey bee colonies in farmlands. Journal of Applied Ecology. 54: 1161–1170.

Requier, F., G. Garcia, G.K.S. Andersson, F. Oddi and L.A. Garibaldi. 2017b. La pérdida global de colonias de la abeja melífera: un mundo de encuestas donde las fronteras persisten. Apicultura sin Fronteras. 92: 13–18.

Requier, F., G.K.S. Andersson, F. Oddi and L.A. Garibaldi. In press. Citizen science in developing countries: how to improve participation. Front Ecol Environ.

Requier, F., G.K.S. Andersson, F. Oddi, G. Garcia and L.A. Garibaldi. 2018a. Perspectives from the survey of honey bee colony losses 2015–16 in Argentina. Bee World. 5: 9–12.

Requier, F., K. Antúnez, C.L. Morales, P. Aldea Sánchez, D. Castilhos, M. Garrido et al. 2018b. Trends in beekeeping and honey bee colony losses in Latin America. Journal of Apicultural Research. 57: 657–662.

Requier, F., L. Garnery, P.L. Kohl, H.K. Njovu, C.W.W. Pirk, R.M. Crewe et al. 2019. The conservation of native honey bees is crucial. Trends in Ecology & Evolution. 34: 789–798.

Ribeiro, M.F., E.M.S. Silva and C.B.S. Lima. 2012. Comparação da utilização de colmeias de abelhas melíferas (*Apis mellifera*) para a polinização em cultivos de melão (*Cucumis melo*) nas regiões de Mossoró (RN) e Salitre (BA). Mensagem Doce. 116: 66. [Online]. http://www.apacame.org.br/mensagemdoce/116/polinizacao11.htm.

Richards, K.W. 1973. Biology of *Bombus polaris* Curtis and *B. hyperboreus* Schonherr at Lake Hazen, Northwest Territorties (Hymenoplera: Bombini). Quaestiones entomologicae. 9: 115–157.

Richards, O.W. 1953. The social insects. MacDonald, London.

Rickli, M., P.A. Diehl and P.M. Guerin. 1994. Cuticle alkanes of honey bee larvae mediate arrestment of bee parasite *Varroa jacobsoni*. Journal of Chemical Ecology. 20: 2437–2453.

Rinderer, T.E., W.C. Rothenbuhler and T.A. Gochnauer. 1974. The influence of pollen on the susceptibility of honey bee-larvae to *Bacillus larvae*. Journal of Invertebrate Pathology. 23: 347–350.

Rinderer, T.E., H.A. Sylvester, M.A. Brown, J.D. Villa, D. Pesante and A.M. Collins. 1986. Field and simple techniques for identifying Africanized and European honey bees. Apidologie. 17: 33–48. doi: 10.1051/apido:19860104.

Rinderer, T.E., H.A. Sylvester, S.M. Buco, V.A. Lancaster, E.W. Herbert, A.M. Collins et al. 1987. Improved simple techniques for Identifying Africanized and European honey bees. Apidologie. 18: 179–196. doi: 10.1051/apido:19870208.

Rinderer, T.E., N. Koeniger, S. Tingek, M. Mardan and G. Koeniger. 1989. A morphological comparison of the cavity dwelling honey bees of Borneo *Apis koschevnikovi* (Buttel-Reepen, 1906) and *Apis cerana* (Fabricius, 1793). Apidologie. 20: 404–411.

Rinderer, T.E., H.V. Daly, H.A. Sylvester, A.M. Collins, S.M. Buco, R.L. Hellmich et al. 1990. Morphometric differences among Africanized and European honey bees and their F1 hybrids (Hymenoptera: Apidae). Annals of the Entomological Society of America. 83: 346–351.

Rinderer, T.E., J.W. Harris, G.J. Hunt and L.I. de Guzman. 2010. Breeding for resistance to *Varroa destructor* in North America. Apidologie. 41: 409–424.

Ritter, W. and D. De Jong. 1984. Reproduction of Varroa jacobsoni O. In Europe, the Middle East and tropical South America. Zeitschrift für Angewandte Entomologie. 98: 55–57.

Rittschof, C.C., C.B. Coombs, M. Frazier, C.M. Grozinger and G.E. Robinson. 2015. Early-life experience affects honey bee aggression and resilience to immune challenge. Scientific reports. 5: 15572. doi: 10.1038/srep15572.

Robin, E.O. 2012. Application of Morphometrics to the Hymenoptera, Particularly Bumble Bees (*Bombus*, Apidae). IntechOpen Limited, London, UK. doi: 10.5772/34745.

Roepke, W. 1930. Beobachtugen an condescend Honigbienen, insbesondere an *Apis dorsata* F. Medendecdeclingen van de Landbouwhoogeschool de Wagningen (Nederland). 34: 1–28.

Roetschi, A., H. Berthoud, R. Kuhn and A. Imdorf. 2008. Infection rate based on quantitative real-time PCR of *Melissococcus plutonius*, the causal agent of European foulbrood, in Honey bee colonies before and after apiary sanitation. Apidologie. 39: 362–371.

Roger, D. 1969. Sur la biologie de *Trigona (Apotrigona) nebulata komiensis* Cock. Biologia gabonica. 5: 151–187.

Rohlf, F.J. 1993. Relative warps analysis and example of its application to mosquito wings. pp. 131–160. *In*: Marcus, L.F., E. Bello and A. Garcia-Valdesas [eds.]. Contributions to Morphometrics. C.S.C.I., Madrid.

Rohlf, F.J. and D.E. Slice. 1990. Extensions of the Procrustes method for the optimal superimposition of landmarks. Systematic Zoology. 39: 40–59. doi: 10.2307/2992207.

Rohlf, F.J. and L.F. Marcus. 1993. A revolution in morphometrics. Trends in Ecology and Evolution. 8: 129–132. doi: 10.1016/0169-5347/90024-j.

Rohlf, F.J. 1996. Morphometric spaces, shape components and the effect of linear transformations. pp. 131–152. *In*: Marcus, L., M. Corti, A. Loy and D. Slice [eds.]. Advances in Morphometrics. Plenum Press, NY, L.

Rooley, A.C. and C.D. Michener. 1969. Observations on nests of stingless honey-bees in Natal (Hymenoptera: Apidae). Journal of the Entomological Society of South Africa. 32: 423–430.

Rortais, A., G. Arnold, M.P. Halm and F. Touffet-Briens. 2005. Modes of honey bees exposure to systemic insecticides: estimated amounts of contaminated pollen and nectar consumed by different categories of bees. Apidologie. 36: 71–83.

Rortais, A., C. Villemant, O. Gargominy, Q. Rome, J. Haxaire, A. Papachristoforou et al. 2010. A new enemy of Honey bees in Europe: the Asian hornet *Vespa velutina*. pp. 11. *In*: Settele, J. (ed.). Atlas of Biodiversity Risks-from Europe to the Globe, from Stories to Maps. Pensoft, Sofia.

Rortais, A., G. Arnold, J.L. Dorne, S.J. More, G. Sperandio, F. Streissl et al. 2017. Risk assessment of pesticides and other stressors in bees: principles, data gaps and perspectives from the European Food Safety Authority. Science of the Total Environment. 587: 524–537.

Roseler, P.F. 1973. Die Anzahl der Spermien im Receptaculum seminis von Hummelkoniginnen (Hym., Apoidea, Bombinae). Apidologie. 4: 267–274.

Rosen, J.G. and K.J. Meginley. 1976. Biology of the bee genus Conanthalictus (Halictidae, Dufoureinae). American Museum Novitates. 2602: 6.

Rosenkranz, P. 1990. Wirtsfaktoren in der Steuerung der Reproduktion der parasitischen Bienen be *Varroa jacobsoni* in Volker von *Apis mellifera*. Ph.D. dissertation, University Tfibingen, Tfibingen.

Rosenkranz, P., P. Aumeier and B. Ziegelmann. 2010. Biology and control of *Varroa destructor*. Journal of Invertebrate Pathology. 103: 96–119.

Rosenzweig, C., J. Elliott, D. Deryng, A.C. Ruane, C. Müller, A. Arneth et al. 2014. Assessing agricultural risks of climate change in the 21st century in a global gridded crop model intercomparison. Proc. Natl. Acad. Sci. 111: 3268–3273.

Rothenbuhler, W.C. 1964. Resistance to American foulbrood in honey bees: I. Differential survival of larvae of different genetic lines. American Zoologist. 4: 111–123.

Roubik, D.W. 1978. Competitive interactions between Neotropical pollinators and africanized honey bees. Science. 201: 1030–1032.

Roubik, D.W. 1980. Foraging behavior of competing Africanized honey bees and stingless bees. Ecology. 61: 836–845.

Roubik, D.W., S.F. Sakagami and I. Kudo. 1985. A note of distribution and nesting of the Himalayan honey bee *Apis laboriosa* Smith (Hymenoptera, Apidae). Journal of the Kansas Entomological Society. 58: 746–749.

Roubik, D.W., J.E. Moreno, C. Vergara and D. Wittmann. 1986. Sporadic food competition with the African honey bee: projected impact on Neotropical social bees. Journal of Tropical Ecology. 2: 97–111.

Roubik, D.W. and H. Wolda. 2001. Do competing honey bees matter? Dynamics and abundance of native bees before and after honey bee invasion. Population Ecology. 43: 53–62.

Roubik, D.W. and R. Villanueva-Guttiérez. 2009. Invasive Africanized honey bee impact on native solitary bees: a pollen resource and trap nest analysis. Biological Journal of the Linnean Society. 98: 152–160.

Roulston, T.H. and K. Goodell. 2011. The role of resources and risks in regulating wild bee populations. Annual Review of Entomology. 56: 293–312.

Rubink, W.L., P. Luévano-Martinez, E.A. Sugden, W.T. Wilson and A.M. Collins. 1996. Subtropical *Apis mellifera* (Hymenoptera: Apidae) swarming dynamics and Africanization rates in north-eastern Mexico and southern Texas. Annals of the Entomological Society of America. 89: 243–251.

Rudenko, E.P. and E.V. Rudenko. 2015. Emerging infectious bee diseases (in Russian). Beekeeping. 5: 32–34.

Rueppell, O., A.M. Hayes, N. Warrit and D.R. Smith. 2011. Population structure of *Apis cerana* in Thailand reflects biogeography and current gene flow rather than *Varroa* mite association. Insectes Sociaux. 58: 445–452.

Ruiz-Matute, A.I., M. Brokl, A.C. Soria, M.L. Sanz and I. Martinez-Castro. 2010. Gas chromatographic-mass spectrometric characterisation of tri- and tetrasaccharides in honey. Food Chemistry. 120: 637–642.

Ruottinen, L., P. Berg, J. Kantanen, T.N. Kristensen and A. Præbel. 2014. Status and Conservation of the Nordic Brown Bee: Final report. NordGen. Nordic Genetic resource center, Norway.

Rust, R.W. 1976. Notes on the biology of North American species of *Panurginus* (Hymenoptera: Andrenidae). The Pan-Pacific Entomologist. 52: 159–166.

Rust, R.W. 1980. The biology of *Ptilothrix bombiformis* (Hymenoptera: Anthophoridae). Journal of the Kansas Entomological Society. 53: 427–436.

Ruth, A.I., E.R. Ruth, H.H. Ruth, M.J. Dale and J.A. Ruth. 1993. The encyclopedia of beekeeping. Fiction and MP Brother, Moscow.

Ruttner, F. 1969. Biometrische Charakterisierung der österreichischen *Carnica*-Biene. Zeitschrift für Bienenforschung. 9: 469–491.

Ruttner, F. 1972. Ergebnisse der Forschung der letzten 25 Jahre auf dem Gebiet der Paarung und der Drohnensammelplätze. Wissenschaftliche Bulletin der Apimondia. International Apiculture Congress, Apimondia.

Ruttner, F. 1977. The problem of the cape bee (*Apis mellifera capensis* Excholtz): parthenogenesissize of populationevolution. Apidologie. 8: 281–294.

Ruttner, F., L. Tassencourt and J. Louveaux. 1978. Biometrical-statistical analysis of the geographic variability of *Apis mellifera* L. Apidologie. 9: 363–381. doi: 10.1051/apido:19780408.

Ruttner, F. and V. Maul. 1983. Experimental analysis of reproductive interspecies isolation of *Apis mellifera* L. and *Apis cerana* Fabricius. Apidologie. 14: 309–327.

Ruttner, F. 1987. The evolution of honey bees. pp. 8–20. *In*: Menzel, R. and A. Mercer [eds.] Neurobiology and Behavior of honey bees. Springer, Berlin.

Ruttner, F. 1988. Biogeography and Taxonomy of Honey bees. Springer-Verlag Heidelberg GmbH, Berlin. doi: 10.1007/978-3-642-72649-1.

Ruttner, F., D. Kauhausen and N. Koeniger. 1989. Position of the red honey bee, *Apis koschevnikovi* (Buttel-Reepen 1906), within the genus *Apis*. Apidologie. 20: 395–404.

Ruttner, F. 1992. Naturgeschichte der Honigbienen. Ehrenwirth Verlag, München.

Ruttner, F. 2006. The Technique of Breeding and Selection of Bees. AST: Astrel, Moscow.

Sadeghi, S., D. Adriaens and H.J. Dumont. 2009. Geometric morphometric analysis of wing shape variation in ten European populations of *Calopteryx splendens* (Harris, 1782) (Zygoptera: Odonata). Odonatologica. 38: 343–360.

Safiullin, R.R. and L.N. Savushkina. 2011. Biological signs of bees, Queens and drones of breed type of the Middle Russian breed "Tatarskiy". Materials of the International scientific and practical conference "Ways of development of beekeeping in Russia through the successful experience of regions, CIS countries and Abroad", Yaroslavl: 108–110.

Safiullin, R.R., N.I. Krivtsov, A.V. Borodachev and L.N. Savushkina. 2011. Selection of the breed type "Tatarskiy" in the Middle Russian breed. Zootechny. 4: 4–6.

Safiullin, R.R., R.G. Nabiullin, A.V. Borodachev and L.N. Savushkina. 2013. Breeding resources of Middle Russian bees of Tatarstan (in Russian). Beekeeping. 3: 8–9.

Sakagami, S. and T. Matsumura. 1980. *Apis laboriosa* in *Himalaya*, the little known world largest honey bee (Hymenoptera, Apidae). Insect Matsumurana 19: 47–77.

Sakagami, S.F. 1960. Ethological peculiarities of the primitive social bees, *Allodape Lepeltier* and allied genera. Insectes Sociaux. 7: 231–249.

Sakagarmi, S.F. and R. Zucchi. 1965. Winterverhalten einer neotropischen Hummel, *Bombus atratus*, innerhalb des Beobachtungskastens. Ein. Beitrag zur Biologie der Hummeln. Journal of the Faculty of Science, Hokkaido University. Series 6, Zoology: 712–762.

Sakagami, S.F. 1976. Specific difference in the bionomic characters of bumble bees. A comparative review. Journal of the Faculty of Science, Hokkaido University. Series 6, Zoology. 20: 390–447.

Sakagami, S.F. and Y. Maeta. 1977. Some presumably presocial habits of Japanese *Ceratina* bees, with notes on various social types in Hymenoptera. Insectes Sociaux. 24: 319–343.

Sakai, L., D. Keene and E. Engvall. 1986. Fibrillin, a new 350-kD glycoprotein, is a component of extracellular microfibrils. Journal of Cell Biology. 103: 2499–2509.

Salmah, S., T. Inoue and S.F. Sakagami. 1990. An analysis of apid bee richness (Apidae) in Central Sumatra. pp. 139–174. *In*: S.F. Sakagami et al. [eds.]. Natural History of Social Wasps and Bees in Equatorial Sumatra. Hokkaido University Press, Hokkaido, Japan.

Sánchez-Bayo, F. and K.A. Wyckhuys. 2019. Worldwide decline of the entomofauna: A review of its drivers. Biological Conservation. 232: 8–27.

Santas, L.A. 1983. *Varroa* disease in Greece and its control with Malathion. In *Varroa jacobsoni* Oud. affecting honey bees: present status and needs. Proceedings of a Meeting of the EC Experts" Group, Wageningen: 73–76.

Santoso, M.A.D., B. Juliandi and R. Raffiudin. 2018. Honey bees species differentiation using geometric morphometric on wing venations. IOP Conference Series: Earth and Environmental Science. 197: 012015. doi: 10.1088/1755-1315/197/1/012015.

Santrac, V., A. Granato and F. Mutinelli. 2010. Detection of *Nosema ceranae* in *Apis mellifera* from Bosnia and Herzegovina. Journal of Apicultural Research. 49: 100–101.

Satta, A., I. Floris and G. Pigliaru. 2004. DataBees: uno strumento informatico per la gestione delle risorse Api e Mieli. APOidea. 1: 25–30.

Savushkina, L.N. and A. Borodachev. 2014. Biological signs of Priokskiy bees (in Russian). Beekeeping. 10: 10–12.

Schachter-Broide, J., J.-P. Dujardin, U. Kitron and R.E. Gürtler. 2004. Spatial structure of *Triatoma infestans* (Hemiptera, Reduviidae) populations from Northwestern Argentina using wing geometric morphometry. Journal of Medical Entomology. 14: 643–649. doi: 10.1603/0022-2585-41.4.643.

Scheloske, H.W. 1974. Untetsuchungen fiber das Vorkommer, die Biologic und den nestbau Der Seidenbiene *Colletes daviesanus* Sm. Zoologische Jahrbücher. Abteilung für Systematik. 101: 153–172.

Schiff, N.M., W.S. Sheppard, G.M. Loper and H. Shimanuki. 1994. Genetic diversity of feral honey bee (Hymenoptera: Apidae) populations in the southern United States. Annals of the Entomological Society of America. 87: 842–848.

Schiff, N.M. and W.S. Sheppard. 1995. Genetic analysis of commercial honey bees (Hymenoptera: Apidae) from the southeastern United States. Journal of Economic Entomology. 88: 1216–1220.

Schmidt, J.O., S.C. Thoenes and M.D. Levin. 1987. Survival of honey-bees, *Apis-mellifera* (Hymenoptera, Apidae) fed various pollen sources. Annals of the Entomological Society of America. 80: 176–183.

Schmidt, L.S., J.O. Schmidt, H. Rao, W.Y. Wang and L.G. Xu. 1995. Feeding preference and survival young worker honey bees fed rape, sesame, sunflower pollen. J. Econ. Entomol. 88: 1591–1595.

Schneider, S.S., L.J. Leamy, L.A. Lewis and G. DeGrandi-Hoffman. 2003. The influence of hybridization between African and European honey bees, *Apis mellifera*, on asymmetries in wing size and shape. Evolution. 57: 2350–2364. doi: 10.1111/j.0014-3820.2003.tb00247.x.

Schneider, S.S., G. DeGrandi-Hoffman and D.R. Smith. 2004. The African honey bee: Factors contributing to a successful biological invasion. Annual Review of Entomology. 49: 351–376.

Schulz, D.J., Z.Y. Huang and G.E. Robinson. 1998. Effects of colony food shortage on behavioral development in honey bees. Behavioral Ecology and Sociobiology. 42: 295–303.

Schulz, D.J., M.J. Vermiglio, Z.Y. Huang and G.E. Robinson. 2002. Effects of colony food storage on social interactions in honey bee colonies. Insectes Sociaux. 49: 50–55.

Seeley, T. 1979. Queen substance dispersal by messenger workers in honeybee colonies. Behavioral Ecology and Sociobiology. 5: 391–415.

Seeley, T.D., D.R. Tarpy, S.R. Griffin, A. Carcione and D.A. Delaney. 2015. A survivor population of wild colonies of European honey bees in the north-eastern United States: investigating its genetic structure. Apidologie. 46: 654–666.

Seitz, N., K.S. Traynor, N. Steinhauer, K. Rennich, M.E. Wilson, J.D. Ellis et al. 2015. A national survey of managed honey bee 2014–2015 annual colony losses in the USA. Journal of Apicultural Research. 54: 292–304.

Settele, J., O. Kudrna, A. Harpke, I. Kühn, C. van Swaay, R. Verovnik et al. 2008. Climatic risk atlas of European butterflies. BioRisk. 1: 1–710.

Shalumova, T. and J.M. Tanski. 2010. 5-(Hy-droxy-meth-yl)furan-2-carbaldehyde. Acta crystallographica. Section E, Structure reports online. 66: o2266. doi: 10.1107/S1600536810031119.

Sharipov, A.Ya. 2019. Assessment of purebred Burzyan bees (in Russian). Beekeeping. 3: 8–9.

Sharov, M. 2018. *Apis mellifera* Far-Eastern honey bee. Rus. Patent #9428. 2018.01.18.

Sharov, M.A. 2015. Breeding far Eastern bees on honey productivity in short rapid honey with lime. Proceedings of the Federal State Budgetary Institution "Scientific Research Institute of Beekeeping": 62–66.

Sharov, M.A. 2018. The honey bees breed "The Ear-Eastern bees" Abstracts of the XXII International Congress of Apislavia. Lab print company, Moscow: 117–119.

Sheppard, W.S. and S.H. Berlocher. 1984. Enzyme polymorphism in *Apis mellifera* from Norway. Journal of Apicultural Research. 23: 64–69.

Sheppard, W.S. 1989. A history of the introduction of honey bee races into the United States. The American Bee Journal. 129: 617–667.

Sheppard, W.S., A.E.E. Soares and D. DeJong. 1991. Hybrid status of honey bee populations near the historic origin of Africanization in Brazil. Apidologie. 22: 643–652.

Sheppard, W.S., M.C. Arias, A. Grech and M.D. Meixner. 1997. *Apis mellifera ruttneri*, a new honey bee subspecies from Malta. Apidologie. 28: 287–293. doi: 10.1051/apido:19970505.

Sheppard, W.S. and D.R. Smith. 2000. Identification of African-derived bees in the Americas: a survey of methods. Annals of the Entomological Society of America. 93: 159–176. doi: 10.1603/0013-8746(2000)093[0159:IOADBI]2.0.CO;2.

Sheppard, W.S. and M.D. Meixner. 2003. *Apis mellifera pomonella*, a new honey bee subspecies from Central Asia. Apidologie. 34: 367–375. doi: 10.1051/apido:2003037.

Shimanuki, H. and D.A. Knox. 2000. Diagnosis of honey bee diseases. Agriculture handbook no. AH690. US Department of Agriculture, Beltsville.

Shinmura, Y., H. Okuyama, T. Kiyoshi, C. Lin, T. Kadowaki and J. Takahashi. 2018. The complete mitochondrial genome and genetic distinction of the Taiwanese honey bee, *Apis cerana* (Hymenoptera: Apidae). Conservation Genetics Resources. 10: 621–626.

Shullia, N.I., R. Raffiudin and B. Juliandi. 2019. The phosphofructokinase and pyruvate kinase genes in *Apis andreniformis* and *Apis cerana indica*: exon intron organisation and evolution. Trop. Life Sci. Res. 30: 87–105.

Shumnyi, V.K., Yu.I. Shokin, N.A. Kolchanov and A.M. Fedotov. 2006. Biodiversity and Ecosystem Dynamics: Information Technologies and Modeling (in Russian). Publishing house of Siberian Branch of the Russian Academy of Sciences, Novosibirsk.

Sihanuntavong, D., S. Sittipraneed and S. Klinbunga. 1999. Mitochondrial DNA diversity and population structure of the honey bee, *Apis cerana*, in Thailand. Journal of Apicultural Research. 38: 211–219.

Simankov, M.K., A.N. Nikitin and A.S. Konovalov. 2017. Monitoring of the morphometric characters of the castes of bees in the Perm region. pp. 73–76. *In*: The role of biodiversity and maintenance of ecosystem homeostasis. Kirov research Institute of the North-East, Kirov.

Simon-Delso, N., G. San Martin, E. Bruneau, L.-A. Minsart, C. Mouret and L. Hautier. 2014. Honey bee Colony Disorder in Crop Areas: The Role of Pesticides and Viruses. PLOS ONE. 9: e103073.

Sinacori, A., T.E. Rinderer, V. Lancaster and W.S. Sheppard. 1998. A morphological and mitochondrial assessment of *Apis mellifera* from Palermo, Italy. Apidologie. 29: 481–490.

Singh, S. 1962. Bee keeping in India. Indian Council of Agrium. Research Publisher, New Delhi.

Sittipraneed, S., D. Sihanuntavong and S. Klinbunga. 2001. Genetic differentiation of the honey bee (*Apis cerana*) in Thailand revealed by polymorphism of a large subunit of mitochondrial ribosomal DNA. Insectes. Soc. 48: 266–272.

Siuda, M., J. Wilde and N. Koeniger. 1996. Further research on honey bee breeding with short postcapping periods. Pszczelnicze Zesyty Naukowe. 40: 135–143.

Sivaram, V. 2012. Status, prospects and strategies for development of organic beekeeping in the South Asian countries. Bangalore University, Department of Botany. [online]. http://www.apiservices. com/articles/us/23-02-12conceptorganicsouthAsia.pdf.

Skirkevicius, A.V. 1986. Pheromone communication of insects. Vilnius: Moxlas.

Slice, D.E. 2007. Geometric morphometrics. Annu. Rev. Anthropol. 36: 261–281.

Smith, D.R. 1991a. African bees in the Americas: Insights from biogeography and genetics. Trends in Ecology and Evolution. 6: 17–21.

Smith, D.R. 1991b. Mitochondrial DNA and honey bee biogeography. pp. 131–176. *In*: Smith, D.R. [ed.]. Diversity in the Genus *Apis*. Westview Press, Boulder, Colorado.

Smith, D.R., M.F. Palopoli, B.R. Taylor, L. Garnery, J.M. Cornuet, M. Solignac et al. 1991. Geographical overlap of two mitochondrial genomes in Spanish honey bees (*Apis mellifera iberica*). Journal of Heredity. 82: 96–100.

Smith, D.R. and T.C. Glenn. 1995. Allozyme polymorphisms in Spanish honey bees (*Apis mellifera iberica*). Journal of Heredity. 86: 12–16.

Smith, D.R. and R.H. Hagen. 1996. The biogeography of *Apis cerana* as revealed by mitochondrial DNA sequence data. Journal of the Kansas Entomological Society. 69: 294–310.

Smith, D.R., B.J. Crespi and F.L. Bookstein. 1997. Fluctuating asymmetry in the honey bee, *Apis mellifera*: effects of ploidy and hybridization. Journal of Evolutionary Biology. 10: 551–574. doi: 10.1007/s000360050041.

Smith, D.R. and W.M. Brown. 1988. Mitochondrial DNA restriction site polymorphism in American and Africanized honey bees (*Apis mellifera*). Experientia. 44: 257–260.

Smith, D.R., L. Villafuerte, G. Otis and M.R. Palmer. 2000. Biogeography of *Apis cerana* F. and *A. nigrocincta* Smith: insights from mtDNA studies. Apidologie. 31: 265–279.

Smith D.R. 2002. Biogeography of Apis cerana: southeast Asia and the Indo-Pakistan subcontinent. pp. 113–114. *In*: Proceedings of the XIV International Congress, IUSSI, Sapporo: 209.

Smith, D.R., M.R. Palmer, G. Otis and M. Damus. 2003. Mitochondrial DNA and AFLP markers support species status of *Apis nigrocincta*. Insect Soc. 50: 185–190.

Smith, D.R., N. Warrit and H.R. Hepburn. 2004. *Apis cerana* from Myanmar (Burma): unusual distribution of mitochondrial lineages. Apidologie. 35: 637–644.

Smith, D.R. 2011. Asian honey bees and mitochondrial DNA. pp. 69–94. *In*: Hepburn, H.R. and S.E. Radloff [eds.]. Honey bees of Asia. Springer-Verlag Berlin Heidelberg. New York, USA.

Smith, F. 1861. Descriptions of new species of hymenopterous insects collected by Mr. A.R. Wallace at Celebes. Proceedings of the Linnean Society of London. 5: 57–93.

Smith, F.G. 1960. Beekeeping in the Tropics. Longmans, Green and Co., London.

Sodhi, N.S., L.P. Koh, B.W. Brook and P.K.L. Ng. 2004. Southeast Asian biodiversity: an impending disaster. Trends in Ecology and Evolution. 19: 654–660.

Soland-Reckeweg, G., G. Heckel, P. Neumann, P. Fluri and L. Excoffier. 2009. Gene flow in admixed populations and implications for the conservation of the Western honey bee, *Apis mellifera*. Journal of Insect Conservation. 13: 317–328.

Solignac, M., D. Vautrin, A. Loiseau, F. Mougel and E. Baudry. 2003. Five hundred and fifty microsatellite markers for the study of the honey bee (*Apis mellifera* L.) genome. Molecular Ecology. Notes. 3: 307–311.

Solovyeva, L.F. 2012. Protect the bees from pesticide poisoning. Plant Protection and Quarantine. 5: 53–54.

Somerville, D. 2003. Small hive beetle in the USA. A report for the RIRDC (Rural Industries Research and Development Corporation). 03/050: 57.

Sommeijer, M.J., F.T. Beuvens and H.J. Verbeek. 1982. Distribution of labour among workers of *Melipona favosa* F.: construction and provisioning of brood cells. Insectes Sociaux. 29: 222–237.

Songram, O., S. Sittipraneed and S. Klinbunga. 2006. Mitochondrial DNA diversity and genetic differentiation of the honey bee (*Apis cerana*) in Thailand. Biochemical Genetics. 44: 256–269.

Spivak, M., D.J.C. Fletcher and M.D. Breed. 1991. The "African" Honey Bee. Westview Press, San Francisco, California.

Spivak, M. and D. Downey. 1998. Field assays for hygienic behaviour in honey bees (Hymenoptera: Apidae). Journal of Economic Entomology. 91: 64–70.

Spivak, M. and G.S. Reuter. 1998. Honey bee hygienic behavior. The American Bee Journal. 138: 283–286.

Spivak, M. and G.S. Reuter. 2001. Resistance to American foulbrood disease by honey bee colonies, *Apis mellifera*, bred for hygienic behavior. Apidologie. 32: 555–565.

Spivak, M. and O. Boecking. 2001. Honey bee resistance to *Varroa* mites. pp. 205–227. *In*: Webster, T.C. and K.S. Delaplane [eds.]. Mites of the honey bee. Dadant and Sons, Inc., Hamilton, IL.

Spivak, M., R. Masterman, R. Ross and K.A. Mesce. 2003. Hygienic behavior in the honey bee (*Apis mellifera* L.) and the modulatory role of octopamine. Journal of Neurobiology. 55: 341–354.

Spleen, A.M., E.J. Lengerich, K. Rennich, D. Caron, R. Rose, J.S. Pettis et al. 2013. A national survey of managed honey bee 2011–2012 winter colony losses in the United States: results from the Bee Informed Partnership. Journal of Apicultural Research. 52: 44–53.

Steffan-Dewenter, I. and T. Tscharntke. 2000. Resource overlap and possible competition between honey bees and wild bees in central Europe. Oecologia. 122: 288–296.

Steffan-Dewenter, I. and C. Westphal. 2008. The interplay of pollinator diversity, pollination services and landscape change. Journal of Applied Ecology. 45: 737–741.

Steinhage, V., B. Kastenholz, S. Schröder and W. Drescher. 1997. A hierarchical approach to classify solitary bees based on image analysis. Mustererkennung 19. DAGM-Symposium, Braunschweig, Informatik Aktuell. Springer, Germany.

Steinhage, V., S. Schröder, K.H. Lampe and A.B. Cremers. 2007. Automated extraction and analysis of morphological features for species identification. pp. 115–129. *In*: MacLeod, N. [ed.]. Automated taxon identification in systematics: theory, approaches and applications. Natural History Museum, London, UK; CRC Press Taylor and Francis Group Boca Raton, FL.

Steinhauer, N.A., K. Rennich, M.E. Wilson, D.M Caron, E.J. Lengerich, J.S. Pettis et al. 2014. A national survey of managed honey bee 2012–2013 annual colony losses in the USA: results from the Bee Informed Partnership. Journal of Apicultural Research. 53: 1–18.

Stephen, W.P. and P.P. Torchio. 1961. Biological observations on *Emphoropsis miserabilis* (Cresson), with comparative notes on other Anthophorids (Hymenoptera: Apoidea). Annals of the Entomological Society of America. 54: 687–692.

Stephen, W.P., G.E. Bohart and P.F. Torchio. 1969. The biology and external morphology of bees with a synopsis of the genera of synopsis of the genera of northwestern America. Agricultural Experimental Station of Oregon Slate University, Corvallis.

Suchail, S., D. Guez and L.P. Belzunces. 2000. Characteristics of imidacloprid toxicity in two *Apis mellifera* subspecies. Environmental Toxicology and Chemistry. 19: 1901–1905. doi: 10.1002/etc.5620190726.

Suchail, S., D. Guez and L.P. Belzunces. 2001. Discrepancy between acute and chronic toxicity induced by imidacloprid and its metabolites in *Apis mellifera*. Environmental Toxicology and Chemistry. 20: 2482–2486.

Suka, T. and H. Tanaka. 2005. New mitochondrial CO1 haplotypes and genetic diversity in the honey bee *Apis koschevnikovi* of Crocker Range Park, Sabah, Malaysia. Journal of Tropical Biology and Conservation. 1: 1–7.

Sumpter, D.J.T. and S.J. Martin. 2004. The dynamics of virus epidemics in *Varroa*-infested honey bee colonies. Journal of Animal Ecology. 73: 51–63.

Swanson, J., B. Torto, S. Kells, K. Mesce, J. Tumlinson and M. Spivak. 2009. Odorants that induce hygienic behavior in honey bees: identification of volatile compounds in chalkbrood-infected honey bee larvae. Journal of Chemical Ecology. 35: 1108–1116. doi: 10.1007/s10886-009-9683-8.

Taber, S. 1980. Bee behavior. The American Bee Journal. 120: 64–266.

Takahashi, J., T. Yoshida, T. Takagi, S. Akimoto, K.S. Woo, S. Deowanish et al. 2007. Geographic variation in the Japanese islands of *Apis cerana japonica* and in *A. cerana* populations bordering its geographic range. Apidologie. 38: 335–340.

Takahashi, J., S. Hadisoesilo, H. Okuyama and H.R. Hepburn. 2018. Analysis of the complete mitochondrial genome of *Apis nigrocincta* (Insecta: Hymenoptera: Apidae) on Sangihe Island in Indonesia. Conserv. Genetic Resour. 10: 755–760.

Takahashi, J., J. Nakamura, M. Sasaki, S. Tingek and S. Akimoto. 2002. New haplotypes for the non-coding region of mitochondrial DNA in cavity-nesting honey bees *Apis koschevnikovi* and *Apis nuluensis*. Apidologie. 33: 25–31.

Takahashi, J., T. Wakamiya, T. Kiyoshi, H. Uchiyama, S. Yajima, K. Kimura et al. 2016. The complete mitochondrial genome of the Japanese honey bee, *Apis cerana japonica* (Insecta: Hymenoptera: Apidae). Mitochondrial DNA Part B: Resources. 1: 156–157.

Takahashi, S., S. Tingek and H. Okuyama. 2017. The complete mitochondrial DNA sequence of endemic honey bee *Apis nuluensis* (Insecta: Hymenoptera: Apidae) inhabiting Mount Kinabalu in Sabah Province, Borneo Island. Mitochondrial DNA Part B: Resources. 2: 585–586.

Tan, H.W., G.H. Liu, X. Dong, R.Q. Lin, H.Q. Song, S.Y. Huang et al. 2011. The complete mitochondrial genome of the Asiatic cavity-nesting honeybee *Apis cerana* (Hymenoptera: Apidae). PLoS One. 6: e23008

Tan, K., N. Warrit and D.R. Smith. 2007. Mitochondrial DNA diversity of Chinese *Apis cerana*. Apidologie. 38: 238–246.

Tan, K., M.X. Yang, S.E. Radloff, H.R. Hepburn, Z.Y. Zhang, L.J. Luo et al. 2008. Dancing to different tunes: heterospecific deciphering of the honey bee waggle dance. Naturwissenchaften. 95: 1165–1168.

Tan, N.Q. and P.T. Binh. 1993. Harmony or conflict? *Apis mellifera* and *Apis cerana* in southern Vietnam. Beekeeping and Development. 32: 4–7.

Tanaka, H., D.W. Roubik, M. Kato, F. Liew and G. Gunsalam. 2001a. Phylogenetic position of *Apis nuluensis* of northern Borneo and phylogeography of *A. cerana* as inferred from mitochondrial DNA sequences. Insectes Sociaux. 48: 44–51. doi: 10.1007/pl00001744.

Tanaka, H., T. Suka, D.W. Roubik and M. Maryati. 2001b. Genetic differentiation among geographic groups of three honey bee species, *Apis cerana, A. koschevnikovi* and *A. dorsata*, in Borneo. Nature and Human Activities. 6: 5–12.

Tanaka, H., T. Suka, S. Kahono, H. Samejima, M. Maryati and D.W. Roubik. 2003. Mitochondrial variation and genetic differentiation in the honey bees (*Apis cerana, A. koschevnikovi* and *A. dorsata*) of Borneo. Tropics. 13: 107–117.

Tarpy, D.R., S. Hatch and D.J.C. Fletcher. 2000. The influence of queen age and quality during queen replacement in honeybee colonies. Animal Behaviour. 59: 97–101.

Teixidó, E., F.J. Santos, L. Puignou and M.T. Galceran. 2006. Analysis of 5-hydroxymethylfurfural in foods by gas chromatography-mass spectrometry. Journal of Chromatography A. 1135: 85–90.

Tentcheva, D., L. Gauthier, N. Zappulla, B. Dainat, F. Cousserans, M.E. Colin et al. 2004. Prevalence and seasonal variations of six bee viruses in *Apis mellifera* L. and *Varroa destructor* mite populations in France. Applied and Environmental Microbiology. 70: 7185–7191.

The Honeybee Genome Sequencing Consortium. 2006. Insights into social insects from the genome of the honeybee *Apis mellifera*. Nature. 443: 931–949.

Thomas, C.D., A. Cameron, R.E. Green, M. Bakkenes, L.J. Beaumont and Y.C. Collingham. 2004. Extinction risk from climate change. Nature. 427: 145–148.

Thomas, G. 1698. An Historical and Geographic Account of the Province and Country of Pensilvania and of West-New-Jersey in America. Baldwin, London.

Thomas, R. and W. Sirivat. 1992. Pollen resource partitioning by *Apis dorsata, A. cerana, A. florea*. Journal of Apicultural Research. 1: 3–7.

Thompson d'Arcy, W. 1992. On growth and form. Dover Publisher, N.Y.

Thompson, H. 2003. Behavior effects of pesticides in bees-their potential for use in risk assessment. Ecotoxicology. 12: 317–330. doi: 10.1023/A:1022575315413.

Thomson, D.M. 2004. Competitive interactions between the invasive European honey bee and native bumble bees. Ecology. 85: 458–470.

Thomson, D.M. 2016. Local bumble bee decline linked to recovery of honey bees, drought effects on floral resources. Ecology Letters. 19: 1247–1255.

Tingek, S., M. Mardan, T.E. Rinderer, N. Koeniger and G. Koeniger. 1988. Rediscovery of *Apis vechti* (Maa, 1953): The Saban Honey Bee. Apidologie. 19: 97–102.

Tingek, S., G. Koeniger and N. Koeniger. 1996. Description of a new cavity nesting species of *Apis* (*Apis nuluensis* n. sp.) from Sabah, Borneo, with notes on its occurrence and reproductive biology. Senckenbergiana Biologica. 76: 115–119.

Tingek, S., M. Mardan, T.E. Rinderer, N. Koeniger and G. Koeniger. 1988. The rediscovery of *Apis vechti* Maa 1953: the Sabah honey bee. Apidologie. 19: 97–102.

Tofilski, A. 2004. DrawWing, a program for numerical description of insect wings. Journal of Insect Science. 4: 1–5. doi: 10.1673/031.004.1701.

Tofilski, A. 2007. Automatic Measurement of Honey bee Wings. pp. 289–298. *In*: MacLeod, N. [ed.]. Automated taxon identification in systematics: theory, approaches and applications. Natural History Museum, London, UK; CRC Press Taylor and Francis Group Boca Raton, FL.

Tofilski, A. 2008. Using geometric morphometrics and standard morphometry to discriminate three honey bee subspecies. Apidologie. 39: 558–563. doi: 10.1051/apido:2008037.

Tomkies, V., J. Flint, G. Johnson, R. Waite, S. Wilkins and C. Danks. 2009. Development and validation of a novel field test kit for European foulbrood. Apidologie. 40: 63–72.

Topolska, G., A. Gajda and A. Hartwig. 2008. Polish honey bee colony-loss during the winter of 2007/2008. Journal of Apicultural Science. 52: 95–104.

Topolska, G., A. Gajda, K. Pohorecka, A. Bober, S. Kasprzak and M. Skubida. 2010. Winter colony losses in Poland. Journal of Apicultural Research. 49: 126–128.

Torchio, P.F. and V.J. Tepedino. 1980. Sex ratio, body size and seasonality in a solitary bee, *Osmia linaria propinqua* Cresson (Hymenoptera: Megachilidae). Evolution (USA). 34: 993–1003.

Tosi, E., M.C. Ciappini, E. Re and H. Lucero 2001. Honey thermal treatment effects on hydroxymethylfurfural content. Food Chemistry. 77: 71–74.

Toth, A.L., S. Kantarovich, A.F. Meisel and G.E. Robinson. 2005. Nutritional status influences socially regulated foraging ontogeny in honey bees. Journal of Experimental Biology. 208: 4641–4649.

Trouiller, J., G. Arnold, B. Chappe, Y. Le Conte and C. Masson. 1992. Semiochemical basis of infestation of honey bee brood by *Varroa jacobsoni*. Journal of Chemical Ecology. 18: 2041–2053.

Trung, L.Q., P.X. Dung and T.X. Ngan. 1996. A scientific note on first report of *Apis laboriosa* F. Smith, 1871 in Vietnam. Apidologie. 27: 487–488.

Tsing, A.L. 2003. Cultivating the wild: Honey-hunting and forest management in Southeast Kalimantan. pp. 24–55. *In*: Zerner, C. [ed.]. Culture and the question of rights. Forests coasts and seas in Southeast Asia. Duke University Press, Durham.

Tsuneki, K. 1970. Bionomics of some species of Megachile, Dasypoda, Colletes and Bombus (Hym.: Apoidae). Etizenia. 48: 1–20.

Underwood, B.A. 1990. Seasonal nesting cycle and migration patterns of the Himalayan honey bee *Apis laboriosa*. National Geographic Research and Exploration. 6: 276–290.

UNEP. 2010. UNEP Emerging Issues: Global Honey Bee Colony Disorder and Other Threats to Insect Pollinators. Pascal Peduzzi, Ron Witt, R. Norberto Fernandez, Mwangi, Theuri.

Uzunov, A., C. Costa, B. Panasiuk, M. Meixner, P. Kryger, F. Hatjina et al. 2014a. Swarming, defensive and hygienic behaviour in honey bee colonies of different genetic origin in a pan-European experiment. Journal of Apicultural Research. 53: 248–260.

Uzunov, A., M.D. Meixner, H. Kiprijanovska, S. Andonov, A. Gregorc, E. Ivanova et al. 2014b. Genetic structure of *Apis mellifera macedonica* in the Balkan Peninsula based on microsatellite DNA polymorphism. Journal of Apicultural Research. 53: 288–295.

Valli, E. and D. Summers. 1988. Honey hunters of Nepal. Harry N. Abrams, Inc., New York.

Van der Sluijs, J.P., V. Amaral-Rogers, L.P. Belzunces, M.B. Van Lexmond, J.M. Bonmatin, M. Chagnon et al. 2015. Conclusions of the Worldwide Integrated Assessment on the risks of neonicotinoids and fipronil to biodiversity and ecosystem functioning.

van der Zee, R. 2010. Colony losses in the Netherlands. Journal of Apicultural Research. 49: 121–123.

van der Zee, R. and L. Pisa. 2010. Wintersterfte 2009–10 en toxische invertsuikersiroop. Netherlands Centre for Bee Research Report 02/2010 [online] http://www.beemonitoring.org/Downloads/Bijensterfte%202009-10_en%20toxische_%20invertsuikersiroop.pdf.

van der Zee, R., L. Pisa, S. Andonov, R. Brodschneider, J.-D. Charrière, R. Chlebo et al. 2012. Managed honey bee colony losses in Canada, China, Europe, Israel and Turkey, for the winters of 2008–9 and 2009–10, Journal of Apicultural Research. 51: 100–114.

van der Zee, R., A. Gray, C. Holzmann, L. Pisa, R. Brodschneider, R. Chlebo et al. 2013. Standard survey methods for estimating colony losses and explanatory risk factors in *Apis mellifera*. Journal of Apicultural Research. 52: 1–35.

van der Zee, R., R. Brodschneider, V. Brusbardis, J.-D. Charrière, R. Chlebo, M.F. Coffey et al. 2014. Results of international standardized beekeeper surveys of colony losses for winter 2012–2013: analysis of winter loss rates and mixed effects modelling of risk factors for winter loss. Journal of Apicultural Research. 53: 19–34.

van der Zee, R., A. Gray, L. Pisa and T. de Rijk. 2015. An observational study of honey bee colony winter losses and their association with *Varroa destructor*, Neonicotinoids and other risk factors. PLOS ONE. 10: e0131611.

Vanbergen, A.J., M.P. Garratt, A.J. Vanbergen, M. Baude, J.C. Biesmeijer, N.F. Britton et al. 2013. Threats to an ecosystem service: Pressures on pollinators. Frontiers in Ecology and the Environment. doi: 10.1890/120126.

Vandame, R. and M.A. Palacio. 2010. Preserved honey bee health in Latin America: a fragile equilibrium due to low-intensity agriculture and beekeeping? Apidologie. 41: 243–255.

vanEngelsdorp, D., R. Underwood, D. Caron and J. Hayes. 2007. An estimate of managed colony losses in the winter of 2006-2007: a report commissioned by the Apiary Inspectors of America. The American Bee Journal. 147: 599–603.

vanEngelsdorp, D., J. Hayes, R.M. Underwood and J. Pettis. 2008. A Survey of Honey Bee Colony Losses in the US, Fall 2007 to Spring 2008. PLOS ONE. 3: e4071.

vanEngelsdorp, D., J.D. Evans, C. Saegerman, C. Mullin, E. Haubruge, B.K. Nguyen et al. 2009. Colony Collapse Disorder: A Descriptive Study. PLOS ONE. 4: e6481.

vanEngelsdorp, D. and M.D. Meixner. 2010. A historical review of managed honey bee populations in Europe and the United States and the factors that may affect them. Journal of Invertebrate Pathology. 103: 80–95.

vanEngelsdorp, D., E. Lenderich, A. Speen, B. Dainat, J. Cresswell, K. Baylis et al. 2013. Standard epidemiological methods to understand and improve *Apis mellifera* health. Journal of Apicultural Research. 52: 1–16.

Vasilev, A.G., I.A. Vasileva and A.O. Shkurikhin. 2018. Geometric morphometrics: from theory to practice (in Russian). KMK Scientific Press, Moscow.

Vejsnaes, F., S.L. Nielsen and P. Kryger. 2010. Factors involved in the recent increase in colony losses in Denmark. Journal of Apicultural Research. 49: 109–110.

Velthuis, H.W., J. Clement and R.A. Morse. 1971. The ovaries of *Apis dorsata* workers and queens form the Philippine. Journal of Apicultural Research. 10: 63–66.

Velthuis, H.W. 1997. The biology of Stingless bees. Utrecht University, the Netherlands and University of Sao Paulo, Brazil.

Vereshchaka, O.A. and N.N. Grankin. 2011. The potential reproduction of the Middle Russian bees (in Russian). Beekeeping. 5: 8–10.

Vergara, C., A. Dietz and A. Perez de Leon. 1993. Female parasitism of European honey bees by Africanized honey bee swarms in Mexico. Journal of Apicultural Research. 32: 34–40.

Verma, L.R. 1990. Beekeeping in Integrated Mountain Development. Oxford and IBH, New Delhi.

Verma, L.R. 1992. Honey bees in Mountain Agriculture. Oxford and IBH, New Delhi.

Verma, L.R. and U. Partap. 1993. The Asian Hive Bee, *Apis cerana*, as a Pollinator in Vegetable Seed Production. ICIMOD, Kathmandu.

Verma, L.R. and U. Partap. 1994. Foraging behaviour of *Apis cerana* on cabbage and cauliflower and its impact on seed production. Journal of Apicultural Research. 33: 231–236.

Verma, L.R., V.K. Mattu and H.V. Daly. 1994. Multivariate morphometrics of the Indian honey bee in the northwest Himalayan region. Apidologie. 25: 203–223.

Verma, L.R. 1998. Conservation and management of Asian hive bee, *Apis cerana*. The 4th Asian Apicultural Association Conference, Kathmandu.

Vidau, C., M. Diogon, J. Aufauvre, R. Fontbonne, B. Vigues, J.L. Brunet et al. 2011. Exposure to sublethal doses of fipronil and thiacloprid highly increases mortality of honey bees previously infected by *Nosema ceranae*. PLOS ONE. 6: e21550.

Vijayakumar, K. and R. Jayaraj. 2013. Geometric morphometry analysis of three species of stingless bees in India. Internatl. Journal of Life Sciences Research. 1: 91–95.

Villaneuva, G.R. and D.W. Roubik. 2004. Why are African honey bees and not European bees invasive? Pollen diet diversity in community experiments. Apidologie. 35: 481–491.

Vinson, S.V. and G.W. Frankie. 1977. Nest of *Cantridini acthyctera* (Hymenoptera: Apoidae: Anthophoridae) in the dry forest of Costa Rica. Journal of the Kansas Entomological Society. 50: 301–311.

Visscher, P.K., R.S. Vetter and F.C. Baptista. 1997. Africanized bees, 1990–1995: Initial rapid invasion has slowed in the US. California Agriculture. 51: 22–25.

Vung, N.N., M.L. Lee, M.Y. Lee, H.K. Kim, E.J. Kang, J.E. Kim et al. 2017. Breeding and selection for resistance to sacbrood virus for *Apis cerana*. Journal of Apiculture. 32: 345–352.

Wafa, A.K., S. Rashad and M.A. Moustafa. 1972. On the nesting habits of *Andrena ovatula* (K) in Egypt. (Hymenoptera, Apoidea). Deutsche Entomologische Zeitschrift. 4: 303–306.

Wahl, O. and K. Ulm. 1983. Influence of pollen feeding and physiological condition on pesticide sensitivity of the honey bee *Apis mellifera carnica*. Oecologia. 59: 106–128. doi: 10.1007/BF00388082.

Wakamiya, T., S. Tingek, H. Okuyama, T. Kiyoshi and J. Takahashi. 2017. The complete mitochondrial genome of the cavity-nesting honey bee, *Apis koschevnikovi* (Insecta: Hymenoptera: Apidae). Mitochondrial DNA Part B Resources. 2: 24–25.

Wallberg, A., F. Han, G. Wellhagen, B. Dahle, M. Kawata, N. Haddad et al. 2014. A worldwide survey of genome sequence variation provides insight into the evolutionary history of the honey bee *Apis mellifera*. Nat. Genet. 46: 1081–1088. doi: 10.1038/ng.3077.

Wang, D.I. and F.E. Moeller. 1970. The division of labor and queen attendance behavior of Nosema-infected worker honey bees. Journal of Economic Entomology. 63: 1539–1541.

Wang, A.R., M.J. Kim, J.Y. Lee, Y.S. Choi, R. Thapa and I. Kim. 2015. The mitochondrial genome of the black dwarf honey bee, Apis andreniformis (Hymenoptera: Apidae). Mitochondrial DNA. 26: 914–916.

Wang, A.R., J.S. Kim, M.J. Kim, H. Kim, Y.S. Choi and I. Kim. 2018. Comparative description of mitochondrial genomes of the honey bee Apis (Hymenoptera: Apidae): four new genome sequences and Apis phylogeny using whole genomes and individual genes. Journal of Apicultural Research. 57: 484–503.

Ward, K., R. Danka and R. Ward. 2008. Comparative performance of two mite-resistant stocks of honey bees (Hymenoptera: Apidae) in Alabama beekeeping operations. Journal of Economic Entomology. 101: 654–659.

Watmough, R.H. 1974. Biology and behaviour of *Caprenter bees* in Southern Africa. Journal of the Entomological Society of South Africa. 37: 261–281.

Weinstock, G.M., G.E. Robinson, R.A. Gibbs, K.C. Worley, J.D. Evans and R. Maleszka et al. 2006. Insights into social insects from the genome of the honey bee *Apis mellifera*. Nature. 443: 931–949. doi: 10.1038/nature05260.

Wen-Cheng, H. and Z. Chong-Yuan. 1985. The relationship between the weight of gueen honey bee at various stages and the number of ovarioles, eggs laid and sealed brood produced. Honey bee Science. 6: 113–116.

Wendorf, H. 2002. Beekeeping development with *Apis mellifera* in the Philippines. Bees for Development Journal. 65: 4–5.

White, J.W. and L.W. Doner. 1980. Honey composition and properties: beekeeping in the United States. Agriculture Handbook. 335: 82–91.

Whitehorn, P.R., S. O'Connor, F.L. Wackers and D. Goulson. 2012. Neonicotinoid pesticide reduces bumble bee colony growth and queen production. Science. 336: 351–352. doi: 10.1126/science.1215025.

Whitfield, C.W., S.K. Behura, S.H. Berlocher, A.G. Clark, J.S. Johnston, W.S. Sheppard et al. 2006a. Thrice out of Africa: Ancient and recent expansions of the honey bee, *Apis mellifera*. Science. 314: 642–645. doi: 10.1126/science.1132772.

Whitfield, C.W., Y. Ben-Shahar, C. Brillet, I. Leoncini, D. Crauser, Y. Le Conte et al. 2006b. Genomic dissection of behavioral maturation in the honey bee. Proceedings of the National Academy of Sciences of the United States of America. 103: 16068–16075.

Wiener, N. 1963. New chapters of cybernetics. Moscow: Soviet radio.

Wilkins, S., M.A. Brown and A.G. Cuthbertson. 2007. The incidence of honey bee pests and diseases in England and Wales. Pest Management Science. 63: 1062–1068.

Willard, L.E., A.M. Hayes, M.A. Wallrichs and O. Rueppell. 2011. Food manipulation in honey bees induces physiological responses at the individual and colony level. Apidologie. 42: 508–518.

Wille, A. and E. Orozco. 1960. The life cycle and behavior of the social bee *Lasioglossum (Dialictus) umbripenne* (Hymenoptera: Halictidae). Revista de Biología Tropical. 17: 199–245.

Wille, A. and E. Orozco. 1970. The life cycle and behavior of the social bee *Lasioglossum (Dialictus) umbripenne* (Hymenoptera: Halictidae). Revista de Biología Tropical. 17: 199–245.

Wille, A. and C.D. Michener. 1973. The nest architecture of stingless bees with special reference to those of Costa Rica (Hymenoptera, Apidae). Revista de Biología Tropical. 21: 279.

Williams, G.R., A. Shafer, R.E. Rogers, D. Shutler and D.T. Stewart. 2008. First detection of *Nosema ceranae*, a microsporidian parasite of European honey bees (*Apis mellifera*) in Canada and central USA. Journal of Invertebrate Pathology. 97: 189–192.

Williams, P.H. 1986. Environmental change and the distributions of British bumble bees (*Bombus* Latr.). Bee World. 67: 50–61.

Williams, P.H. and J.L. Osborne. 2009. Bumble bee vulnerability and conservation world-wide. Apidologie. 40: 367–387.

Williamson, S.M., D.D. Baker and G.A. Wright. 2012. Acute exposure to a sublethal dose of imidacloprid and coumaphos enhances olfactory learning and memory in the honey bee *Apis mellifera*. Invertebrate Neuroscience. 13: 63–70. doi: 10.1007/s10158-012-0144-7.

Willis, L.G., M.L. Winston and B.M. Honda. 1992. Phylogenetic relationships in the honey bee (genus *Apis*) as determined by the sequence of the cytochrome oxidase II region of mitochondrial DNA. Molecular Phylogenetics and Evolution. 1: 169–178. doi: 10.1016/1055-7903900013-7.

Wilms, W., V.L. Imperatriz-Fonseca and W. Engels. 1996. Resource partitioning between highly eusocial bees and possible impact of the introduced Africanized honey bee on native stingless bees in the Brazilian Atlantic rainforest. Studies on Neotropical Fauna and Environment. 31: 137–51.

Wilson, E.O. 1971. The Insect Societies. Harvard University Press, Massachusetts, USA.

Wilson-Rich, N., M. Spivak, N.H. Fefferman and P.T. Starks. 2009. Genetic, individual and group facilitation of disease resistance in insect societies. Annual Review of Entomology. 54: 405–423.

Winfree, R., R. Aguilar, D.P. Vazquez, G. LeBuhn and M.A. Aizen. 2009. A meta-analysis of bees" responses to anthropogenic disturbance. Ecology. 90: 2068–2076.

Winston, M.L., J. Dropkin and O.R. Taylor. 1981. Demography and life history characteristics of two honey bee races (*Apis mellifera*). Oecologia. 48: 407–413.

Winston, M.L., O.R. Taylor and G.W. Otis. 1983. Some differences between temperate European and tropical African and South American honey bees. Bee World. 64: 12–21.

Winston, M.L. 1992. The biology and management of Africanized honey bees. Annual Review of Entomology. 37: 173–193.

Winston, M.L. 1994. The biology of the honey bee, B-Beauvechain, Nauwelaerts Ed.

Wojcik, V.A., L.A. Morandin, L.D. Adams and K.E. Rourke. 2018. Floral Resource Competition Between Honey Bees and Wild Bees: Is There Clear Evidence and Can We Guide Management and Conservation? Environ. Entomol. 47: 822–833.

Wongsiri, S., C. Lekprayoon, R. Thapa, K. Thirakupt, T.E. Rinderer, H.A. Sylvester et al. 1997. Comparative biology of Apis andreniformis and *Apis florea* in Thailand. Bee World. 78: 23–35.

Woodard, S.H., B.J. Fischman, A. Venkat, M.E. Hudson, K. Varala and S.A. Cameron. 2011. Genes involved in convergent evolution of eusociality in bees. Proceedings of the National Academy of Sciences of the United States of America. 108: 7472–7477.

Woodrow, A.W. and E.C. Holst. 1942. The mechanism of colony resistance to American foulbrood. Journal of Economic Entomology. 35: 327–330.

Woyke, J. 1966. Wovon hangt die Zahl der Spermien in der Samenblase der auf naturlichem Wege begatteten Koniginnen ab. Zeit Der Bienenforscher. 8: 236–247.

Woyke, J. 1976. Brood-rearing efficiency and absconding in *Indian honeybees*. Journal of Apicultural Research. 15: 133–143.

Woyke, J. 1977. Cannibalism and brood-rearing efficiency in the honeybees. Journal of Apicultural Research. 16: 84–94.

Woyke, J. 1980a. Effect of sex allele homo-heterozygosity on honey bee colony populations and on their honey production.1. Favorable development conditions and unrestricted queens. Journal of Apicultural Research. 19: 51–63.

Woyke, J. 1980b. Evidence and Action of Cannibalism Substance in *Apis cerana indica*. Journal of Apicultural Research. 19: 6–16.

Woyke, J. 1984. Correlations and interactions between population, length of worker life and honey production by honey bees in a temperate region. Journal of Apicultural Research. 23: 148–156. doi: 10.1080/00218839.1984.11100624.

Wu, Y. and B. Kuang. 1987. Two species of small honey bee—a study of the genus *Micrapis*. Bee World. 68: 153–155.

Yang, J., J. Xu, J. Wu, X. Zhang and S. He. 2019. The complete mitogenome of wild honey bee *Apis dorsata* (Hymenoptera: Apidae) from South-Western China. Mitochondrial DNA Part B Resources. 4: 231–232.

Yokoi, K., H. Uchiyama, T. Wakamiya, M. Yoshiyama, J.I. Takahashi, T. Nomura et al. 2018. The draft genome sequence of the Japanese honey bee, *Apis cerana japonica* (Hymenoptera: Apidae). European Journal of Entomology. 115: 650–657. doi: 10.14411/eje.2018.064.

Yusuf, A., C.W.W. Pirk and R.M. Crewe. 2015. Mandibular gland pheromone contents in workers and queens of *Apis mellifera adansonii*. Apidologie. 5: 559–572.

Zelditch, M.L., D.L. Swiderski, H.D. Sheets and W.L. Fink. 2004a. Geometric Morphometrics For Biologists: A Primer. Elsevier Academic, New York, USA. doi: 10.1016/b978-012778460-1/50007-7.

Zelditch, M.L., D.L. Swiderski and H.D. Sheets. 2004b. A Practical Companion to Geometric Morphometrics for Biologists: Running analyses in freely-available software.

Zeuner, F.E. and F.J. Manning. 1976. A monograph of fossil bees (Hymenoptera: Apoidea). Bulletin of the British Museum (Natural History). 27: 155–254.

Zhang, J.F. 1990. New fossil species of Apoidea (Insect: Hymenoptera). Acta Zootaxon Sinica. 15: 83–91.

Zhen-Ming, J., Y. Guanhang, H. Shuangxiu, L. Shikui and R. Zaijin. 1992. The advancement of apicultural science and technology in China. pp 133–147. *In*: Verma, L.R. [ed.]. Honey bees in mountain agriculture. Oxford and IBH, New Delhi.

Zhao, W., K. Tan, D. Zhou, M. Wang, C. Cheng, Z. Yu, Y. Miao and S. He. 2014. Phylogeographic analysis of Apis cerana populations on Hainan Island and southern mainland China, based on mitochondrial DNA sequences. Apidologie. 45: 21–33.

Zolina, G.D. and A.G. Mannapov. 2017. Breed type "Moskovskiy" recorded in the state register of selection achievements (in Russian). Beekeeping. 8: 6–8.

Zucchi, R., S.F. Sakagami and J.M.F. Camargo. 1969. Biological observations on a Neotropical parasocial bee, *Eulkma nigrita* with a review on the biology of Euglossinae (Hymenoptera: Apidae). A comparative study. Journal of the Faculty of Science, Hokkaido University. Series 6, Zoology. 17: 271–380.

Index

Color Plate Section

Chapter 2

(A)   (B)

Figure 2.7: he morphological characteristics of *Apis nigrocincta* from Parigi Moutong, Central Sulawesi: yellowish (A) clypeus and (B) hind legs.

(A)   (B)   (C)

Figure 2.8: *Apis dorsata binghami* in South Sulawesi: (A) worker honey bee, (B) nesting in a *Litsea mappacea* (Lauraceae) tree approximately 1 m above the ground, (C) nesting in a *Artocarpus sericocarpus* (Fam. Moraceae) tree at 32.6 m above the ground (Photograph: Muh. Teguh Nagir).

Chapter 2

(A)　　　　　　　　　(B)　　　　　　　(C)

Figure 2.10: Asian honey bee *Apis andreniformis*. (A) nest encircled with twigs and worker bees, (B) queen bee, and (C) drone bee (Photograph: Rika Raffiudin).

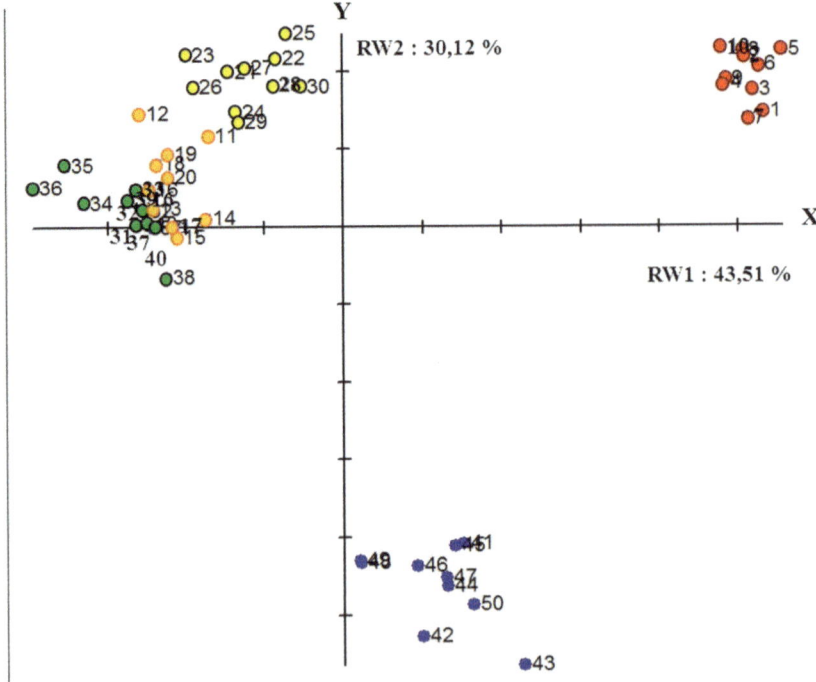

Figure 2.11: The Relative Warp (RW) ordination plot of five honey bee species: geometric morphometric wing venations. 1–10, red = *A. andreniformis*; 11–20, orange = *A. cerana*; 21–30; yellow = *A. koschevnikovi*; 31–40, green = *A. mellifera* and 41–50, blue = *A. dorsata* (Santoso et al. 2018).

Chapter 6

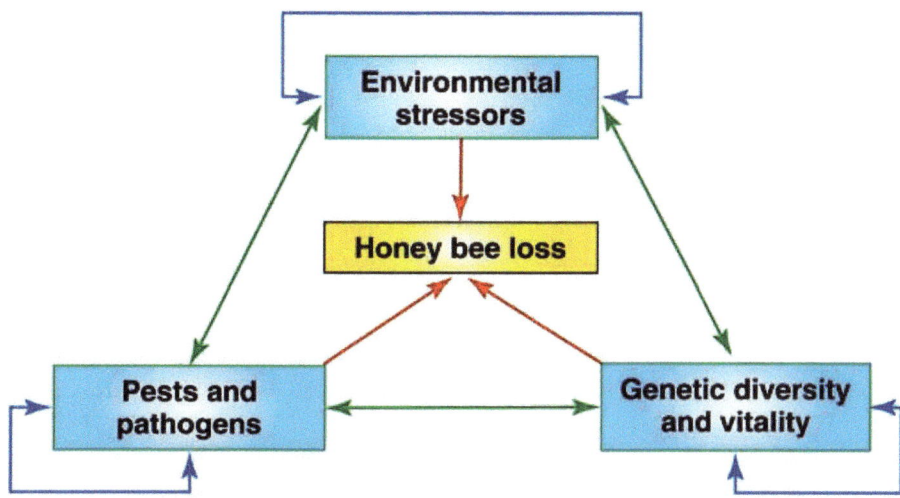

Figure 6.1: Interactions among multiple drivers of honey bee loss. Blue boxes represent the three main groups of drivers associated with honey bee loss; red arrows represent direct pressures on honey bees from drivers; green arrows represent interactions between drivers; black arrows represent interactions within drivers. (Retrieved from Potts et al. 2010- TRENDS in Ecology and Evolution).

Chapter 6

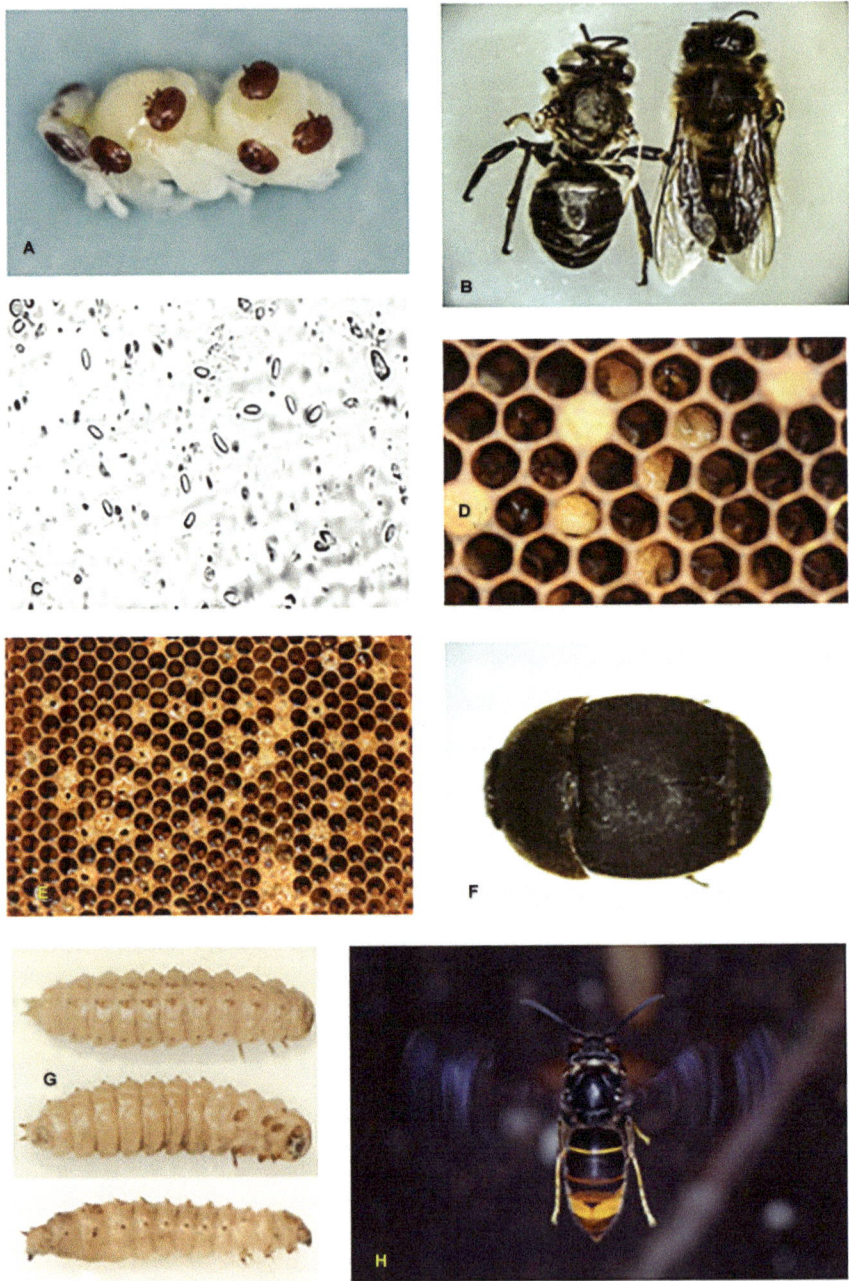

Figure 6.4: (A) The *Varroa* mite females feeding on a honey bee pupa. (B) A bee with deformed wings and shortened abdomen due to the DWV infection. (C) *Nosema ceranae* spores seen in 400x magnification. (D) A brood comb with twisted, diseased larvae attacked with *Melissococcus plutonius*. (E) a brood comb with brood cappings clearly punctured and scattered brood pattern. (F) The small hive beetle imago. (G) larva. (H) The yellow-legged hornet (Photos A, B, C, D, E, F, G: A. Gajda, Photo H: Per Kryger).

Chapter 6

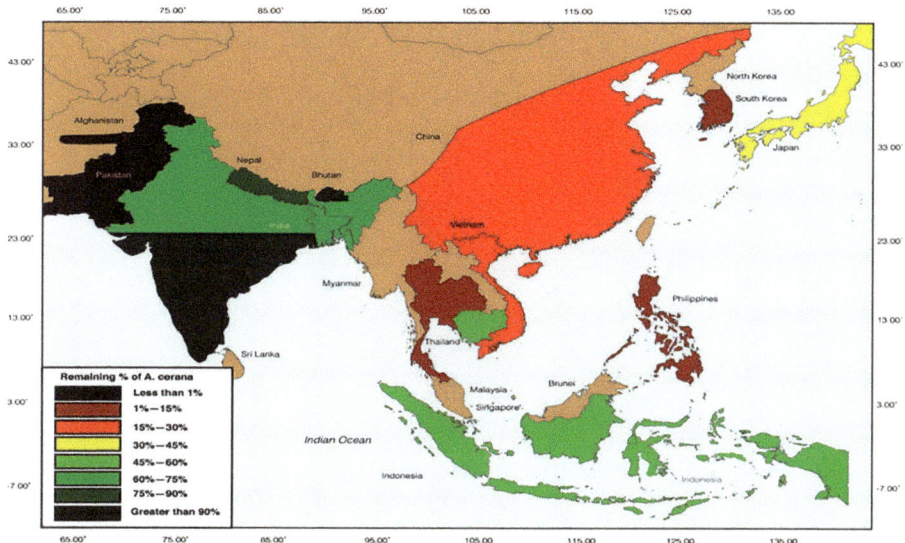

Figure 6.8: Map of the estimated remaining percentage of *A. cerana* compared with *A. mellifera* based on 2014–15 of 31 apiculturists in 16 Asian countries. Due to lack of data about wild honey bee species, only the commercially kept species (*A. cerana* + *A. mellifera*) could be taken into account. The sum of both species represents 100%, while the map details the proportion of *A. cerana* still extant (Source: Jones and Bienefeld 2016).

Chapter 9

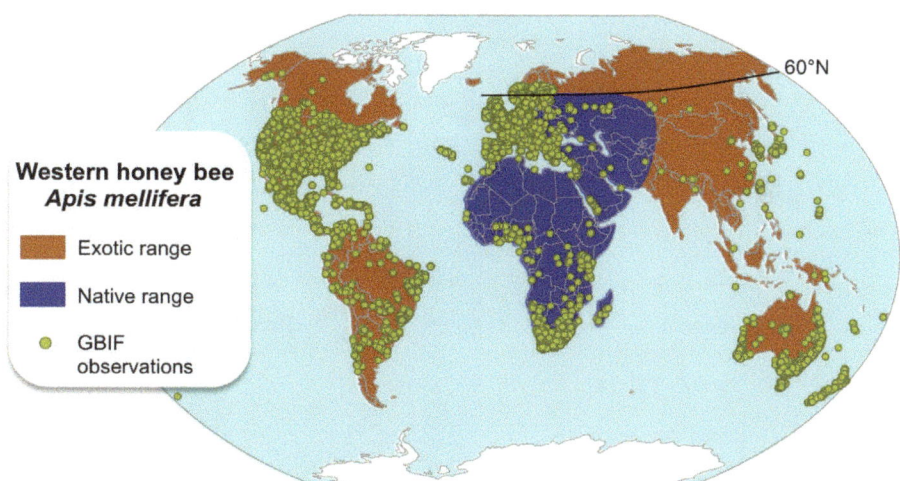

Figure 9.1: The worldwide expansion of the western honey bee *Apis mellifera*. In Europe, the native range of the species is limited around the 60°N (Ruttner 1988). The Eastern limits of the native range include the Middle East, Kyrgyzstan, Western China, and parts of Kazakhstan (Ruttner 1988). The GBIF dataset (in dots, GBIF Occurrence Download 2018a) shows the human-mediated introduction of the species throughout the rest of the world, and are plotted using the maps and rgdal R packages (R Core Team 2018).

Chapter 9

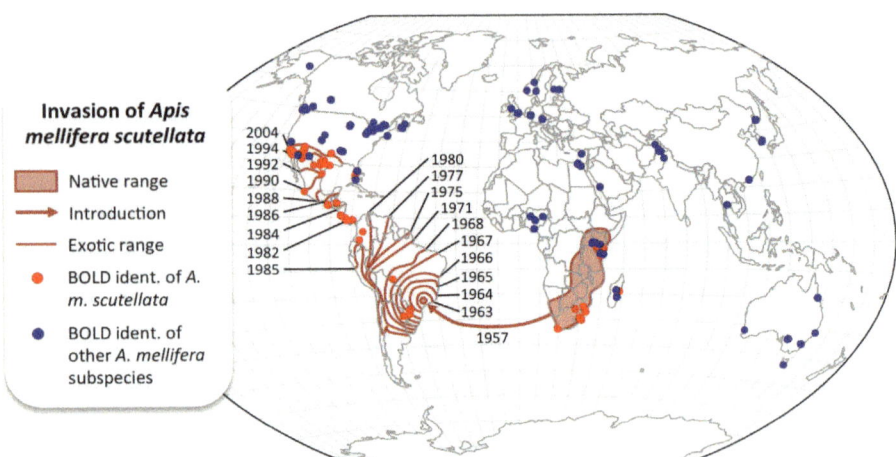

Figure 9.2: Introduction and invasion of the African honey bee *Apis mellifera scutellata* in the Americas. The native range of *A. m. scutellata* covers the southwestern of Africa, from Tanzania to South Africa (Ruttner 1988). The trends of the exotic range expansion are based on former studies (Spivak et al. 1991, Visscher et al. 1997, Moritz et al. 2005, Kono and Kohn 2015, Rangel et al. 2016). The BOLD dataset (in dots, Barcode of Life Database 2018) confirms the human-mediated wide distribution of *A. m. scutellata* in the Americas, and are plotted using the maps and rgdal R packages (R Core Team 2018).

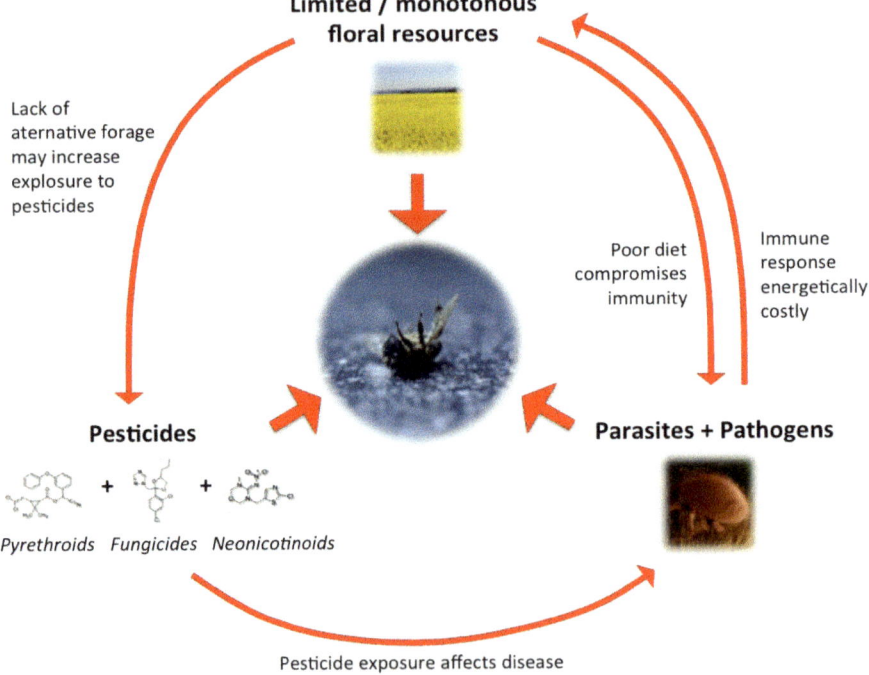

Figure 9.4: The decline of bee pollinators—a consequence of multiple stresses. Redraw from Goulson et al. (2015).

Chapter 9

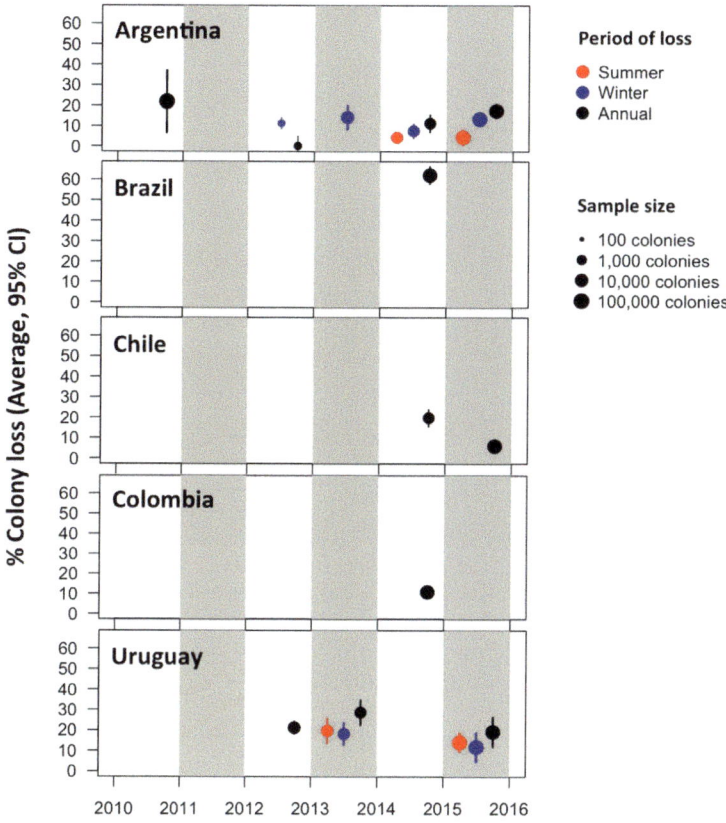

Figure 9.5: Synthesis of unpublished data of honey bee colony losses in Argentina, Brazil, Chile, Colombia, and Uruguay. The loss rates are calculated as the average value per beekeeper (95% Confidence Interval) following vanEngelsdorp et al. (2013). Redraw from Requier et al. (2018b), where more details on the surveys' methods are available.

Chapter 10

Figure 10.1: The original fossil specimen of a worker of *Apis nearctica*. It is the holotype of the species, i.e., the specimen designated as the type of this species by the authors who named and described the species. It is in the collection of the California Academy of Sciences. From Engel et al. (2009).